AF896125

AUSTRALIA'S Fossil Heritage

AUSTRALIA'S Fossil Heritage

A Catalogue of Important Australian Fossil Sites

The Australian Heritage Council

PUBLISHING

National Library of Australia Cataloguing-in-Publication entry

Australian Heritage Council.

Australia's fossil heritage : a catalogue of important Australian fossil sites / by the Australian Heritage Council.

9780643101777 (pbk.)
9780643101784 (epdf)
9780643102309 (epub)

Includes bibliographical references and index.

Paleontology – Australia
Fossils – Australia.
Natural history – Australia.
Cultural property – Protection – Australia

560.450994

Published by
CSIRO PUBLISHING
36 Gardiner Road, Clayton VIC 3168
Private Bag 10, Clayton South VIC 3169
Australia

Telephone: [+613] 9545 8555
Local call: 1300 788 000 (Australia only)
Fax: +61 3 9662 7555
Email: csiropublishing@csiro.au
Web site: www.publishing.csiro.au

Contributors

New South Wales: Yong-Yi Zhen

Northern Territory: Peter Murray and Dirk Megirian

Queensland: Alex Cook, Robyne Leven (Lark Quarry entry)

South Australia: Dennis Rice with Jim Gehling, Liz Reed (Naracoorte entry)

Tasmania: Clive Calver, Max Banks, Patrick Bender, Jim Jago, Greg Jordan, Patrick Quilty, Andrew Rozefelds

Victoria: John Long, Tom Rich, David Holloway, Erich Fitzgerald

Western Australia: Mikael Siversson

Plant fossil sites: Raymond J. Carpenter and Robert S. Hill

Department of Sustainability, Environment, Water, Population and Communities: Kate O'Callaghan, Leah Schwartz, Rachel Sanderson, Stephanie Russo, Tessa Bird, Ruth Donovan, Robyne Leven, Tania Laity and Kirsty Douglas

Editors: Alex Cook, John Magee, Karen Roberts, Kirsty Douglas, Kate O'Callaghan and Rachel Sanderson

Cover image: Peter Schouten. Images in this catalogue by acclaimed palaeoartist Peter Schouten are inspired by Henry Thomas de la Beche's *Duria Antiquior or a more ancient Dorset* (1830), an early and very famous dioramic representation of ancient life, a Jurassic scene based on fossils found on the Dorset coast. De la Beche's images helped to establish a tradition of pictorial representation of scenes from the deep past ably realised by Schouten in his series of evocative dioramas.

Set in Adobe Century Schoolbook 9/12 and Century Gothic
Edited by Elaine Cochrane
Cover and text design by Andrew Weatherill
Typeset by Andrew Weatherill
Index by Russell Brooks
Printed by Ingram Lightning Source

Mar26_RP_IPS

Foreword

The Australian Heritage Council was formed in 2004 as an independent expert body to advise the Minister for the Environment on the listing, protection, promotion and management of heritage places.

As Chair of the Australian Heritage Council I have the privilege of overseeing Australia's most prestigious heritage list, the National Heritage List. The National Heritage List captures critical moments in our development as a nation. It reflects the lives and times of many exceptional Australians; it also tells the stories of those places that reveal the richness of Australia's extraordinarily diverse natural heritage. The National Heritage List was established to list places of outstanding heritage significance to Australia and includes natural, historic and Indigenous places.

I am honoured to introduce the publication of this inventory of Australia's fossil heritage, which will make a valuable addition to our understanding of Australia's natural heritage.

Australia has a wealth of fossil sites that includes some of the oldest fossils in the world. The information we learn from fossil sites helps us to piece together a national story about how Australia's plants, animals and environments have transformed over many millions of years. Australia's fossil history contributes to an understanding of past and present biodiversity and situates Australian biota and ecosystems in a global perspective.

This book makes a valuable contribution to cataloguing many of our most important fossil sites, and serves as a useful resource to draw from in determining those fossil sites that are of outstanding significance to Australia. What do I mean by 'outstanding'? I mean sites that have fundamentally shifted our understanding of the development of Australia's flora and fauna, or sites that illustrate incredible diversity of species, or act as a catalogue of different species over a great length of time, or sites that house exquisitely preserved specimens.

For example, the Ediacara fossil sites in South Australia record the first discovery of the fossilised remains of an entire community of soft-bodied creatures in such abundance anywhere in the world. Before their discoverer Reg Sprigg unearthed and interpreted the sites in 1946, scientists believed that only organisms with hard parts, such as shells or skeletons, could be preserved in the fossil record. As the oldest known evidence for complex multicellular organisms, these sites in the Ediacara Hills gave scientists a new understanding of the evolution of life on Earth, as well as a better understanding of how soft tissue can be preserved in the fossil record. In 2004, the International Union of Geological Sciences ratified the Ediacaran Period, the first new geological period in 120 years, and the only one named for a southern hemisphere site. The Ediacara fossil site in South Australia was added to the National Heritage List in 2007.

The Yea flora fossil site in Victoria [the Silurian Baragwanathia site] is an example of how a seemingly insignificant fossil site can reveal an important national story. Although first discovered in 1875, the significance of the Yea flora fossils was not realised until 60 years later, when they were studied by pioneering botanist Dr Isabel Cookson in 1935. Dr Cookson identified the remains as ancient vascular land plants, about 415 million years old and the oldest of their kind in the world. Her findings were internationally significant as they suggested that not only did complex land plants develop much earlier than previously thought, but that they also first evolved in the southern hemisphere. The Yea flora fossil site was added to the National Heritage List in 2007.

One of the ways the Council determines outstanding significance is to compare and contrast sites with similar attributes in order to show which select few sites best illustrate an understanding of the evolution of our flora and fauna. This book lists a relatively large number of sites. Some of the most outstanding sites are already included in the National Heritage List, and some more may be included in the future. I hope that this publication will serve as a useful resource for students and scientists to understand our important fossil heritage, and more broadly increase our understanding and appreciation of Australia's National Heritage List.

Professor Carmen Lawrence
Chair, Australian Heritage Council

Contents

Foreword v
Acknowledgements ix
Introduction xi

1. **New South Wales** **1**
Fennel Bay (Kurrur-Kurran) fossil forest 1
Lake George 3
Lightning Ridge 4
Belmont insect beds 5
Canowindra fossil fish site 8
Cuddie Springs 9
Taemas–Cavan–Wee Jasper 12
Cliefden Caves – Belubula River Valley 13
Talbragar fossil beds 14
Wellington Caves 15

2. **Northern Territory** **19**
Alcoota vertebrate fossil beds 19
Bauhinia Downs flora, Balbarini 21
Bitter Springs Formation, Ross River Homestead 21
Bullock Creek vertebrate fossil site 23
Ellery Creek Cambrian–Ordovician trace fossil locality 24
Cape Van Diemen flora 25
Ellery Creek Section 25
Mt Skinner fauna 28
Mt Watt and Mt Charlotte 29

3. **Queensland** **31**
Walsh River and Elizabeth Creek 31
Fletcherview – Big Bend 'Leichhardt Cliffs' Devonian reef complex 32
Mount Morgan Jurassic dinosaur footprint site 34
Rewan, The Crater 34
Black Mountain Cambrian–Ordovician Boundary Section 35
Chinchilla Rifle Range 36
Chatsworth Limestone echinoderm and trilobite localities 37
Redbank Plains Formation 38
Mount Crosby Triassic insect locality 38
Jack Hills Gorge and Fish Hill, Broken River Province 40
Riversleigh 41
Homevale Station 44
Dinmore Quarry 46
Lark Quarry Dinosaur Stampede National Monument (Winton Formation) 46
Late Cretaceous (Cenomanian) Winton Formation 50

4. **South Australia** **51**
Ediacara sites, Nilpena, Flinders Ranges 51
Emu Bay (Cape d'Estaing to Big Gully), Kangaroo Island 55
Myponga Beach, Fleurieu Peninsula 56
Ajax Mine archaeocyathid locality 57
Stuart, Poole and Nelly Creek fossil plant sites 58
Moon Plain Cretaceous fossil site, Coober Pedy 60

Lake Palankarinna, Lake Eyre and Tirari Desert 61
Green Waterhole Cave 63
Lake Frome Basin, Namba Formation 65
Lake Ngapakaldi Leaf Locality 65
Lake Callabonna 67
Naracoorte Caves 69

5. **Tasmania** **71**
Christmas Hills 71
Darwin Meteorite Crater 72
Ordovician faunal succession of the Florentine Valley 72
Fossil Cliffs, Maria Island 74
Early Oligocene macroflora sites of north-western Tasmania 77
Lune River Jurassic plant site 78
Eastern shore of Macquarie Harbour 79
The Shoreline near Point Hibbs, Western Tasmania 81
Table Cape – Fossil Bluff, Wynyard 82
Triassic vertebrate faunas, Knocklofty Formation 84

6. **Victoria** **89**
Mt Howitt fish fossil site 89
Genoa River Devonian tetrapod site 91
Otway and Strzelecki Ranges 93
Hines Quarry Diprotodon site, Bacchus Marsh 96
Grange Burn Pliocene mammal site, Hamilton 97
Victorian Ordovician graptolite province 98
Silurian Baragwanathia site, Yea 100
Torquay Oligocene whale and invertebrate site 101
Hamilton Miocene–Pliocene invertebrate sites 105
Early Cretaceous (Aptian) Koonwarra site 106
Middle Eocene Anglesea site, Eastern View Formation 107
Latrobe Valley coal measures 109

7. **Western Australia** **111**
Devonian 'Great Barrier Reef' and Gogo fish fossil site, Fitzroy Crossing 111
Devil's Lair, Cape Leeuwin–Cape Naturaliste region 116
Tumblagooda Sandstone, Kalbarri and Murchison River Gorge trace fossils 117
West Dale, Darling Plateau 118
Roe Calcarenite invertebrate fauna, Roe Plains 120
Callytharra Formation, Callytharra Springs (Permian invertebrates and foraminifera) 121
Broome Sandstone dinosaur trackways and macroflora, Broome 121
Miria Marl, Giralia Anticline 122
Warrawoona Group (North Pole and Strelley Pool sites), Proterozoic early life fauna 123
Cundlego Formation Permian invertebrate fauna, Gascoyne Junction 124

Appendix A: List of sites by type 125
Appendix B: Geological timeline 127
Glossary 129
References and further reading 151
Index 183

This catalogue is compiled from reports by a number of contributors from each state and the Northern Territory. Thankyou to Yong-Yi Zhen (New South Wales/ACT chapter), Peter Murray and Dirk Megirian (Northern Territory chapter), Alex Cook (Queensland chapter), Dennis Rice, Jim Gehling (with Liz Reed, who contributed the Naracoorte entry) (South Australia chapter), Clive Calver, Max Banks, Patrick Bender, Jim Jago, Greg Jordan, Patrick Quilty and Andrew Rozefelds (Tasmania chapter), John Long, Tom Rich, David Holloway, Erich Fitzgerald (Victoria chapter), and Mikael Siversson (Western Australia chapter), and to Raymond J. Carpenter and Robert S. Hill whose reports on significant plant fossil sites were used throughout this publication. The state and territory museums were particularly helpful in providing the assistance and expertise of their palaeontologists, which was critical to the success of this project.

Thankyou to Alex Cook, Kirsty Douglas, John Magee, Karen Roberts, Kate O'Callaghan and Rachel Sanderson who played a vital role in transforming a series of diverse reports from numerous authors into a single manuscript. John Magee also prepared the glossary.

Publication of this book has been made possible by the financial support and personnel provided by the Australian Government Department of Sustainability, Environment, Population, Water and the Arts. Jane Ambrose, Terry Bailey, Tessa Bird, Jennifer Carter, Ruth Donovan, Kirsty Douglas, Kate O'Callaghan, Stephanie Russo, Rachel Sanderson and Leah Schwartz provided expertise and management that enabled the initial reports to be commissioned and subsequently transformed into a book. Tania Laity provided invaluable creative and technical assistance with locality maps. The project was first envisaged by Cameron Slatyer, and Robyne Leven's excellent work on the Lark Quarry National Heritage assessment in 2004 underpins that entry.

Peter Schouten, whose artworks feature throughout the publication, created beautiful paintings that bring to life the stories told in the pages of this book. Peter understood the need to create paintings that would invoke the imagination of the audience, drawing them into the stories and places of our past.

The following individuals and organisations are gratefully acknowledged for granting permission to reproduce images, as well as for providing images: The National Library of Australia; Australian Museum; Richard Barwick and The Age of Fishes Museum; Sarine Locke and ABC Rural; Gavin Young; Anne Musser; Steve Bourne; the Centre for Learning Innovation, New South Wales Department of Education and Training; Gavin Dally and the Museum and Art Gallery of the Northern Territory; the Museum Board of South Australia; Catherine Rogers; John A. Long; and Museum Victoria.

A particular thanks to all of the academic and amateur palaeontologists and fossil hunters whose vocation and enthusiasm has led to the uncovering of Australia's rich fossil heritage.

The National Heritage List was created in January 2004 to recognise places of outstanding natural or cultural heritage value to the nation. National heritage encompasses those places that reveal the richness of Australia's extraordinarily diverse natural heritage. Australia's wealth of exceptional fossil sites is one aspect of this natural heritage that has as yet been little explored. While a small number of fossil sites have risen to public prominence, there are many lesser-known sites that may be of outstanding heritage value.

The Australian Heritage Council is establishing a comparative framework to aid in the assessment of Australian fossil heritage sites for nomination to the National Heritage List. As part of this process, in 2006 palaeontologists were engaged to compile lists of outstanding fossil sites and to document their characteristics and importance.

Palaeontologists from each state museum and the Northern Territory museum selected ten fossil sites. Each representative also chose three exceptional fossil sites from anywhere in Australia for comparative analysis. Reports collect data for each site including:

- scientific importance and research potential
- fossil significance
- taphonomy and site condition
- comparison with other sites
- stories and dioramic potential
- key references.

Contributors took different approaches to the guidelines, which has led to some variation in the detail or emphasis provided by each state, and between different sites. The reports have been edited for clarity, and compiled in this publication to present a valuable catalogue of fossil sites with heritage importance in Australia.

The evaluation of each state's fossil sites was based on the evidence available at the time the reports were prepared. The discussions thus reflect current knowledge in 2006–07, and are not a definitive statement of comparative value. Several contributors observed that their selection of sites reflects both sample bias and their areas of expertise. It is also important to distinguish between heritage value and scientific value – for instance, some sites that have little potential for future research may nonetheless have significant heritage value, due to their former role in furthering understanding of the evolutionary history of Australian biota. Other sites have risen to prominence during the gestation of the catalogue, like the Nullarbor Caves mammal localities, and do not appear in detail despite their growing importance.

Technology and interpretations have also changed: as a prominent example, in 2010, the late Dirk Megirian and colleagues Gavin Prideaux, Peter Murray and Neil Smit published an Australian Land Mammal Age schema (ALMA), the first formal division of Australian Cenozoic time based on the mammalian fauna (Megirian *et al.* 2010).

It is hoped that this document will provide a useful national reference to the wonderful fossil sites it catalogues, and give some understanding of how the heritage values of such sites might be assessed.

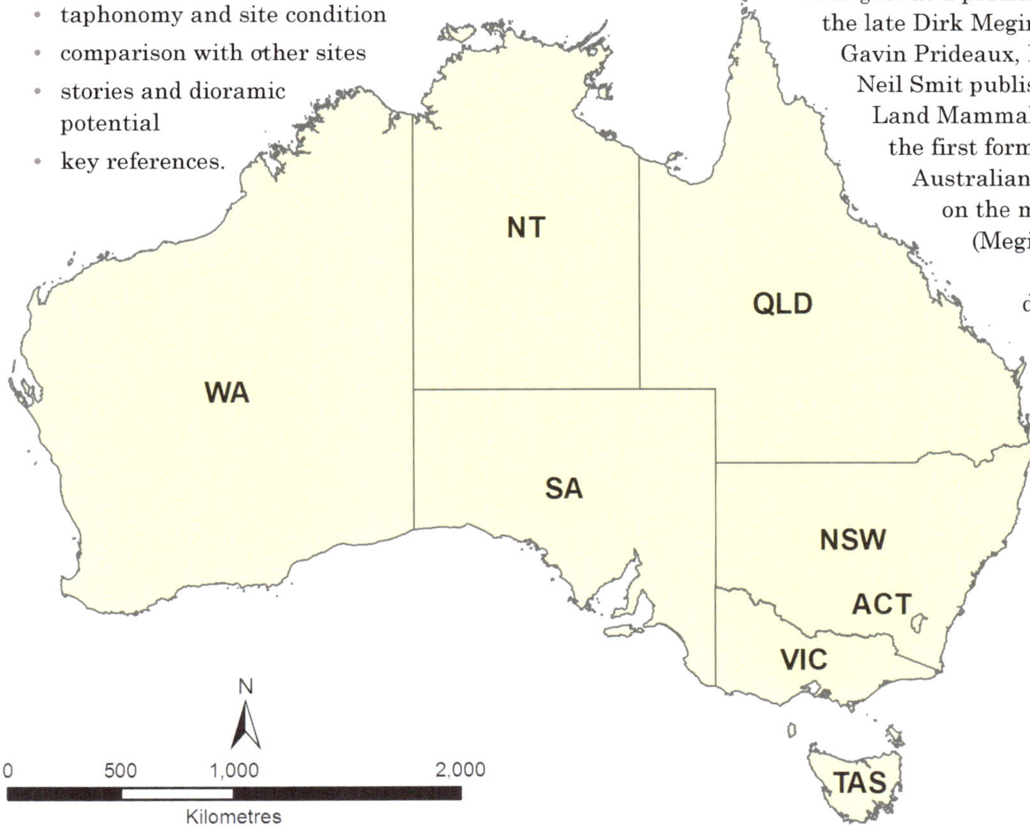

Australia © SEWPaC

New South Wales

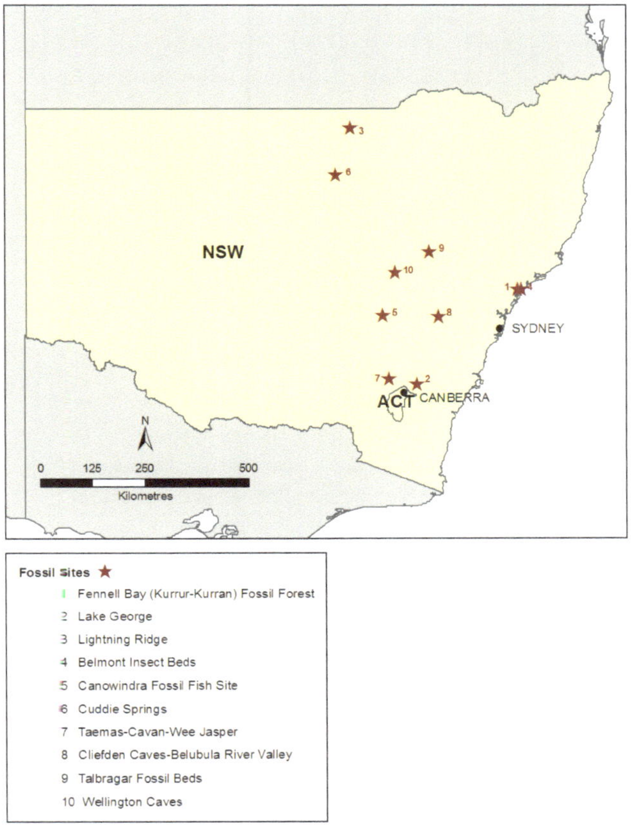

New South Wales locality map © SEWPaC

Fennell Bay (Kurrur-Kurran) fossil forest

Overview

Fennell Bay (Kurrur-Kurran) is a late Permian fossil forest site located in Lake Macquarie. It is approximately 6 hectares in area and extends approximately 150 metres into the lake.

Scientific importance and research potential

The Fennell Bay (Kurrur-Kurran) fossil site is the largest and best preserved Permian *in situ* fossil forest known in New South Wales. It was the subject of early scientific work by the Reverend William Branwhite Clarke and became the first site of geological significance to be accorded official recognition and protection in that state when it was gazetted on 17 September 1904. The site has been well documented, and is known for the quality of fossil preservation. No recent research has been undertaken and it is considered to have low research potential.

Fossil significance

Petrified tree stumps of about 0.3–0.5 metres in diameter (probably species of glossopterids, an extinct group of seed ferns) are preserved in apparent growth position, which is a rare phenomenon. They have been

Figure 1.1 Fossilised tree trunk at Fennell Bay. Photograph John Houldsworth © SEWPaC

converted into chalcedony, frequently with excellent preservation of growth rings and wood structure. Clarke estimated that there were at least 500 trees on this site (Clarke 1885, article signed and dated 1842). However, as a result of souveniring of specimens by collectors, the number of petrified trees has declined significantly over the years. Percival (1985) reported approximately 30–40 trees left.

Glossopterids are a diverse group of plants that thrived throughout the cool, temperate forests of the southern hemisphere during the Permian period (299–251 million years ago). Glossopterids are named for their large, tongue- or strap-shaped leaves and are often interpreted as deciduous. It has been suggested that they once dominated the vegetation on the southern landmass called Gondwana in the same way that eucalypts dominate the Australian vegetation today. They grew into woody trees and probably also shrubs of all sizes, suiting them to a range of different habitats and niches (White 1994). As glossopterids lived in wet and swampy areas like the Fennell Bay (Kurrur-Kurran) site, their roots had specialised internal structures to cope with the conditions.

Taphonomy and condition of the site/fossils

The chert-like sediments in which the fossil forest is embedded are referred to as the Eleebana Formation of the Moon Island Beach Subgroup, which forms the uppermost part of the Newcastle Coal Measures (Late Permian). Stratigraphically, the site lies 9–15 metres below the Great Northern Seam (of the Newcastle Coal Measures) and about 30 metres above the Upper Pilot Seam. The area is subject to slight tidal fluctuation and the rocky lake bottom from which the trees protrude is largely covered by a thick deposit of fine silt, supporting a seaweed carpet. Some small mangroves have become established in the lake adjacent to the muddy shore.

Comparison with similar sites in Australia

The Cooyal locality near Mudgee (NSW) may host the best Permian impression flora in Australia, and the Homevale Station site (Qld), now within Homevale National Park, has exceptional fossil preservation. These sites both show excellent two-dimensional preservation. While glossopterid gymnosperms dominate most Australian Permian coal-bearing

Figure 1.2 Lake George as painted by Louis Frank © National Library of Australia

sediments, they rarely show the three-dimensional *in situ* preservation and demonstration of growth direction of the Fennell Bay forest.

Stories

The most important story associated with this site is that of the early discovery and documentation by the Reverend W.B. Clarke (Branagan and Valance 2008; Clarke 1885) and T.W. Edgeworth David (David 1907), two of the most distinguished Australian pioneer geologists. It is also linked to stories of fossil theft and vandalism as demonstrated by the severe loss of fossil trees since the first documentation by Clarke.

Lake George

Overview

Lake George is a Pleistocene and Holocene microfossil site. It is about 2900 hectares in extent, approximately 2 kilometres south of Collector and 40 kilometres north-east of Canberra.

Figure 1.3 Field trip at the Lake George quaternary site © SEWPaC

Scientific importance and research potential

The site has been well documented and is regarded internationally as an important scientific site. It has been the subject of solid scientific work (e.g. Coventry 1976; Singh *et al.* 1981a, 1981b; Singh and Geissler 1985) and continues to have research potential.

Fossil significance

The site contains the longest relatively continuous Pleistocene and Holocene continental sequence of pollen, spore, algal and charcoal particles known in Australia. These fossil records provide crucial information on vegetation changes (particularly plant migrations and extinctions) and climate changes during glacial–interglacial cycles. This information is extremely important as it enables reconstruction of the Pleistocene–Holocene biogeographic, environmental and climate history of the wider region.

Taphonomy and condition of the site/fossils

Palaeomagnetic data indicates that the lake was already in existence during the Middle Miocene (15–11 Ma). It is a closed lake that owes its origin to the disruption of a pre-existing drainage system by post-Palaeozoic (less than 250 Ma) warping or faulting. Sedimentary records of Lake George go back 7–4 Ma. These sediments consist of clays and sandy clays which are weakly calcareous and low in both organic and inorganic carbon content.

Comparison with similar sites in Australia

Palaeoenvironmental sequences that describe the vegetation of the last glacial maximum (LGM) and of the subsequent climatic amelioration are relatively rare in Australian region (Black *et al.* 2006). The Lake George microfossil site is one of the oldest continuous vegetation and fire history records in Australia, and is also significant globally. The site has provided:

- vegetation history and a sequence of lake level changes, which have been studied comparatively against known global climatic changes
- microfossil evidence that enables the reconstruction of the climatic history of the Lake George area over the last 350 000 years
- charcoal records that document a continuous fire history (including evidence of early human activity and occupation) for the Lake George area.

Cores from a number of sites in the Atherton Tableland (Qld) also preserve important Late Quaternary pollen records, most notably Lynch's Crater, which has recently revealed a complete pollen record of the last 230 000 years (Kershaw *et al.* 2007). However, the Lake George pollen record of more than 350 000 years is accompanied by palaeomagnetic data and sedimentary data that extend the regional geological record back to the Middle Miocene.

Stories

The site is linked to the story of how the environment evolved in south-eastern Australia during the Quaternary Period (2.6 Ma to the present). Together with marine cores and records from other sites in Australia, it is critical to articulating climate and biological changes across the continent immediately before and during the period of human occupation. The pioneering work of Gurdip Singh and colleagues in the now defunct Department of Biogeography and Geomorphology at the Australian National University in interpreting the Lake George pollen and charcoal records is an important entry into the history of Australian Quaternary studies.

Lightning Ridge

Overview

Lightning Ridge is an Early Cretaceous, opalised fossil site. It is located along Castlereagh Highway, Walgett Shire, in north-western New South Wales, and is renowned throughout the world for its deposits of precious black opal.

Scientific importance and research potential

Lightning Ridge is a fascinating fossil site, not only because the fossils are preserved as opals. It preserves a diverse assemblage of vertebrate, invertebrate and plant fossils all together, which is very rare. Some of the most important discoveries at Lightning Ridge include the early mammals *Steropodon galmani* (Archer *et al*. 1985) and *Kollikodon ritchei* (Flannery *et al*. 1995). These two species have only been found at Lightning Ridge and are extremely old relatives of the living monotremes (e.g. platypus). There has been solid scientific work done on the site, and some fossils are published (see Smith and Smith 1999). However, most are yet to be documented and there remains a high potential for further discoveries and research.

Fossil significance

Lightning Ridge is an important site to study and reconstruct the palaeoecology, palaeoenvironment, and palaeoclimate of this region during the Early Cretaceous (approximately 110 Ma). It has produced a wide range of fossil groups including plants (e.g. tree trunks, plant stems, pine cones and seed pods), microfossils, invertebrates (mostly freshwater molluscs) and vertebrates (bones, claws and teeth), as well as trace fossils (dinosaur footprints and worm burrows). Vertebrates represented at Lightning Ridge include dinosaurs (e.g. small sauropods, theropods and ornithopods), pterosaurs, plesiosaurs, turtles, crocodiles, fishes, lungfish, birds and mammals. The mammals found at Lightning Ridge are very early relatives of the living monotremes (e.g. platypus) and are very important for what they reveal about mammal evolution. The preservation, regularly precious opal, of the Lightning Ridge fossils adds to their significance, rarity and popular appeal (Smith and Smith 1999).

Taphonomy and condition of the site/fossils

The fossil-bearing sediments (Finch claystone) are generally between 0.25–2 metres in thickness, but can reach up to 5 metres thickness. The Finch claystone occurs within the Wallangulla Sandstone Member of the Griman Creek Formation and has been dated using pollen data to approximately 110 Ma. The Formation crops out or lies at shallow depths over thousands of square kilometres in north-western New South Wales, southern Queensland and central Australia. The sandstone and mudstone was once sediment deposited in a large ancient delta where plants, aquatic life and occasionally the bones and teeth of animals were preserved.

Commonly the bones are found isolated and scattered, but in rare cases semi-articulated skeletons are found. Fossils are generally preserved as non-precious opal, called potch, and when recovered are fragmentary and broken because they have been damaged by mining machinery. Specimens recovered direct from opal-bearing sediments in mines are in better condition, but are often isolated and the process is much slower. Most specimens are a combination of pseudomorph and replacement. The outstanding quality of preservation is partly because the sediment is extremely fine-grained and is an ideal natural casting medium.

Comparison with similar sites in Australia

Lightning Ridge was the first site of discovery of early mammals in Australia. Cretaceous mammal fossils are extremely rare in Australia and the only other sites that have produced them are in Victoria (Otway and Strzelecki Ranges). Compared to other opalised fossil sites in Australia (e.g. Coober Pedy in South Australia), Lightning Ridge is the only opal field in Australia that has preserved terrestrial organisms, and at such a high diversity.

Stories

Lightning Ridge has strong links to the story of the early evolution of mammals, as it is the site of discovery of extremely rare early mammal fossils that have not been found anywhere else. It has significant dioramic potential, particularly as it provides evidence of the coexistence of small monotremes with dinosaurs, small crocodiles and turtles.

The human history of the site as a mining district is also a source of many important stories. Opal was found at Wallangulla (later named Lightning Ridge) in late 1900 by Jack Murray, who became the first to mine the area in the following year. The first mineshaft was sunk around 1903 and shortly after came the first opal rush (1905). The first Council of Walgett Shire formed in 1906. By 1909, about 1200 people, mostly single men, lived on the Three Mile Field (an independent community to Old Town), and the population on the Wallangulla Opal fields became larger than Walgett. In the same year, a local newspaper reported the finding of fossils from opal fields, and an opalised fossil lungfish was documented in 1914. In the mid 1980s, the town of Lightning Ridge began to boom as new prospecting and extraction techniques were applied. Lightning Ridge is now the only significant source of black opal in the world.

Figure 1.4 Opal mining at Lightning Ridge © National Library of Australia

Opal mining continues today, and the town has a population of 6000 people, with over 5000 separate mining claims in the area. Opal mining is the town's primary industry, and the region produces the world's finest black opals. A visitor to the area can experience the thrill of mining opals, making it a tourist attraction for opal mining and opal mining history.

The site is associated with Robert Etheridge Junior, a distinguished Australian palaeontologist, who was the first Curator in Palaeontology and later became the Director of the Australian Museum. In 1917, he described a primitive crocodile from the Lightning Ridge area.

Belmont insect beds

Overview

The Belmont insect beds are a series of late Permian insect fossil sites. They occur at Lake Macquarie, near Newcastle, approximately 145 kilometres north of Sydney, New South Wales.

Scientific importance and research potential

The Belmont insect beds have been well documented, and are the only extensive Permian age fossil insect sites found in Australia. The subject of numerous publications since 1898, including ongoing research, they have great future research potential (e.g. Beattie 2007; Jones and Chen 2000; Mitchell 1925; Rasnitsyn and Aristov 2002; Riek 1953, 1968, 1970a, 1970b; Tillyard 1918, 1926).

Fossil significance

The main *Konservat-Lagerstätten* is one of the most prolific sites and has exceptionally high insect content, both in abundance and diversity. The Belmont insect beds are the type locality for over 126 insect species and two species of conchostracans, and are thus extremely important. One hundred and forty-five insect species (belonging to 97 genera) have been described, including about 60 Homoptera (cicadas, leafhoppers, aphids), about 15 Psocoptera (related to book lice), a number of species of Neuroptera (lacewings), one species of Plecoptera (stoneflies), one species of Glosselytrodea (an extinct order of winged insects related to grasshoppers and crickets), one species of Odonata (dragonflies), and a number of new species of Coleoptera (beetles). Belmont is currently the only location at which all six archostemate beetle families have been found.

These insect remains are associated with plentiful fossil conchostracans (branchiopod crustaceans that are related to water fleas), represented by two species. Fish scales are plentiful in some localities, and partial fish carcasses occur at the Warner's Bay site. Fossil fish are very rare in the Newcastle Coal Measures. Associated plant remains include *Glossopteris*, *Phyllotheca*, and *Noeggerathiopsis* with occasional *Annularia*, *Sphenopteris*, woody gymnosperm trees, ferns, and horsetails that constitute the flora of the coal seams in the Newcastle Coal Measures, and other southern hemisphere Permian coals.

The Belmont insect beds sites offer many possibilities for palaeoenvironmental reconstruction, and have strong potential for further research. It is a very interesting and puzzling question why thousands of insects were densely accumulated in this tuffaceous chert unit, when there is no infaunal evidence such as burrows, bioturbation or anything similar.

Taphonomy and condition of the site/fossils

The fossil insect horizon is relatively thin (the average is approximately 0.3 metres for most sites but can reach 2 metres). However, it is an extensive unit of Late Permian age (about 250 Ma), and consists of hard, fine-grained tuffaceous chert (grey to black when fresh). It lies approximately 20 metres below the bottom of the economically exploited Fassifern Coal Seam in the upper Newcastle Coal Measures. Outcrops of the fossil insect bed occur for about 3.2 kilometres along the Violet Town ridge. Its lateral extent has never been traced due to housing estates and industrial developments in the surrounding areas, but it is believed to extend at least 9.6 kilometres in a 1.6-kilometre-wide belt that tends in a north-west to south-east direction.

The fine grain size of the tuffaceous chert bed has facilitated the detailed preservation of even the venation in the many insect wings found at the site. Insect fossils are predominantly wings, but also include portions of body. Site to site, this insect bed is underlain by a 4.6–5.5-metre layer of varied facies that includes fine sandstone, shale, silty sandstone or coal. Abundant fossil wood, including sections of fossilised tree trunks, is found at Insect Reserve site. Underlying the sequence directly below the insect bed is a conglomerate that contains directional cross-bedding. The insect beds were probably deposited in ephemeral relict water bodies that were present beside the Belmont conglomerate braided channel system that drained south-westerly, close to a coastal plain. It appears that the Belmont fish remains were the carcasses of fish trapped in overbank ponds after flooding and draining of the main channels.

Comparison with similar sites in Australia

The Belmont insect beds are the oldest and one of the most prolific groups of insect fossil sites known in Australia. Comparable sites include one that has produced two insect wings from strata slightly older within the Newcastle Coal Measures (Tillyard 1918), and another in Tasmania that has produced a single specimen of Late Carboniferous age (Riek 1973, 1976). The Belmont sites are an extraordinarily rich part of Australia's palaeontological heritage, having produced the majority of known Permian insect species from Australia.

Stories

The first fossil insect from Belmont were discovered by Mr John Mitchell in 1898. Mitchell traced the source of the chert back to a small quarry near the John Darling Colliery site, on the east side of Sydney Road, about 2 kilometres north of Belmont. Specimens were then sent to the entomologist and geologist Robert John (Robin) Tillyard, who was awarded his PhD in 1917; from that year until his accidental death in January 1937, Tillyard published 12 papers on the Belmont insects.

In 1918, Tillyard made his first visit with John Mitchell to the Belmont insect beds. By 1922, Mitchell had collected 52 fossil insect specimens from Belmont. He sent more specimens to Tillyard in the early 1920s, and Tillyard then arranged another dig with Mitchell in 1923. About 50 specimens were obtained from this site between 1917 and 1923. In 1925, a new exposure of the insect seam was discovered at Warner's Bay, along the Esplanade road bank on the landward or eastern side.

John Mitchell died in 1928, and in 1930 the Australian Museum purchased his large fossil collection, which included a large number of Belmont and Warner's Bay insect fossils. In the years up to 1937, Tillyard made a number of fossil collecting trips to Belmont, collecting both at the John Darling site and at the Warner's Bay site. Tillyard arranged with the Shire Engineer to dig a load out of the road-cutting at Warner's Bay late in 1936.

T. Pincombe and M. Stanley continued to collect from 1937 and through the war years. In 1942, an engineer, Mr O.L. Knight, started collecting at Pincombe's Outcrop, Warner's Bay, and Belmont (Knight 1950). In the period 1942–44 he collected and donated 242 specimens to the Australian Museum. Stanley donated his entire collection to the Australian Museum in 1943, over 1000 specimens of insects, conchostraca and plant fossils.

Dr John William Evans, Director of the Australian Museum, and Mr Harold Fletcher, Curator of Palaeontology at the Australian Museum, were responsible for setting aside an area of 18.7 hectares (Public Reserve R.81914 for the Preservation of Fossils) at Tingira Heights. They made a number of trips together in the late 1950s and 1960s to collect fossil insect specimens at Belmont, and later were joined by Dr Alex Ritchie.

The Australian Museum collection in Sydney consists of approximately 1500 Belmont insect specimens. The British Museum of Natural History collection in London consists of approximately 1000 Belmont insect specimens.

The Belmont insect beds have important associations with several noteworthy scientists, including Mitchell, Tillyard, Evans, Consett Davis and Edgar Riek.

Figure 1.5 Fish fossils at Canowindra.
Photographer Mark Mohell © SEWPaC

Canowindra fossil fish site

Overview

The Canowindra fossil fish site is a Late Devonian (360 Ma) fish fossil site, located, as the name suggests, near Canowindra. The original roadside locality on Gooloogong Road (now renamed Fish Fossil Drive) was temporarily reburied in 1993 to protect this unique fossil layer from vandalism. The site is about 1 hectare in area and is 10 kilometres west of Canowindra.

Scientific importance and research potential

The Canowindra fossil fish site has been well documented since 1990, with solid scientific work undertaken by Johanson and others (e.g. Johanson 1997a, 1997b, 1997c, 1998; Johanson and Ahlberg 1997; Thomson 1973). It is known for its extraordinary abundance and high diversity of fossils. Although most of the Canowindra fauna has now been described, the site still has potential for future research. Much of the mass-kill deposit still lies buried at the original site on Fish Fossil Drive. Almost certainly many thousands of fish specimens remain buried at Canowindra, and these probably include other taxa new to science (see also Ritchie and Johanson 2006).

Fossil significance

The fossils are abundant at the main roadside site and most are complete, articulated specimens that are extremely well preserved. Eight fish species described from the Canowindra fish fauna are represented by over 3700 mostly complete fish specimens. The fauna is dominated by two species of antiarch placoderms (armoured fishes), *Bothriolepis* and *Remigolepis* (97% of the fauna). *Groenlandaspis*, an arthrodiran placoderm, is much rarer with only about 50 specimens recovered. The sarcopterygians (lobe-finned fishes) are even rarer, with the remains of approximately 30 individuals so far noted. Sarcopterygians are represented by the genera *Canowindra*, *Cabonnichthys*, *Gooloogongia* and *Mandageria*, all of which are only known from this site. The dipnoan genus *Soederberghia*, which is also present at Canowindra, is also known from East Greenland, Pennsylvania (USA) and from near Forbes (NSW). Further investigation is expected to improve the understanding of the palaeoecology of the Canowindra fauna and the processes involved in their burial and preservation as fossils.

Taphonomy and condition of the site/fossils

All of the fossil specimens from the main site are preserved on a single bedding plane of a sandstone layer called the Late Devonian Mandagery Sandstone. Canowindra is a unique site with a preserved fauna that represents part of an ancient fish community that had been trapped and killed in a catastrophic event (mass-kill). Incoming sediments later buried these fishes quickly, with minor disturbance to the fish skeletons. The fishes are well preserved and show little evidence of post-mortem disarticulation. They are crowded together on the rock surface and there is no sign that any decay, predation or scavenging occurred before burial. Only fishes have been found, with plants and invertebrates conspicuously absent. The rock directly overlying the fish fauna is a reddish, fine to medium grained quartz sandstone that lacks strong evidence of cross-bedding.

Comparison with similar sites in Australia

Canowindra is a spectacularly rich, unique, well preserved Late Devonian fossil fish assemblage and is one of the best worldwide. There are four other known Devonian fish sites in Australia that preserve articulated skeletons (Long 1996). One of these, Wee Jasper, is located in New South Wales; two, Mt Howitt and Buchan, are located in Victoria; and the last, the Gogo fish fossil sites of the Lennard Shelf, are found in the Kimberley region of Western Australia. Preservation at these sites varies, with Canowindra and Mt Howitt preserving fish in two dimensions, while Buchan, Wee Jasper and Gogo preserve fish in three dimensions.

Two-dimensional preservation at Mt Howitt consists mainly of impressions of the bones left behind in the mudstone – where the bone is preserved it is usually badly weathered. Negative preparations are used to study the Mt Howitt fish fauna by dissolving the weathered bone and making latex casts of the

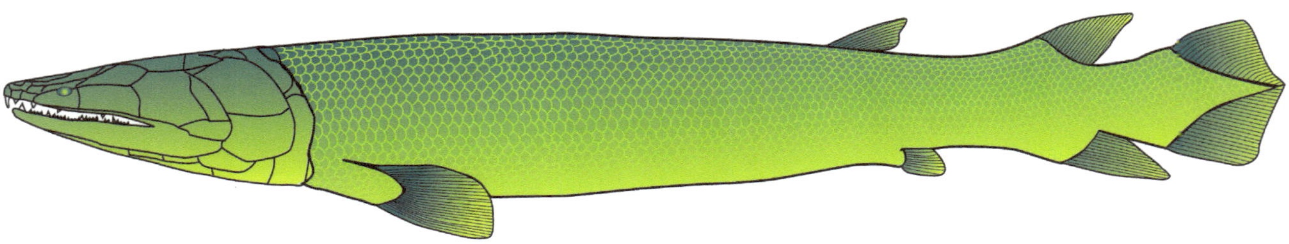

Figure 1.6 Reconstruction of *Mandageria*, the largest of the Devonian fishes from the fish fossils at Canowindra. Illustrated by Richard Barwick © Richard Barwick and The Age of Fishes Museum, Canowindra

impressions. Canowindra's preservation in contrast is mostly natural moulds filled with hard sandstone casts, some of which preserve internal structure of braincases and gills. There are at least eight species of fish found in Canowindra, but the fauna is overwhelmingly dominated by just two species. Fine detail is retained in many of the Canowindra specimens. The Canowindra specimens are unique in Australia for the evidence they preserve of mass mortality events caused by cyclical, drought-like conditions in the Devonian.

In contrast to the freshwater settings of Canowindra and Mt Howitt, the sites at Wee Jasper, Buchan and Gogo represent Devonian coral reefs. At all three of these sites the fish fauna is preserved in limestone rather than shale, which results in a different type of preservation. Buchan's fossil fish are rare and, as the most fossiliferous part of the facies has been mined for many years, the deposit may be all but removed. Wee Jasper, by contrast, has nearly as diverse a fish fauna as Gogo and apparently similar preservation (Long 1996). However, the evidence of the behaviour and evolution of freshwater fish provided at Mt Howitt and Canowindra is absent from these sites.

Stories

Canowindra is an important site for education, with stories of its unique taphonomy, scientific significance and the history of discovery. The site can be linked to the story of the early evolution of fishes, and has dioramic potential as the location of a mass mortality event featuring a Devonian billabong and drought.

Its accidental discovery in 1956, uncovered during roadworks, is also historically significant. A large rock slab covered with strange impressions was unearthed and moved to the side of the road. A local beekeeper, William Simpson, identified the impressions as fossils, realised their importance, and notified Harold Fletcher, then the Australian Museum's Curator of Palaeontology. The slab was saved, transported to Sydney, and placed on permanent public display in the Australian Museum.

Dr Alex Ritchie, Fletcher's successor as Curator of Palaeontology at the Australian Museum, visited Canowindra several times between 1973 and 1990 trying to relocate the original site. The search intensified after 1980 when Ritchie discovered two small specimens of the genus *Groenlandaspis*, a Late Devonian placoderm, while cleaning and re-examining the original 1956 Canowindra slab. This was a fish he had earlier discovered, and reconstructed, from Devonian rocks in Antarctica.

In 1993, with the support of the Cabonne Shire Council and the local community, Ritchie relocated and later excavated the Canowindra fossil site. From 1993 to 1999, volunteers helped Australian Museum scientists to clean and cast the fossils. Over 200 fish-covered slabs with a total of approximately 70 tonnes (individual slabs may weigh over 2 tonnes) that were recovered in the 1993 excavation are stored at Canowindra showground. A selection of the most important slabs and fossil specimens are on display at the Age of Fishes Museum in Canowindra, which was built by the local council and the community as a tourist and educational centre in central west New South Wales.

Cuddie Springs

Overview

Cuddie Springs is a Pleistocene megafauna and archaeological site. It is about 2 hectares in area and is 35 kilometres west-north-west of Carinda, in central north New South Wales.

Scientific importance and research potential

Cuddie Springs has been well documented, and continues to be the subject of active research and excavation by a team involving researchers from University of Sydney, University of New South Wales, Australian Museum, University of Wollongong, and other institutions in Australia and overseas (e.g. Field *et al.* 2002; Field and Fullagar 2001; Trueman *et al.* 2005). It is one of the most important and contentious sites in New South Wales for its combination of vertebrate palaeontology and archaeology, and it has a high potential for future research.

Fossil significance

Over 20 species of megafauna have been recognised from this site, including two species of *Diprotodon*, four species of an extinct short-faced kangaroo (*Sthenurus*), 12 species of a range of herbivorous marsupial genera, seven species of reptiles and five species of birds, including a giant flightless bird (*Genyornis newtoni*). The site is unusual and contentious because of the apparent association of fossils and archaeological objects, including stone tools and stone blades. Cuddie Springs is one of the best sites in the semi-arid zone of Australia for palaeoenvironmental and climatic reconstruction of the Late Pleistocene through multidisciplinary research.

Taphonomy and condition of the site/fossils

The fossils of extinct Pleistocene megafauna are preserved at Cuddie Springs in apparent association with stone tools, charcoal, and evidence of other human activities. They are found within a shallow clay pan about 160 metres in diameter in lacustrine clay

Figure 1.7 Reconstuction of megafauna assemblage at Cuddie Springs. Image Peter Schouten © SEWPaC

sediments deposited on the middle of an ephemeral lake floor. The site is situated in flat, open woodland surrounded by very low relief Quaternary alluvial and aeolian deposits. It extends about 2 kilometres east–west and about 3 kilometres north–south.

Comparison with similar sites in Australia

Cuddie Springs is one of a number of important Australian Pleistocene fossil sites including (among others) Lancefield Swamp (Vic), Lake Callabonna and Naracoorte Caves (SA), the Darling Downs (Qld), the Wellington Caves (NSW) and Devil's Lair (WA), which have played significant roles in more than a century of debate about the cause or causes of the extinction of the Australian Pleistocene megafauna.

The Cuddie Springs deposits are important to this debate for the evidence they provide of an association of extinct megafauna with humans. Stratified deposits (36 000–28 000 ka) provide an indication of environmental and climatic change leading up to the last glacial maximum, and may shed light on the causes and timing of the extinction of the Pleistocene megafauna. Although it may contain fewer individual animals than a site like Lake Callabonna, and span a shorter time period than the extensive Wellington and Naracoorte cave deposits, demonstrating less species diversity, none of the other iconic early megafauna sites include Cuddie's controversial association with the presence of humanity.

Stories

Fossils were first discovered at Cuddie Springs in 1876, when Mr Yeomans, the property owner, sank a well in the centre of the claypan. Some of Yeomans' collection was sent to London and studied by the distinguished palaeontologist Sir Richard Owen. In 1914 W.W. Froggatt also displayed some specimens of this early collection of bones before the Linnean Society of New South Wales. The Australian Museum made an excavation in 1933. Important discoveries, including significant archaeology, have been made more recently.

Cuddie Springs is an important Indigenous site, with evidence detailing human activities extending back to 30 000 years BP. It is also the subject of a dreamtime story (Langloh Parker 1905) and thus has ongoing cultural significance for Aboriginal people.

Taemas–Cavan–Wee Jasper

Overview

Taemas–Cavan–Wee Jasper (Burrinjuck Reservoir) is an Early Devonian fossil invertebrate and fish site on the Goodradigbee River in southern New South Wales. The site includes two areas about 13 kilometres apart.

Scientific importance and research potential

Taemas–Cavan–Wee Jasper is regarded as one of the best Early Devonian successions in Australia. There has been solid taxonomic work on brachiopods, trilobites, conodonts, rugose corals and fishes from this site (e.g. Campbell and Barwick 2000; Chatterton 1971, 1973; Edgecombe 1998; Gorter and Young 1981; Jackson *et al.* 1970; Lindley 2001, 2002; Young 1984). Many invertebrate groups show high potential for future research, particularly stromatoporoids, tabulate corals, bryozoans, and algae.

The exquisite three-dimensional preservation of extinct placoderms, including internal braincase structures and eyes, allows insight into the nature of ancestral gnathostomes (jawed vertebrates) and the relationship of placoderms to the two major extant groups of jawed fish – osteichthyans and chondrichthyans (Young 2010).

Fossil significance

Taemas–Cavan–Wee Jasper is one of the most important sites for the reconstruction of the depositional environment of the carbonate shelves that developed along eastern Australia in the Early Devonian (416–398 Ma). The area is extremely diverse and abundant in fossil fauna, particularly brachiopods, tabulate corals and stromatoporoids. Over 150 species are represented, including more than 46 brachiopod species, 25 conodont species, over 25 rugose coral species, 16 trilobite species, and over 12 fish species, including the first discovery of the braincase of a fossil lungfish, and fossil placoderm eyeballs.

Taphonomy and condition of the site/fossils

The site is extensive Early Devonian limestone with small scale reefs (bioherms and patch reefs) and stratiform faunal accumulations (biostromes) distributed in the Murrumbidgee Valley. The organisms were deposited in a tropical marine shelf setting. The Murrumbidgee Group, which overlies the Sugarloaf Creek Formation, consists of three units with a total maximum thickness of about 840 metres. It comprises two fossiliferous limestone units

Figure 1.8
Gastropod fossil from Wee Jasper. Photograph Sarina Locke
© Sarina Locke and the Australian Broadcasting Commission

(Cavan Limestone at the base and Taemas Limestone at the top) separated by a siliceous clastic unit (Majurgong Formation). The specimens are well exposed in outcrop.

Comparison with similar sites in Australia

This is one of the most celebrated geological terrains of the continent, and the best Early Devonian marine shelf succession in Australia. It is the type locality for more than 70 species, including more than 34 species (or subspecies) of brachiopods, 12 species of trilobites, 12 rugose coral species, over 12 fish species, and one species of myriapod. The site contains diverse and abundant invertebrate and vertebrate faunas. It is of similar age and fauna to the Jack Hill Gorge and Fish Hill sites (Qld).

Like the Gogo fish fossil sites of the Kimberley (WA), Taemas–Cavan–Wee Jasper preserves beautiful three-dimensional fish fossils. It includes the oldest detailed evidence for the vertebrate brain and eye, and preserves unique cartilaginous brain cases (Young 2008, 2010). However, it is significantly older than the Late Devonian Gogo fossils and the Kimberley Devonian barrier reef system of the Lennard Shelf. Although considerably less extensive than the Kimberley Devonian reefs, Taemas–Cavan–Wee Jasper nonetheless is one of the best exposures of a Devonian tropical reef in the world, and preserves the world's oldest known coral fish assemblage.

Stories

Taemas–Cavan–Wee Jasper has been important to the work of influential palaeontologists of the nineteenth and twentieth centuries. Limestones were first noted in 1824. Sir Thomas Mitchell (1838) collected coral and stromatoporoid fossils from the area and suggested a Silurian age for the succession. The Reverend William Branwhite Clarke, Anglican clergyman and pioneering Australian geologist, was Curator of the Australian Museum from 1841 to 1843. Clarke visited and collected from the Taemas–Cavan–Wee Jasper area around 1844. He sent 71 fossils to Cambridge University for examination and these were later described by Professor L.G. de Koninck in the *Mémoires de la Société Royale des Sciences de Liège* in 1876, as reported in the *Geological Magazine* in 1877. Since then the diverse Devonian faunas have been studied by numerous palaeontologists and research students, including Robert Etheridge Jr., Dorothy Hill, Brian Chatterton, Alan Pedder, Ken Campbell and Gavin Young. Professor T.W. Edgeworth David's earliest paper was on the Taemas-Cavan – Wee Jasper karst.

Wee Jasper fossils were used in the 1940s to develop the acetic acid preparation technique that is now standard practice for extracting vertebrate fossils from calcareous rock matrix (G. Young pers. comm. 2010).

The site has dioramic potential for showcasing the diverse marine fauna that was present during the Early Devonian in the area that is now New South Wales. Fossil horizons document environmental change from a deep volcanic lake through tropical reef to river and lake deposits. The site also provides important evidence for early evolution of fishes and evidence of early land-dwelling animals (Young 2008, 2010).

Cliefden Caves – Belubula River Valley

Overview

Cliefden Caves – Belubula River Valley is a Late Ordovician marine invertebrate fossil site about 20 kilometres east of Canowindra, New South Wales. The site is part of an arcuate belt of limestone adjacent to the south side of the Belubula River.

Scientific importance and research potential

The Cliefden Caves – Belubula River Valley site is one of the best documented Late Ordovician successions in Australia. Geologists around the world regard it as a superb example of an Ordovician island faunal assemblage. The site contains well documented faunas including brachiopods, corals, stromatoporoids, sponges, trilobites, and conodonts. Several detailed taxonomic and biostratigraphic studies have been made (e.g. Percival 1991; Rigby and Webby 1988; Stevens 1950; Webby *et al.* 1997; Zhen and Webby 1995). However, there remains substantial potential for further faunal analysis.

Fossil significance

The site is unusual because of the well exposed, diverse faunas showing the progression from nearshore to deep water marine environments. It contains abundant examples of some of the earliest shell beds in the geological record and the earliest rugose corals known. More than 180 species have been recorded from the Cliefden Caves Limestone Group and Malongulli Formation including brachiopods (over 47 species), trilobites (12 species), corals (over 27 species), stromatoporoids (16 species), sponges (36 species), and conodonts (36 species from Cliefden Caves Limestone Group and 48 species from Malongulli Formation). Eight genera and 12 species of radiolarians have also been described from the Malongulli Formation, including the rare *Etymalbaillella yennienii*, recorded for the first time outside China (Noble and Webby 2009).

Figure 1.9 Fossilised gastropods and molluscs from the Cliefden Caves – Belubula River Valley region. *Fossils in NSW*. Photograph © Centre for Learning Innovation, New South Wales Department of Education and Training

Taphonomy and condition of the site/fossils

The Cliefden Caves Limestone Group is a 460-metre-thick carbonate unit deposited on an island shelf with abundant corals, brachiopods and stromatoporoids. At its maximum development it formed extensive biostromes. This carbonate unit is overlain by another unit composed of shales and siltstones that formed in deeper water (the Malongulli Formation) and is 380 metres thick. The Malongulli Formation consists of a fauna dominated by graptolites, trilobites, and siliceous sponges.

Comparison with similar sites in Australia

The Ordovician, especially the Middle to Late Ordovician, was a critical time in the evolution of marine organisms, and records a second major marine radiation after the important Cambrian radiation (Kakuwa and Webb 2007). As such, Ordovician fossil localities with good stratigraphic resolution can provide critically important data, both biostratigraphic and evolutionary.

The Cliefden Caves – Belubula River Valley sites contain the best exposures of Late Ordovician island marine invertebrate fossil assemblages in Australia. The well documented faunas and biostratigraphy serve as the reference sections of the Late Ordovician in Australia and as a window to Late Ordovician island arc biota. The sites include the type localities for over 100 species (including brachiopods, corals, stromatoporoids, trilobites, conodonts and sponges) and the Stratotype Section of the three conodont biozones within the Eastonian stage of the Late Ordovician.

In comparison, the superb Ellery Creek assemblage in the MacDonnell Ranges of the Northern Territory contains a well studied and near-complete sequence of Cambrian to Ordovician trace fossils. This sequence is unrivalled in Australia for biostratigraphic correlation (see p. 24). The Middle to lowest Upper Ordovician succession of interbedded radiolarian chert and shales exposed at Seal Creek in eastern Victoria contains a diverse, abundant, and well preserved assemblage of trace fossils of organisms that inhabited pelagic, deep ocean sediments on an abandoned section of a submarine fan complex. The Seal Creek trace fossils are useful for clarifying the timing and extent of the Ordovician radiation in benthic faunas of the deep pelagic realm, and the cherts at Seal Creek are exceptional in that trace fossils are clearly evident in outcrop as well as on cut and polished surfaces. Virtually all the chert beds at Seal Creek are extensively bioturbated (Kakuwa and Webb 2007).

Both the Ellery and Seal Creek trace fossils are older than the body fossils of the Cliefden Caves – Belubula River Valley sites. Within the Cliefden Caves – Belabula River Valley sites, the radiolarian species of the Malongulli Formation occur across different depositional environments and from a range of geographically widespread localities, suggesting they will become biostratigraphically important faunal markers with correlation potential within the Upper Ordovician.

Stories

The Cliefden Caves – Belubula River Valley offers an excellent example of the development of a carbonate rim fringing an island shelf in the volcanic island setting, providing insight into Late Ordovician island marine ecosystems.

The first limestone deposit found in New South Wales in 1815 is located at the southern end of this site (Oxley 1820; Stevens 1950). The site includes the first fossiliferous Ordovician limestone found in New South Wales (Stevens 1950).

Talbragar fossil beds

Overview

Talbragar is a Jurassic fossil fish and plant site north-east of Gulgong. It covers an area of approximately four hectares. Discovered by Mr A. Lowe of Wilbertree in 1889, the site was publicised by New South Wales

Mines Department Geological Surveyor William Anderson the same year (White 1981).

Scientific importance and research potential

The Talbragar fossil beds are the best Jurassic fossil site in New South Wales. The site contains many well documented fish and a profusion of plant fossils (e.g. Bean 2006; Dulhunty and Eadies 1969; Wade 1941, 1953; White 1981, 1994; Woodward *et al.* 1895). Many distinguished Australian scientists have published on the Talbragar fossils, including William Anderson, Robert Etheridge Jr, T.W. Edgeworth David and Mary E. White. There is high potential for discovery of new species, particularly species of insects and other invertebrates, but also the possibility of finding reptiles and birds.

Fossil significance

The fossil assemblage provides a very complete picture of a particular flora from the environs of a small lake. The plant fossils in particular are prolific and exceptionally well preserved. Twenty-five species are represented from this site, including fossil plants (16 species), fossil fish (eight species) and fossil insects (one species), all of which are unique to the locality. The assemblage represents Kauri pine forest with podocarp conifers and cycadophyte understorey with *Pentoxylon* (a possible ancestor of the monocotyledonous angiosperm *Pandanus*). Seed ferns, tree ferns and ground ferns are also represented. Multidisciplinary research may help to reconstruct the depositional environment and palaeoclimate of this Jurassic lake site, which at the time was situated at high latitudes on the Gondwanan landmass.

Taphonomy and condition of the site/fossils

The Talbragar fossil site is very unusual for its excellent and beautiful preservation of Jurassic plants and fishes. The fossil-bearing rocks (siltstone or cherty mudstone) are a part of the Purlawaugh Formation of Jurassic age (approximately 167–190 Ma) and occur as loose-floating blocks in soil over an area of about 300 metres by 80 metres. Stratigraphically, they are overlain by remnants of Pilliga Sandstone and underlain by Triassic Narrabeen Sandstone. They are weathered into rectangular blocks exhibiting concentric iron-stained bands. Recent excavation revealed the *in situ* successions of these beds. There are three fossil bearing beds with a thickness of 17 centimetres for the upper layer, 30 centimetres of the mid layer, and 30 centimetres of the lower layer, each separated by weathered shales. Fish fossils are more common in the upper and mid layers. These shales and siltstones are thought to have been the erosional remnant of sediments from a small freshwater lake. Plants are preserved as white impressions on ochre-coloured silicified shale.

Comparison with similar sites in Australia

Although there are a number of important Palaeozoic fish fossil sites in Australia, well-preserved Mesozoic sites are more rare. Talbragar is the only Jurassic fish site in New South Wales, and is the most significant Jurassic terrestrial fossil deposit in Australia. The fossils are extremely well preserved. It is the type locality of 25 species including 16 plants, eight fishes, and one insect (with more under investigation).

Stories

Following their discovery by Lowe in 1889, the Geological Survey of New South Wales immediately commissioned a professional museum collector, Charles Cullen, to collect from the site (White 1981). He assembled a superb representative collection, probably much larger than what remains *in situ*. The collection is in the State Mineral and Fossil Reference Collection (NSW Department of Mineral Resources). The collection is not only significant for its scientific value, but is a testament to the foresight and decisive action in the public interest by the Geological Survey of New South Wales. The site has been excavated for over 100 years, and a huge amount of rock has been removed. The Cullen collection ensured that the loss to science was limited (White 1994). A recent authorised excavation in 2006 by a postgraduate student of the Australian National University has resulted in a large collection and a detailed stratigraphical section being collected. It was from this site that the first rocks of Jurassic age in New South Wales were recognised. The Australian Museum has featured Talbragar fossils in its displays.

Because of the profusion of fossils, their excellent preservation and sequence, dioramic potential is high for reconstructing lacustrine environments of the Jurassic Period.

Wellington Caves

Overview

The Wellington Caves is a Pliocene–Pleistocene vertebrate fossil site near Wellington, approximately 360 kilometres north-west of Sydney in central New South Wales.

Scientific importance and research potential

The fossil sites of the Wellington Caves have been well documented (e.g. Dawson 1982, 1985, 1999; Dawson and Augee 1997; Dawson *et al.* 1999; Hand *et al.* 1988) but the site has strong future research potential; for example, four new taxa are currently being described and many more are possible. Current research activities are led by palaeontologists from the University of New South Wales, the Australian Museum, and other institutions. The Wellington Caves is the only known vertebrate fossil site in Australia that has the potential to reveal sequential faunal change over a period of four million years from the Early Pliocene Epoch up to the Late Pleistocene Epoch.

The fossil faunas of the Wellington Caves attracted wide international attention during the nineteenth century. It is a site with great historical significance and is the type locality for many established fossil mammal species in Australia. It is the source of large collections that are housed in several museums in Australia and overseas. Major collections exist in the Australian Museum, Sydney; the Natural History Museum, London; University of California Museum of Paleontology, Berkeley; the Field Museum, Chicago; the American Museum of Natural History, New York; and lesser collections in many other museums.

Fossils from the Wellington Caves sites made an important contribution to the development of Charles Darwin's theory of evolution and are closely associated with well-known nineteenth century naturalists such as Darwin, Charles Lyell, and especially with Sir Richard Owen, arguably the greatest comparative anatomist of his era. Collections were made and studied by several past directors or curators of fossils at the Australian Museum, including Gerard Krefft, E.P. Ramsay and Charles Anderson. During the latter half of the twentieth century Wellington fossils have been studied by such eminent scientists as R.A. Stirton (University of California), Richard Tedford (American Museum of Natural History), William Ride (Australian National University) and Michael Archer (University of New South Wales).

Fossil significance

The site is very productive and contains abundant and diverse faunas with many fossils yet to be excavated. The age of the fossil assemblages ranges from approximately 30 000 years through to approximately four million years. Over 60 marsupial species have been identified, including at least 35 extinct species. The extinct species include *Thylacoleo* (marsupial lions), *Diprotodon*, giant kangaroos, 7-metre-long goannas (*Megalania*), other reptiles, and birds. Wellington Caves is the type locality for 11 species of marsupial megafauna, including the iconic *Diprotodon optatum*. Recent studies indicate that the Wellington Caves deposit is important in understanding the evolution of modern kangaroos and wallabies (e.g. Dawson 1982, 1985; Dawson and Augee 1997). The site shows strong potential for the reconstruction of the palaeoecosystem of the region during the Pliocene–Pleistocene.

Taphonomy and condition of the site/fossils

A Palaeozoic (Early Devonian, Garra Formation) limestone has cave systems developed in low hills, at an elevation of about 50 metres above the present streams. There are at least five natural caves, which have been expanded and greatly disturbed by phosphate mining and fossil collecting. Fossils are preserved in the Pliocene–Pleistocene sediments deposited in the cave system. These were deposited in various stages during cave development and contain several assemblages of slightly different ages.

Comparison with similar sites in Australia

The Wellington Caves site is one of the first well documented vertebrate fossil sites in Australia and holds great historical significance for the role it played in the discovery of vertebrate fossils (particularly extinct Pleistocene megafauna) in Australia, and for its importance in the scientific career of Sir Richard Owen. It has been studied for approximately 180 years and is the type locality for over 25 species of fossil mammals.

The Wellington Caves site is strongly comparable to the Naracoorte World Heritage fossil site in South Australia and the newly discovered mid-Pleistocene cave fauna from the Nullarbor Plain (WA). The Pliocene component of the Wellington Caves is comparable with the Chinchilla Local Fauna from Chinchilla (Qld), the Bow Local Fauna from Bow, near Merriwa (NSW) and, most importantly, the Hamilton Local Fauna (Vic) and the Bluff Downs Local Fauna (Qld). Spanning the Pliocene–Pleistocene, the Wellington Caves faunas cover a much longer period than all these sites apart from Naracoorte. Further, the historical intellectual/scientific significance of the Wellington Caves is unmatched by other Pliocene–Pleistocene sites in Australia.

Stories

The dioramic potential for Wellington Caves is primarily focused on its historical story and its influence on the development of scientific theories and hypotheses regarding vertebrate evolution in Australia during the nineteenth century. It is significant as the first place of discovery of marsupial fossils in Australia (in 1830) and contains the largest deposit of Pliocene–Pleistocene mammal fossils in Australia.

The site made an important contribution to the development of Darwin's theory of evolution of the fossils he found in South America. It seems that Darwin was previously aware of the similar significance of the Wellington Caves fossils and formulated the law of succession of species based on description in Charles Lyell's *Principles of Geology*, which quotes William Clift's identifications of dasyures, wombats and kangaroos from the Wellington Caves.

The first European record of the Caves consists of a series of sketches by landscape artist Augustus Earle in 1826. Hamilton Hume might be the first to give a written description of the caves, in a diary entry of 1828. Two years later (1830), George Ranken, a local magistrate, found piles of fossil bones in the Bone Cave, when he accidentally fell into the cave entrance. Many of these fossil bones were of enormous size and could not be matched with any known Australian animal. In the same year he returned with his friend Major Thomas Mitchell, the Surveyor General of New South Wales, and collected more than 1000 specimens with a huge variety of bones from the caves that appear to have acted as a natural trap for the fauna. Mitchell then sent a large collection of fossils from Wellington Caves to Europe for examinations by Georges Cuvier and Owen. Cuvier may never have seen the samples, though he commented on some of them. As a major part of his life work, Sir Richard Owen was progressively able to identify a number of large extinct marsupials and birds first from the Wellington Caves and then from a number of other important nineteenth century mammal fossil sites. His involvement continued until 1888 when, aged 84, he published his last paper on Wellington Caves fossils.

Established in 1829, the Australian Museum soon became actively involved in the collecting and studies of the Wellington Caves fossils. Two curators of the Australian Museum, Gerard Krefft and E.P. Ramsay, made extensive collections from the Wellington Caves. In 2000, the Australian Museum presented a major exhibition with fossil bones from the Wellington Caves.

Since soon after their discovery, the Wellington Caves have been a steady source of information about

Figure 1.10 Major Mitchell sketching the entrance of the caves in Wellington Valley, New South Wales.
Painting William Romaine Govett 1843 © National Library of Australia

ancient geology and faunas, although collapses and other geological phenomena have splintered and scattered skeletons and, tragically, phosphate mining from 1913–1971 has rendered priceless palaeontological evidence into fertiliser for farmers.

The site has much potential as an educational and tourism resource.

Northern Territory

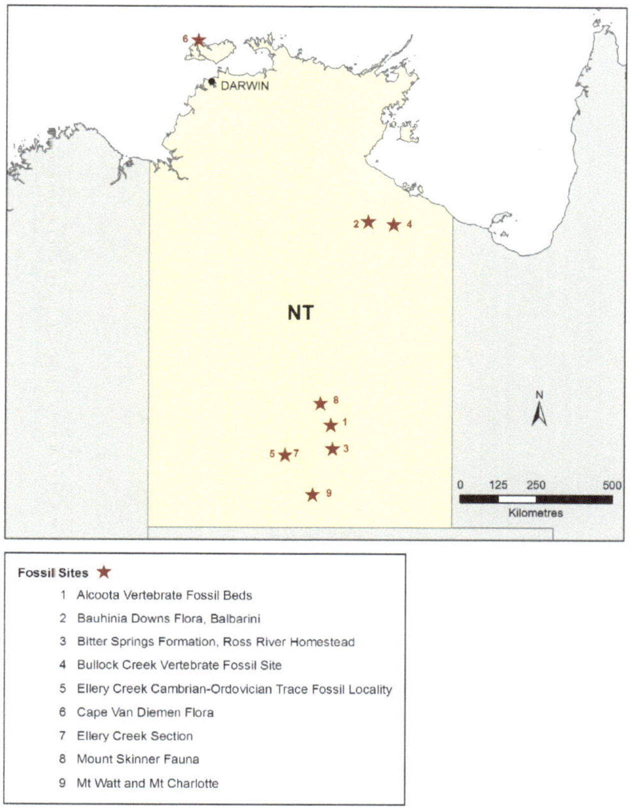

Northern Territory locality map © SEWPaC

Alcoota vertebrate fossil beds

Overview

The Alcoota vertebrate fossil beds are unique, rich assemblages of Late Miocene marsupial, bird and reptilian fossils (Alcoota and Ongeva Local Faunas), located near Alice Springs in the Northern Territory. Under Northern Territory regulations, they are a listed Heritage Place, 'Alcoota scientific reserve', which is 48 hectares in area.

Scientific importance and research potential

The site has been relatively well studied (e.g. Murray *et al.* 2000; Newsome and Rochow 1964; Woodburne 1967a, 1967b), but productive research is ongoing and new discoveries continue to be made. Published results have largely stood up to testing and re-examination. The site is very unusual for its great abundance of fossil material, and the period of geological time represented, which is extremely poorly represented elsewhere in Australia. The investigation by Michael O. Woodburne (1967b) stands as one of the key investigations carried out by a team led by Ruben A. Stirton (University of California, Berkeley). Earlier in the century, the work of the Stirton team established marsupial palaeontology as a serious academic discipline in Australia (Tedford 1991), and Woodburne's (1967b) Alcoota contributions to the project are fine examples of rigorous, field-based, vertebrate palaeontological research.

Fossil significance

Alcoota is a unique assemblage with some rare species that have only been recorded from Alcoota. It has produced the biochronological index fossils for the Late Miocene of the Australian land mammal biochronological scheme (Murray *et al.* 1993, 2000; Woodburne *et al.* 1985). Alcoota provides a very good representation of large herbivorous vertebrates that enables complete skeletal reconstructions of many species. In many instances, species found at Alcoota are the most completely preserved representatives of their families. Small carnivorous species are less well represented, but continuous collecting has yielded representative material.

The potential for palaeoenvironmental reconstruction at Alcoota is excellent. Sedimentological and palaeontological data provide the earliest evidence of significant aridification of the Australian continent leading to the prevailing dry climatic regime of the continent.

Taphonomy and condition of the site/fossils

The site contains disarticulated, jumbled skeletons in unconsolidated, poorly sorted, sandy to gravelly fluvial channel deposits. The majority of bones are complete, though heavily fractured by post-depositional expansion and contraction of the host sediments. Extraction of fossils is very time-consuming and requires technical skill. Access to fossil beds requires removal of about 1 metre of overburden. Bones cannot be left exposed to the elements or they will be eroded and disintegrate in the field.

Comparison with similar sites in Australia

Alcoota is the land mammal reference standard for the Late Miocene in Australia. A couple of vertebrate fossil sites of approximately similar age in Victoria and New South Wales have produced only a few, fragmentary specimens of a small number of taxa, many of which cannot be identified to species. It remains the most important Late Miocene vertebrate fossil site in Australia.

Figure 2.1 Baru and Diprotodon skulls at Alcoota. Photograph © of the Museum and Art Gallery of the Northern Territory

Alcoota occupies an intermediate temporal position between, for example, the Riversleigh (Qld) and Naracoorte (SA) World Heritage sites, and it is younger than the Lake Eyre Basin Etadunna and Wipajiri Formation assemblages (SA) that stand as the reference standards for the base of the Australian land mammal biochronological scheme (a framework of relative ages for land mammals based on biocorrelation between vertebrate faunas).

Stories

The assemblages record a drought event surrounding a waterbody that resulted in mass mortality. This was a dramatic event that marked the onset of a major climate shift: from relatively equable, mesic (wet or humid) conditions in the Early and Middle Miocene to much drier, warmer, strongly seasonal conditions with increasingly unreliable rainfall (relatively xeric conditions) that characterised the continental interior of the Pliocene through to the present day. In concert with the effects of this global climate change, Australia's fauna underwent a marked turnover from assemblages typical of the Miocene (dominated by a diversity of large marsupial browsers and giant flightless birds) to assemblages characteristic of the Pliocene and Pleistocene (major radiation of, and dominance by, kangaroos and large grazers). The Alcoota assemblage represents a key transitional vertebrate fauna during this time, which is otherwise poorly represented in Australia.

Alcoota is the subject of displays in the Museum of Central Australia (Alice Springs). It is featured in popular published accounts of the origins and evolution of Australia's unique fauna (e.g. Vickers-Rich and Rich 1993b) and has been featured in television documentaries (ABC and Discovery Channel productions). Using Alcoota material, the Museum and Art Gallery of the Northern Territory staff have been able to assemble complete composite skeletons of several species, with sufficient material available for several additional specimens.

The pioneering investigation of the site by Woodburne is of historical significance as a key element in the Australian land mammal biochronological scheme for the Australian continent first articulated by Stirton and colleagues.

The site was known to Aboriginal people and pastoralists at the time of the first scientific investigations. Aboriginal residents on Alcoota pastoral station recognise it as a significant site, but not one of secret or sacred cultural significance. The protection of the site as a scientific reserve was with the support of Aboriginal people on Alcoota.

The Alcoota fossil reserve also contains the type section of the Waite Formation (Geoscience Australia stratigraphic number 19212).

Bauhinia Downs flora, Balbarini

Overview

The Bauhinia Downs flora site is an Early Cretaceous fossil plant site that includes relictual taxa such as giant clubmoss (*Cyclomeia*), which are usually associated with earlier geological periods.

Early Cretaceous strata across the northern part of the Northern Territory and adjacent parts of western Queensland were originally described under informal stratigraphic nomenclature as the Mullaman Beds by Skwarko (1966). Following nomenclatural revision, the name Mullaman Beds now remains applicable only to Skwarko's 'inland belt', a succession of relatively poorly studied, mostly non-marine strata extending across the base of the northern part of the Northern Territory. Apparently equivalent strata from further east in Queensland (the Lees Sandstone) contain a similar flora.

The site nominated here as representative is on Balbarini Station (Northern Territory), from which several previously unrecorded taxa have been recovered and lodged in the collections of the Museum and Art Gallery of the Northern Territory.

Scientific importance and research potential

The Mullaman Beds inland belt and Bauhinia Downs flora site have been competently examined and reported on by several authors in both published and unpublished scientific reports (e.g. Skwarko 1966; White 1961, 1966) and in popular publications (White 1994). They retain considerable research potential in palaeoecological reconstruction, systematics and evolution. The work that has been done is of high quality but primarily taxonomic, leaving scope for further research on broader ecological questions.

Fossil significance

The flora is generally representative of Early Cretaceous floras, but among them are relictual species that are generally associated with earlier geological periods. For example, giant clubmosses (*Cyclomeia*) are numerous and well preserved at Bauhinia Downs, and this is the latest known existence of the group in Australia. Preservation of the plants is exceptionally good, and includes three-dimensional specimens preserved *in situ*, which is rare.

Taphonomy and condition of the site/fossils

The site contains fractured, flat-lying quartz sandstone and pebble conglomerates, mostly *in situ*. During the Early Cretaceous the site was probably a riparian in nature, supporting a plant community that was inundated by periodic flooding. Specimens are abundant and easy to examine.

Comparison with similar sites in Australia

Bauhinia Downs flora is among the better preserved Australian Mesozoic flora. The Bauhinia Downs flora is reasonably well dated by marine intercalations to the Neocomian Stage of the Early Cretaceous (approximately 145–130 Ma). Comparable fossil plant assemblages of this age are known from the Winton Formation of Queensland (McLoughlin *et al.* 1995) and the Dampier Peninsula coast in Western Australia.

Stories

The site has a high dioramic potential (Retallack 1975, 1977) as it provides a strong example of Mesozoic Gondwanan vegetation. The site is important in demonstrating two scientific ideas:

- long survival of giant clubmosses
- the scientific value of complete fossil plant specimens (the ability to associate the leaves, fruits and flowers to a single species).

Bitter Springs Formation, Ross River Homestead

Overview

The Bitter Springs Formation is a Neoproterozoic (approximately 900 Ma) microfossil and cyanobacterial bioherm (stromatolite) site. The fossiliferous chert outcrops extensively and/or is recorded in the subsurface of the Amadeus Basin in the south-western Northern Territory and adjacent parts of South and Western Australia. The formation takes its name from Bitter Springs in the Northern Territory, adjacent to Ross River Homestead, close to the type section of the formation, and the premier stromatolite and microfossil localities (Fortowski *et al.* 1988).

Scientific importance and research potential

The site has been extensively studied, and has been subject to high quality research by renowned specialists, both Australian and international (e.g. Barghoorn and Schopf 1965; Oehler 1976; Schopf 1968a, 1968b; Schopf and Blacic 1971). The Bitter Springs Formation was initially famous for its well preserved individual cells of cyanobacteria, which demonstrated some of the earliest life-forms known at the time. Bitter Springs Formation stromatolites are among the best preserved biohermic reefs and its silicified cellular material continues to yield new information.

Fossil significance

The formation has produced a diverse assemblage of at least 30 types of microfossils, particularly cyanobacteria. The significance of these fossils is partially historical as they helped to provided unequivocal evidence that diverse life-forms existed in rocks this old. The site provides testimony to the geological antiquity of cyanobacteria and their persistence through time. The *in situ* organosedimentary structures called bioherms also demonstrate the reef-building capacity of cyanobacteria even in their early history, and the structure of the reef is obvious in outcrop (Barghoorn and Schopf 1965; Schopf 1968a).

The fossil assemblage is very representative of Pre-Cambrian stromatolite biotas, yet it is one of the most diverse stromatolitic deposits in Australia (Oehler 1976). The site has good potential for palaeoenvironmental reconstruction.

Taphonomy and condition of the site/fossils

The site is exceptionally well preserved, from microscopic single cells preserved in three dimensions through to large-scale reef structures. Taphonomically, it is an indurated marine reef formation that preserves microscopic detail.

Comparison with similar sites in Australia

Bitter Springs is historically important as the location of the first documented Proterozoic cyanobacterial microfossils. It is one of the best preserved ancient bioherm formations and its diversity of taxa continue to be useful as a comparative baseline (e.g. Oehler 1976; Schopf 1968a, 1968b; Schopf and Blacic 1971). It is older than, but comparable in species composition and diversity to, the earliest succession of the Ellery Creek Section (NT). It is also comparable to the Officer Basin fossiliferous chert assemblages located in Western Australia (Cotter 1997).

Bitter Springs is very representative of Neoproterozoic microfossil and biohermal localities. It is readily reconstructable as a graphic or a three-dimensional model, and is comparable to modern bioherms such as those currently growing at Shark Bay, Western Australia.

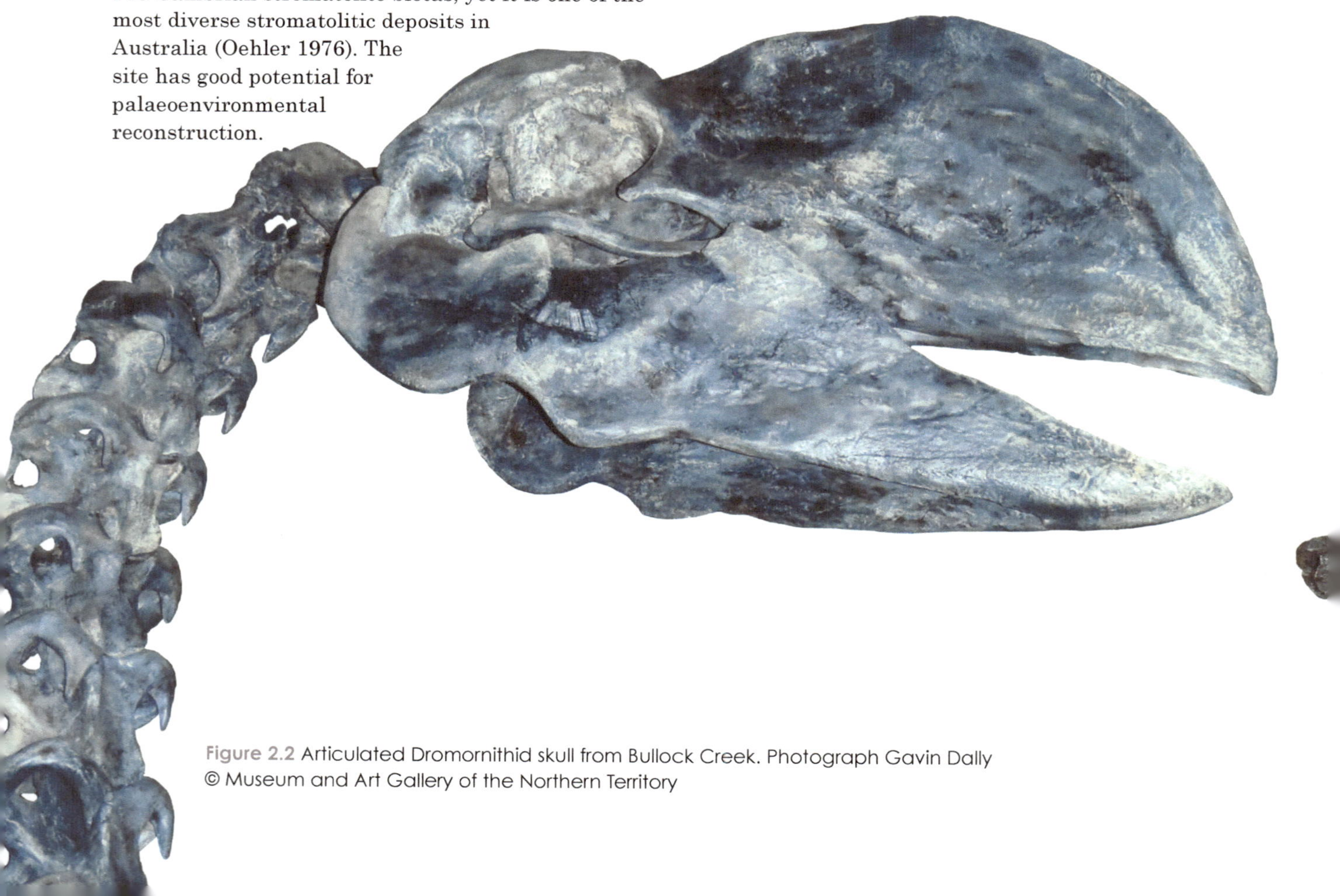

Figure 2.2 Articulated Dromornithid skull from Bullock Creek. Photograph Gavin Dally © Museum and Art Gallery of the Northern Territory

Stories

This site illustrates the story of how stromatolite formations grow, and provides evidence on how the oceans and atmosphere became oxygenated during the Proterozoic. The locality provides an excellent *in situ* representation of what the marine, atmospheric and climatic conditions were like approximately 900 Ma in the Neoproterozoic.

Bullock Creek vertebrate fossil site

Overview

Bullock Creek vertebrate fossil site (Bullock Creek Local Fauna) is a rich assemblage of Middle Miocene marsupial, bird and reptile fossils (Figures 2.2, 2.3). It is a listed Northern Territory Heritage Place.

Scientific importance and research potential

The site has been relatively well studied (e.g. Clemens and Plane 1974; Murray and Megirian 1992; Plane and Gatehouse 1968; Woodburne *et al.* 1985). Research at the site is ongoing, and new discoveries continue to be made. There had been occasional further collecting by a number of researchers up to about 1986, followed by an ongoing program of systematic collecting conducted by the Northern Territory Museum. Results have been published in numerous refereed scientific articles and have largely stood up to testing and re-examination. The site is recognised as an important vertebrate reference assemblage in the Australian land mammal biochronological scheme for the Middle Miocene (Murray and Megirian 1992; Murray *et al.* 2000; Woodburne *et al.* 1985).

Fossil significance

The Bullock Creek local fauna has become established as the reference assemblage for the Middle Miocene of the Australian land mammal biochronological scheme (Murray *et al.* 2000; Woodburne *et al.* 1985). It contains remains of relatively rare species as well as common species; the latter are most valuable in providing biocorrelative information between sites of comparable age, particularly those from Riversleigh (Qld). It is noteworthy for its preservation of relatively rare cranial material (Fortowski *et al.* 1988: 158).

The site has excellent potential for palaeoenvironmental reconstruction. The sedimentological and palaeontological data provide good regional palaeoclimatic proxies for the Middle Miocene of the Northern Territory.

Taphonomy and condition of the site/fossils

The vertebrate fossils consist of mostly disarticulated skeletal elements. Some skeletons are intact and others show predation damage from crocodiles; all are preserved in freshwater calcimudstones and calcareous sandstones and conglomerates. Fossils occur in localised concentrations on low rubbly outcrops. Specimens are extracted using dilute acetic acid and/or by mechanical techniques. Preservation is excellent, including of microscopic histological detail.

Figure 2.3 Diprotodon skull from Bullock Creek. Photograph Gavin Dally © Museum and Art Gallery of the Northern Territory

Comparison with similar sites in Australia

The Bullock Creek local fauna is the land mammal reference standard for the Middle Miocene (Murray and Megirian 1992; Murray *et al.* 2000). It occupies an intermediate temporal position between the Late Miocene Alcoota site and the Late Oligocene–Early Miocene Lake Eyre Basin Etadunna and Wipajiri Formation assemblages (SA). Assemblages of comparable age are known from Riversleigh (Qld), particularly the Faunal Zone C sites and Encore Site (Travouillon *et al.* 2006).

Stories

Bullock Creek local fauna has excellent dioramic potential. It is the subject of displays in the Museum and Art Gallery of the Northern Territory. The assemblage appears to have been substantially accumulated by the large crocodilian predator *Baru darrowi*. Skeletal remains of other taxa commonly show tooth puncture marks and other predator damage.

Bullock Creek is also featured in popular published accounts of the origins and evolution of Australia's unique fauna (e.g. Vickers-Rich and Rich 1993b), and has been featured in television documentaries (ABC productions).

The historical story associated with discovery of the Bullock Creek vertebrate fossil site revolves around M.D. Plane (who with Gatehouse first reported on the discovery of vertebrate fossils at Bullock Creek), Ruben A. Stirton (University of California, Berkeley) and colleagues. The work of the Stirton team established marsupial palaeontology as a serious academic discipline in Australia (Plane and Gatehouse 1968; Tedford 1991; Rich *et al.* 1993).

Ellery Creek Cambrian–Ordovician trace fossil locality

Overview

Ellery Creek is a Cambrian–Ordovician trace fossil locality. It is located in the Ellery Creek section, Pacoota Formation, West MacDonnell Ranges.

Scientific importance and research potential

There has been one detailed study of this site (Haines 1982). References are often made to the presence of trace fossils in the Larapinta Group, though Haines's (1982) study is the only comprehensive formal analysis. The importance of trace fossils in biochronology, palaeoecology and facies recognition has been slow to develop. Haines's work was a pioneering study that demonstrated the usefulness of trace fossils in stratigraphic and facies analysis. Much systematic work remains to be done.

Fossil significance

Trace fossils are not rare in the world, but localities in which there are many ichnospecies preserved in long, continuous successions with clearly definable facies associations are uncommon, making this site unusual (Crimes 1970). The fossils are highly representative of global Cambrian–Ordovician shallow marine ichnofaunas (e.g. Crooks and Gatehouse 1980). The majority of known ichnospecies – *Skolithos*, *Diplocraterion*, *Corophioides*, *Arenicolites*, *Rusophycus* and *Cruziana* – are represented in the Pacoota Formation. The site has high value for potential palaeoenvironmental reconstruction and the majority of the Pacoota ichnofossil assemblages provide useful information on depth, currents and marine floor zonation.

Taphonomy and condition of the site/fossils

The Pacoota Formation ichnofossils are beautifully preserved. The taphonomic interpretation varies with the behaviour of the organism and the facies in which they occur. There are many soft-bodied burrowing organisms known only from their infilled burrows. Elsewhere, delicate and intricate trails or tracks indicate rapid though gentle burial by fine sediment.

Comparison with similar sites in Australia

This site is important because it is one of the few well studied localities containing finely preserved and abundant material within a long, continuous sedimentary sequence. There are many localities on a global scale with the same or similar trace fossils, but few have been examined in such a long and detailed stratigraphic context.

Stories

The attractiveness of the markings and mystery surrounding trace fossils makes them one of the most interesting and easily displayed palaeontological structures. Trace fossils also provide one of the most useful ways to demonstrate environmental change through time, particularly that of sea levels, based on the trace fossils of certain types of organisms. Trace fossils are frequently used in the absence of actual fossil remains to fill the gaps in the fossil record. Detailed studies of ichnofossils provide information on the habits, behaviour and locomotion of organisms otherwise found as inert, incomplete and distorted fossil remains.

The study of trace fossils also occupies an interesting place in the history of palaeontology. Though common and abundant, ichnofossils have been relegated to the fringes of palaeontological investigation until fairly recently. Improved methods and analytical techniques are now bringing ichnofossil studies to the fore.

There is no known direct Indigenous influence or stories relating to the widespread and very conspicuous markings on the rock surface. Some large, vertical assemblages of trace fossils in the MacDonnell Ranges have a vague resemblance to rock engravings and it is therefore surprising that some significance has not been attributed to them.

The abundance and good preservation of the Pacoota and Stairway Sandstone ichnofossils was recognised by Watt (1896) of the Horn Expedition.

Cape Van Diemen flora

Overview

The Cape Van Diemen flora is a Palaeogene plant fossil site The fossils are found in sandstone coastal cliff exposures on the eastern side of Cape Van Diemen, Melville Island, Northern Territory.

Scientific importance and research potential

The Cape Van Diemen flora is scientifically important as the only Palaeogene (Palaeocene–Eocene) macroflora known in northern Australia. While there have been preliminary, high quality descriptive studies undertaken (Pole 1998b; Pole and Bowman 1996; White 1974), there remains a high potential for further research at this site. There was an intensive collecting trip conducted in 1996, with previous collections made in 1994 and 1974. The material is well preserved and abundant, though not securely dated. Considerable scope exists for stratigraphic and sedimentological analysis in relation to dating and taphonomy.

Fossil significance

Fossil macrofloras (leaves, seed, flowers, etc.) are extremely rare for the early part of the Cenozoic in Australia and particularly for the northern half of the continent. The flora represents a semi-deciduous scleromorphic formation indicative of strong seasonality with a warm, periodically dry climate. The fossil leaves are highly representative of a type of flora known in the region today with some genera and species closely related to living species (*Grevillea*, *Brachychiton*, *Dilcbeia*, *Smilax*). The site offers very good potential for palaeoenvironmental reconstruction. Fossils are abundant and taxonomically diverse with many living analogues to draw from.

Taphonomy and condition of the site/fossils

The abundant and well preserved fossils were buried by seasonal inundation of a flood plain, incorporating forest floor leaf litter into fine alluvial sediment.

Comparison with similar sites in Australia

The Cape Van Diemen flora is unique in location and age for the northern half of Australia. The assemblage indicates that diverse scleromorphic floras were present in Australia during the Palaeogene, complicating the prevailing interpretation of a pan-Australian rainforest.

Stories

The site has excellent and important dioramic potential for showcasing a northern assemblage of non-rainforest plants during the Palaeogene. The close similarity of the Cape Van Diemen flora to modern 'dry rainforest' floras of north Australia allows accurate reconstruction and modelling.

Ellery Creek Section

Overview

Ellery Creek Section is a Late Proterozoic–Permian microfossil site of exceptional historical and evolutionary significance that spans a 7800-metre stratigraphic section. The northern part of the section is in West MacDonnell National Park, Ellery Gorge (west of Alice Springs), where it is exposed along an 8-kilometre stretch of the gorge. The southern part is on adjacent Aboriginal land. Ellery Creek is a type section of the Amadeus Basin and it provides evidence for approximately 540 million years of geological and palaeontological evolution in central Australia.

Scientific importance and research potential

The Ellery Creek Section incorporates a nearly complete sequence of sediments deposited in the northern Amadeus Basin that spans approximately 850–310 Ma. Parts of the site have been rigorously studied by Australian and international experts and there are a number of high quality Geological Survey reports dealing with the formations exposed in the section (e.g. Schopf 1968a; Schopf and Blacic 1971; Walter *et al.* 1979). The site is regarded as extremely important within micropalaeontology, especially the Loves Creek Member of the Bitter Springs Formation within the 'Ellery Creek Section' (equivalent to geological monument 48 of Fortowski *et al.* 1988). It has both biostratigraphic and palaeontological research potential.

Fossil significance

The fossil assemblage contains a diverse composition of microfossils, particularly cyanobacteria, but also demonstrates a nearly continuous succession of life-

Figure 2.4 Reconstruction of the Mt Watt and Mt Charlotte fish fauna, which existed around 470 million years ago. Painting Peter Schouten © SEWPaC

forms from cyanobacteria to the vertebrate placoderm fish, such as *Bothriolepis*. The site spans the Late Proterozoic through to the Permian (approximately 810–310 Ma) and has excellent potential for demonstrating successions in climate and environment as well as macroevolutionary events.

Some of the algal microfossils exhibit apparent alternation of diploid and haploid generations, which implies eukaryotic sexual reproduction and genetic recombination. The evolution of genetic recombination was a major macroevolutionary development and was a prerequisite to the 'Cambrian explosion' of multicellular life.

Taphonomy and condition of the site/fossils

Taphonomy and preservation varies according to the mode of formation: numerous states and conditions are exemplified, and contacts between units are easily observed. Strata are tilted near vertical so that true sediment thicknesses are clearly apparent and are exposed horizontally (instead of vertically) in a continuous 7800-metre section. Representative fossils from all strata spanning approximately 540 million years may be seen in the creek bed and adjacent outcrop. Rocks and fossils of geological ages preceding the Ellery Creek Section can be seen nearby in the Hermannsburg Sandstone and Brewer Conglomerate. It is exceptional (and possibly unique) to be able to see such complete geological sections in such a short physical distance. The fossil succession is well mapped and colour aerial photos are readily available.

Comparison with similar sites in Australia

As a stratigraphic section, it is comparable to that exposed in the Grand Canyon of North America, but is much more accessible. This site provides a continuous sequence through time that follows on from the fossiliferous carbonate cherts of the Bitter Springs Formation (p. 21).

Stories

The site has good potential as a supplement to individual dioramas and provides fossil material that covers half a billion years of evolution. This begins with a graphic demonstration of more than 100 million years of Neoproterozoic glaciation and ends dramatically with the Alice Springs orogeny at 310 million years, pushing up hills that produced the Brewer Conglomerate. The Ellery Creek Section is used in the Museum of Central Australia (Alice Springs) as the organising principle for a series of small dioramas depicting faunas and environments of successive periods of geological time.

Mt Skinner fauna

Overview

The Tops Member of the Central Mount Stuart Formation at Mt Skinner contains fossils of rare, multicellular, soft-bodied organisms from the Ediacaran Period in the Late Proterozoic (663–542 Ma). Mt Skinner is approximately 170 kilometres north of Alice Springs, Northern Territory.

Scientific importance and research potential

The site has not been thoroughly researched palaeontologically, but the stratigraphic occurrence of the fossils is well documented (e.g. Wade 1969, Walter *et al.* 1989). Mt Skinner is of historical significance in the development of the understanding of the evolution of multicellular life. It has research potential and is of continued interest as one of relatively few Ediacaran sites in Australia.

Fossil significance

The Mt Skinner fauna includes a number of soft-bodied multicellular organisms, e.g. *Skinnera brooksi* and *Hallidaya brueri* (Wade 1969) plus trace fossils of similar organisms. Sites from the Ediacaran Period containing body fossils are rare and there are still undescribed taxa from the Mt Skinner fauna (Haines 2005). The presence of undescribed and unnamed taxa means that the full significance of the site cannot yet be assessed and shows that its scientific potential is yet to be fully realised. The site has excellent potential for palaeoenvironmental reconstruction.

Taphonomy and condition of the site/fossils

The Tops Member consists of fluvial to deltaic sediments composed of immature sandstone, arkose, conglomerate, mudstone and minor carbonate. The fossils are confined to a restricted interval, at most several tens of metres thick (Haines 2005). Body and trace fossils are present: photographs in Thompson (1991) indicate that at least some examples came from fine sandstone or siltstone facies. The geological and tectonic setting is well understood.

Comparison with similar sites in Australia

The Ediacara type locality and the first place where Proterozoic body fossils of metazoans (multicellular animals) were discovered is Ediacara in the Flinders Ranges of South Australia (Fedonkin *et al.* 2007). As such, the historic importance of that site is unsurpassable. The Mt Skinner Fauna is related to those from Ediacara, but may represent an earlier timeframe and/or a different biozone (Waggoner 2003).

Although there are some other Ediacaran localities across Australia (Wells 1996, Waggoner 2003), the central and South Australian localities appear to be the most significant (Waggoner 2003). Mt Skinner is the type locality for *Skinnera* (Wade 1969), which is of biocorrelative value in comparing the relative age of similar sites.

Stories

The site has good dioramic potential. Once the fauna is better documented, it may prove to be an exemplary stage in the evolution of Ediacaran faunas and has great potential to showcase the evolution of early life-forms.

Mt Watt and Mt Charlotte

Overview

The Mt Watt and Mt Charlotte fish fossil localities have produced natural moulds of the oldest known southern hemisphere vertebrates: jawless fishes (Agnatha). They are found on the indurated surface of the Stairway Sandstone, which is early Middle Ordovician in age (approximately 470 Ma). The Stairway Sandstone is south of Alice Springs, Northern Territory.

The site is very unusual in geography and in stratigraphy: both localities are isolated outliers of fossiliferous Ordovician sandstone in the form of mesas. They are distinctive on the horizon among otherwise rounded hills of similar size, but these two represent different geological ages. The assemblages represent a specific facies or zone of fossils not widely identified in the Stairway Sandstone.

Scientific importance and research potential

The Mt Watt and Mt Charlotte Local Fauna contain fossils of the oldest known vertebrates in the southern hemisphere: bony, jawless fishes of the genera *Arandaspis* and *Porophoraspis* (Long 1993a; Ritchie 1985; Ritchie and Gilbert-Tomlinson 1977). This fossil evidence comes in the form of natural moulds of the scales and dermal plates that formed the exterior of these bony fishes.

Mt Watt and Mt Charlotte have not been extensively collected, but the fauna is known reasonably well. The site s research potential lies in continued collecting for rare early vertebrate fossils and studies of their structure and evolution. Continuing research may lead to the refinement and confirmation of age determination, as well as to the correlative recognition of similar facies elsewhere in the Stairway Sandstone.

Fossil significance

Very rare fossils are found at this site, in particular scales and impressions of head shields of at least two species of early jawless fishes (*Arandaspis* and *Porophoraspis*). These are the oldest vertebrates found in Australia and the southern hemisphere. *Arandaspis* is an extremely important taxon and is related to similar-aged heterostracans (*Anatolepis*) in the northern hemisphere. In addition to fishes, there is abundant evidence of invertebrates such as molluscs, brachiopods and trilobites, as well as trace fossils (particularly *Cruziana* and *Diplocraterion*). Trilobites of the genus *Asaphus* and *Aportophyla*-like brachiopods are indicative of a shallow marine fauna. The site has high potential for palaeoenvironmental reconstruction.

Taphonomy and condition of the site/fossils

The sediments of the Stairway Sandstone were laid down in a shallow epicontinental sea. Elements of fish skeletons were dissembled and transported short distances after decomposition of the organisms (Ritchie and Gilbert-Tomlinson 1977).

Comparison with similar sites in Australia

The Mt Watt and Mt Charlotte local fauna is unique as it has produced the oldest vertebrate fauna in Australia and the southern hemisphere. Most fossil sites in Australia that have produced bony fishes are younger in age (Devonian, Permian or Mesozoic).

Stories

The Mt Watt and Mt Charlotte bony, jawless fishes are a major contribution to our knowledge of the origin, structure, evolution and distribution of the earliest vertebrates. They help to document very early vertebrate evolution in the shallow epicontinental seaways of the southern hemisphere.

The site has excellent dioramic potential: invertebrate assemblage and palaeohabitat are well defined and species are large and distinctive. The Museum of Central Australia, Alice Springs, has produced a small diorama of the fauna, containing models of the fishes, and characteristic invertebrates such as nautiloids, brachiopods and trilobites.

Mt Watt was mapped by Charles Winnecke and fossils were collected by James Watt of the Horn Expedition of 1896. Winnecke named the fossiliferous mesa after Watt. However, no vertebrate remains were found at the time. Tate (1896) recognised Mt Watt as an important invertebrate fossil locality.

Queensland

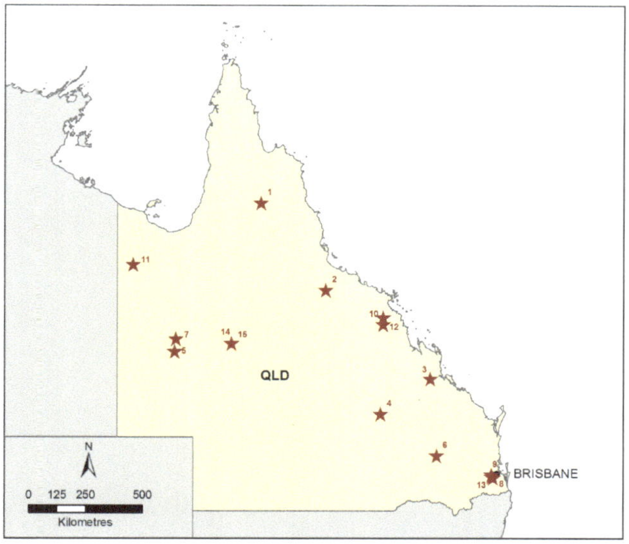

Fossil Sites ★

1 Walsh River and Elizabeth Creek
2 Fletcherview - Big Bend 'Leichhardt Cliffs' Devonian Reef Complex
3 Mount Morgan Jurassic Dinosaur Footprint Site
4 Rewan, The Crater
5 Black Mountain Cambrian-Ordovician Boundary section
6 Chinchilla Rifle Range
7 Chatsworth Limestone Echinoderm and Trilobite Localities
8 Redbank Plains Formation
9 Mount Crosby Triassic Insect Locality
10 Jack Hills Gorge and Fish Hill, Broken River Province
11 Riversleigh
12 Homevale Station
13 Dinmore Quarry
14 Lark Quarry Dinosaur Stampede National Monument (Winton Formation)
15 Late Cretaceous (Cenomian) Winton Formation

Queensland locality map © SEWPaC

Walsh River and Elizabeth Creek

Site overview

Walsh River Cliffs and Elizabeth Creek contain mixed Early Cretaceous marine invertebrate and vertebrate sites. The two localities are located in Wrotham Park north-west of Chillagoe in northern Queensland.

For the last 134 years, the Wrotham Park area, on a highly seasonal and high-flow tropical river, has yielded many marine faunas from the Early Cretaceous Period, when the Great Artesian Basin was inundated. The sites contain an intensely abundant and diverse faunal record, including ammonites, belemnites, crustaceans, bivalves, gastropods, fishes, plesiosaurian remains, echinoderms (comatulids), ichthyosaurs and stromatolites, as well as a microfauna and microflora. Fortuitous finds of terrestrial taxa include coniferales, pterosaur fragments and abundant large fossil logs. The Elizabeth Creek section represents a younger sequence stratigraphically than Walsh River Cliffs, with a slightly different ammonite fauna.

The stratigraphic unit represented is the Doncaster Formation. This unit was originally defined as the Blackdown Formation on the Walsh River, but it correlates with the Doncaster Formation within the rest of the Great Artesian Basin. Essentially mud-dominant, with minor nodular green-grey limestones, intraclastic conglomeratic units and rarer greensand and coarse labile sandstones, the unit represents maximum flooding of the Australian craton during the period 115–110 Ma. The sequence contains a number of sedimentological features including hardgrounds, cool-water stromatolites, intraclastic horizons, and burrowed and bored nodule horizons.

Scientific importance and research potential

This is the best exposed Early Cretaceous (Late Aptian) marine sequence in Australia. It contains key index taxa documenting the Late Aptian marine flooding event which covered over one-third of the continent. The site has been moderately well studied over the last century, including recent studies of stratigraphy and detailed palaeontology (e.g. Etheridge 1920; Jack and Etheridge 1892; McHenry *et al.* 2005). It is the site of recent collections and excavations of marine reptiles, in particular of an 85% complete plesiosaur.

Walsh River and Elizabeth Creek have good research potential, as continual renewal of outcrop by severe monsoonal erosion events reveals new material constantly and it will continue to yield into the foreseeable future. Research work planned includes documentation of the stromatolitic horizons (Cook, unpublished), marine reptile faunas (McHenry, Cook, Kear and Morell pers. comm.) and other opportunistic finds such as pterosaur material (Salisbury pers. comm.).

The robustness of the scientific work undertaken so far is modest. Ammonite faunas require revision in detail, as do other mollusc faunas.

Fossil significance

Walsh River and Elizabeth Creek contain an excellent benthic and nektic assemblage from the Early Cretaceous (Late Aptian), without parallel anywhere in the Great Artesian Basin or elsewhere in Australia. Some taxa found at the site are rare. Several taxa are endemic but most are represented by poorer quality specimens throughout the Great Artesian

Basin. The site offers extremely high potential for palaeoenvironmental reconstruction, including use of stable-isotope geothermometry and other geochemical environmental assessment tools. Palaeoenvironmental reconstruction has been the subject of several informal works and representations in museum displays and books.

Taphonomy and condition of the site/fossils

Fossils within nodules are in excellent condition and there have been near-complete vertebrate fossils recovered. Ammonites are of high commercial grade and together with crustaceans have been collected commercially for a number of years. This has placed the site under some stress from exploitation, ameliorated by a restricted access regime implemented by the property managers.

Comparison with similar sites in Australia

This site contains the best example of Cretaceous marine faunas in Australia, which far exceeds the diffuse Cretaceous sites in the northern Eromanga Basin and Surat Basin. Fossils are better exposed and more diverse than in deposits of similar age in the opal fields in South Australia. For diversity and richness of ammonite species, the Miria Marl of Western Australia's Giralia Range is unparalleled but from a much younger Cretaceous assemblage, and the Wrotham Park sites contain a much more extensive vertebrate record than the Miria Marl.

Stories

The site provides a complete benthic and nektic ecosystem record for its time, and hence can be used to reconstruct the cool-water Aptian faunas and ecosystem. It gives a dioramic insight into the marine fauna of the largest inland sea to flood continental Australia. Mesozoic speciation was engendered by rapid fluctuations in sea level. This is true of the Late Aptian stage in northern Australia, where a sea-level maximum induced speciation in the shallow continental basin (Oosting 2004). This is reflected in a vast diversity of both benthic and nektic invertebrate fossil forms. Of note are the diverse heteromorph ammonites belonging to the *Australiceras-Tropaeum* clade, which produced bizarre and somewhat plastic morphologies during this time. This indicates a genus-level group in flux due to the rapid change in niche and broad ecospace availability. The presence in modest numbers of plesiosaurian remains complements occurrences elsewhere and is at a key speciation time for the group. Large pliosaurids, especially *Kronosaurus*, are known from this horizon, as are a number of unnamed plesiosaurs, including specimens with significant gut contents demonstrating diverse feeding strategies (McHenry *et al.* 2005).

The site was first discovered by the 1872 Hann Expedition to Cape York, which set out to map the Telegraph route (Hann 1873a, 1873b). The Hann expedition found the Palmer River Goldfield approximately one week after finding these Cretaceous sites. Original material from the expedition was sent to Robert Etheridge in London. *En route* it was caught up in a shipwreck, then recovered, before being sent to London, described, and returned to Queensland to the Geological Survey and the Queensland Museum.

Fletcherview – Big Bend 'Leichhardt Cliffs' Devonian reef complex

Site overview

The Fletcherview – Big Bend 'Leichhardt Cliffs' Devonian reef complex is approximately 25 kilometres north-west of Charters Towers, Queensland. The fossil reef complex is exposed as low cliffs and inclined bedding surfaces along the Burdekin River. These extend from the Big Bend Environmental Park downstream past Big Rocks to Fletcherview (Little Rocks) and downstream a further 2 kilometres to the Leichhardt Cliffs, on Burdekin Downs Station.

The Leichhardt Cliffs sites contain millions of fossil organisms representing the various ecologies associated with a nearshore reef complex that developed in the Burdekin Basin in the Middle Devonian (Givetian, approximately 392–386 Ma). The exposures along the Burdekin River show at least 12 different environments (see Cook 1995), ranging from rocky headland communities through muddy and sandy embayments and lagoons to fringing reef complex and offshore shoals. Each ecosystem has a distinct fauna which is well exposed on the immense bedding planes and cliff exposures. The area has long been recognised as a highly accessible example of palaeontological and geological heritage, with protection afforded to the Big Bend Section by the Dalrymple Shire Council as an environmental and recreation reserve.

It is also the site of the best exposure of the youngest basalt flow in northern Australia. The Toomba Flow (approximately 13 500 years BP) flowed west to east along the Burdekin River and plicated the river channel (Whitehead and Stephenson 1998). This is a particularly long basalt flow and was coeval with human habitation in the area. Excellent exposures of the flow are preserved at Big Bend and Fletcherview Station along the Lolworth Creek and Burdekin River banks.

Scientific importance and research potential

Both sites have had extensive and high quality research on their faunas, sedimentology and stratigraphy based on regional and detailed studies for over 150 years. Initial studies of the fauna formed part of the reconnaissance work of Jack and Etheridge (1892). Studies of coral taxa and ecology by Hill (1942b), Zhen (1991), Zhen and Jell (1996); stromatoporoid ecology and taxonomy (Cook 1999), gastropod taxonomy (Cook 1993b, 1997; Etheridge 1917; Heidecker 1959), conodonts (Talent and Mawson 1994), bivalves (Cook 1993a; Heidecker 1959) and a crinoid (Jell *et al.* 1988) have resulted in a moderately well known fauna that is still incompletely documented.

Studies have identified over 35 coral taxa (Zhen and Jell 1996), 25 species of gastropods (Cook 1997; Heidecker 1959), 24 species of stromatoporoid (Cook 1995, 1999), at least 10 tabulate coral species, and a variety of undocumented brachiopods, bivalves, nautiloids, chaetetids and other faunal elements. Major exposure of bedding planes allows accurate assessment of detailed synecology and community structure within the reef complex. It is one of the very few fringing reefs well exposed within eastern Australia and is the only major, well exposed, easily accessible Givetian-age reef complex in the country.

The site is actively used for teaching and research. James Cook University currently uses it as an undergraduate site, and the Queensland Museum is investigating its detailed palaeontology. Significant reef faunas still remain unstudied, such as tabulate corals, algae, nautiloids and microfaunas. Brachiopods require significant work but are known to contain at least 15 taxa.

Fossil significance

Leichhardt Cliffs is unusual for its excellent exposure and its age. It contains the best representation of fringing reef complexes from the Middle Devonian (Givetian) and has excellent potential for palaeoenvironmental reconstruction. While the individual taxa found at the site are not rare, the assemblages are better exposed in these locations than anywhere else for material and ecologies of this age. Several endemic taxa have been recognised, but they may be present in other unstudied localities in adjacent geological provinces.

Taphonomy and condition of the site/fossils

Some taphonomic work by Cook (1995, 1999) showed a variety of preservational styles, from *in situ* reef mounds and banks to *in situ* and toppled coral thickets. Most taxa show little alteration and have limited diagenetic overprinting of their microfabric. Some areas show patchy silicified preservation. In gross scale, the large number of wide bedding surfaces for some ecological communities makes this a very accessible and exceptionally well preserved fossil sequence.

Comparison with similar sites in Australia

Reef complexes elsewhere are either less well exposed or of a different age. The Leichhardt Cliffs sites are less extensive than the classic sequences in the Canning Basin of Western Australia, but the Western Australian sequences are of Late Devonian age rather than Middle Devonian and are generally far less accessible than most of the Leichhardt Cliffs exposures. Its state of preservation is much better than the Middle Devonian sequence in the adjacent Broken River Province. It shows considerably less alteration and better exposure than carbonate sequences in New South Wales (Early Devonian of the Wee Jasper and Cavan areas).

Leichhardt Cliffs is of great importance because of its non-extractive potential, its nature as an *in situ* preserved reef complex, and its relevance to the environment and cultural life associated with the modern reefs of the Queensland coast.

Stories

Reef complexes existed in a number of intervals in Earth history. One of these events occurred during the Devonian when reef complexes grew in the north-east of the continent between 400–380 Ma. Three hundred and eighty million years ago, fringing fossil reefs grew in a wide sheltered bay known as the Burdekin Basin. Rock headlands, muddy lagoons, sandy creek outfalls and a barrier reef complex deposited a mosaic of limestones and other sedimentary rocks, and these record several million years of reef growth as sea level rose.

Evolutionary pathways were promoted by continental connections with other parts of Gondwana during the Devonian, as evidenced by the strong affinity of Chinese and European reef faunas to those in the Burdekin Basin. Stromatoporoid and other reef-builder speciation was possible only through favourable shelf and climatic conditions on the north-eastern margin of Australia.

Ludwig Leichhardt visited here in 1844 during his expedition to northern Australia, and collected fossils that were the first described from Queensland (*Cyathophyllum leichhardtii*). His journals note the limestone cliffs on the river and their relationship to the overlying basalt flow (Leichhardt 1847).

The sites have been used as the basis for a display within the Museum of Tropical Queensland Ancient Seas Display. They detail the story of fringing reef ecology during the Middle Devonian, the zonation of animals in the reef complex, and the effects of increased sedimentation and rising sea level on the reef itself. They have excellent dioramic potential when juxtaposed with the earlier east coast Devonian reefs and the spectacular Late Devonian Canning Basin sequences.

The visual impact of the sites is stunning, as they are exposed in both lateral and vertical extent on the banks of one of the largest rivers in Queensland. The site has both historic and scientific importance and complements and expands the story of the geological history of reef development nationally and internationally.

Mount Morgan Jurassic dinosaur footprint site

Site overview

Over 300 footprints of dinosaurs are preserved as positive impressions on the ceiling of the clay caverns in Mount Morgan, central-eastern Queensland. These represent concentrations of Early Jurassic (approximately 190 Ma) dinosaur footprints and trackways, and are the only abundant record of dinosaurs for the Early Jurassic in Australia. The site is an important local tourist attraction. Several trackways are represented in the mines, with the large and small footprints belonging to at least 13 footprint taxa. Most are considered theropod dinosaurs; others are ascribed to early ornithopods.

Scientific importance and research potential

Initial reporting was undertaken by Staines (1954) and Bartholomai (1966), with additional illustrated material by Molnar (1991). Stratigraphy and age constraints were delimited by Playford and Cornelius (1967) and additional sedimentologic information was provided by Neudert (1986). More recently, detailed work undertaken by Cook *et al.* (2002) and Saini (2005) has documented the footprint types and their importance. Significant additional research could be facilitated by in-depth documentation of the ichnotaxa, which requires substantial replication of the material. Sedimentology and age are well established. The site is not regarded as canonical within the discipline.

Fossil significance

The fossils found at the site are very rare. Other fossil footprints of this age are represented as single prints or limited collections of prints (Thulborn 1994). The wide exposure upon the ceiling of the mine is unusual but not unique. However, this is the only representation of a substantial dinosaur assemblage in the Early Jurassic of Australia. No dinosaur skeletal remains are known and the other footprint occurrences are minimal and isolated. Cook *et al.* (2002) and Saini (2005) have both provided environmental reconstructions for use in the tourism industry.

Taphonomy and condition of the site/fossils

The site is exposed as panels in different parts of the caverns. Some are preserved with an ironstone diagenetic overprint, which has made them the most robust. However, others are exposed directly in the siltstones which contain abundant pyrite and hence are under threat. Some areas of the caverns are subject to geotechnical problems, but major areas are relatively stable. Preservation is within the top 50 centimetres of the siltstone unit and the footprints are in positive hyporelief.

Comparison with similar sites in Australia

Footprints from Carnarvon Gorge (Thulborn 1994) and the Ipswich basin (Queensland Museum Collections) are small in number. Lark Quarry has more abundant footprints preserved and is of major significance due to its concentration, and the diverse Western Australian Cretaceous tracks preserved on Dampier Peninsula are critically important for reconstructing Australia's dinosaur faunas and Mesozoic terrestrial ecologies, but Mount Morgan is equally significant due to the rarity in Australia of evidence for dinosaurs of this age. The Triassic–Jurassic boundary represents an extinction event for some archosaurs that changed the theropod to ornithopod ratios. Evidence of dinosaurs in Australia during the Early Jurassic is scarce, and these footprints represent the only abundant record of their presence and behaviour, making this Queensland site extremely important.

Rewan, The Crater

Site overview

The Crater is an Early Triassic amphibian and fish fossil site close to Carnarvon National Park at Rewan Station, Bowen Basin, in central-eastern Queensland. The Crater contains small areas of outcrop of the Early Triassic Arcadia Formation. The site has produced the remains of well preserved amphibians, fishes, reptiles and insects from the Early Triassic (approximately 235 Ma). This is the most diverse assemblage known from this time in Australia.

Scientific importance and research potential

This is the most diverse Triassic fauna known on the continent. It is the type locality for a number of key amphibian taxa and is cited in baseline data on extinction recovery studies and palaeobiological studies worldwide. The research undertaken on this site has been excellent and of global importance. It includes significant reconnaissance-style palaeontology on the amphibian elements of the fauna and a major review of the assemblage by Northwood (1999). A detailed survey of the site (referred to by Northwood) includes a metre-scale topographic survey and 10-metre scale horizontal data. Groups examined include lungfish (A. Kemp 1991, 1993), fish (Northwood 1999; Turner 1982), amphibia and reptiles (Bartholomai 1979; Bartholomai and Howie 1970; Thulborn 1979, 1983; Warren 1980, 1981, 1985a, 1985b, 1991; Warren and Black 1985; Warren and Hutchinson 1988, 1990a, 1990b; Yates 1996, 2000), coprolites (Northwood 2005) and procolophonids (Hamley and Damiani pers comm.). The site has extremely high research potential for continued elucidation of Early Triassic aquatic and terrestrial faunas.

Fossil significance

Amphibian specimens and related fauna of this age are extremely rare, which makes this site the most important (and best known) site of its kind in Australia. The quality and quantity of material will allow significant investigations into the palaeoecology of the fauna. The site is also unusual for the high number of specimens it contains, their high diversity, and unusual faunal content. It is the most diverse Early Triassic fauna site known in Australia.

Taphonomy and condition of the site/fossils

The material weathers out of red-brown shales within the Arcadia Formation, but consists of partially articulated and articulated material. Skull material is uncommon but is in excellent condition when found.

Comparison with similar sites in Australia

The site is the best Triassic fauna site in Australia and contains the highest diversity for its age. In terms of site aesthetics, it lacks the impact of non-extractive sites and its value lies in the constituent fauna so far sampled. Thus on the basis of on-site preservation it is of lower quality than other Mesozoic sites such as Mount Morgan, Fletcherview, Lark Quarry (Qld). The site is more abundant and diverse than the Blina Shale, Canning Basin (WA), which is the closest comparable site. It is of higher quality than the Duckworth Creek, Mt Crosby, Redbank Plains and Chatsworth Triassic sites (Qld).

Stories

This site tells the story of life around shallow ponds and streams approximately 235 million years ago, the strange amphibians and reptiles that lived in and around the area, and the ecosystem in which they flourished. The fauna from this site provides one of the few insights into the terrestrial and freshwater biota of the continent following the Permian extinction event. The recovery phase and subsequent radiation of amphibia and reptiles saw the development of many amphibian groups and the rise of the reptiles represented herein. The dioramic potential of the site is currently limited without significant reconstruction and interpretation.

Black Mountain Cambrian–Ordovician Boundary Section

Site overview

This site, which includes Mt Unbunmaroo (Black Mountain) Cambrian–Ordovician Section in the Chatsworth and Ninmaroo Limestones, is located to the north-east of the small town of Boulia in western Queensland.

The type and classic section, exposed on the mountain, consists of a highly fossiliferous Late Cambrian to Early Ordovician limestone succession that has been used for international and national studies and research on the Early Palaeozoic. The sequence contains abundant trilobites, nautiloids, gastropods, monoplacophorans, polyplacophorans, rostroconchs, conodonts, bivalves, worms, echinoderms and stromatolites. The site was a potential location (unsuccessful) for the Cambrian–Ordovician Global Stratotype (Druce *et al.* 1982). The nautiloids are equal to the oldest in the world and consist of new taxa of protactinoceratids (Wade pers comm.). The site tells the story of the rise of modern animal groups and is of particular use in geological correlation.

Scientific importance and research potential

Significant studies of the site and its fauna have been summarised by Shergold *et al.* (1991). Work on the specific faunas include conodonts (Druce and Jones 1971; Nicoll 1991; Nicoll and Shergold 1991); trilobites (Shergold 1972, 1975, 1982); polyplacophorans (Runnegar *et al.* 1979); and rostroconchs (Pojeta *et al.* 1977).

It is regarded as a nationally and internationally important standard in chrono, magneto and biostratigraphy for the Cambrian–Ordovician boundary, and is often cited in the geological literature. Examples of this work include Druce *et al.* (1982); Ripperdan *et al.* (1992); Ripperdan (1992); Ripperdan and Kirschvink (1992); Shergold and Nicoll (1992) and Anderson *et al.* (2004).

Extremely high levels of lithostratigraphic and biostratigraphic studies combined with detailed metre-scale biostratigraphy make this sequence among the most studied on the continent. Significant collections of nautiloids made by Wade (held in the Queensland Museum) and of gastropods (Shergold *et al.* 1991) will provide important information on the earliest radiation of mollusc clades. However, the nautiloid studies were not completed by Wade prior to her death, although unpublished manuscripts were recovered. The palaeontological work on the conodont faunas is extremely robust; the trilobite studies may require review but are very sound. Studies of the echinoderm fauna are good but not comprehensive. The locality stands as a typical outcrop area for the terminal Cambrian Datsonian stage and will benefit from additional research.

Fossil significance

While the fossils found at this site are not rare organisms in isolation, as a biostratigraphic assemblage the material is of significance. Such a well preserved laterally and vertically extensive section and sequence of faunas is unusual. The site provides an excellent representation of Late Cambrian and Early Ordovician faunas, which are preserved in more disparate localities elsewhere. The site's potential for palaeoenvironmental reconstruction is good, but as yet no synthetic attempt at palaeoenvironmental reconstruction has been attempted.

Taphonomy and condition of the site/fossils

Much of the site contains accumulations of silicified replacements of calcitic and aragonitic shelly fauna, with unreplaced phosphatic fossils. Each unit varies, but notable accumulations are disarticulated and current-swept echinoderm debris units and a mix of benthic and nektic fauna.

Comparison with similar sites in Australia

The site represents a fossiliferous sequence that is laterally and temporally extensive and crosses a major geological time boundary. Its level of exposure, and the quality and quantity of key organisms are unrivalled for its time period. It is less important than non-extractive sites such as Lark Quarry and Mount Morgan (Qld), but more important to protect than extractive 'pit' and diffuse faunal sites such as Mount Crosby and Rewan (Qld). It is comparable in stratigraphic importance to the Mesozoic–Cenozoic boundary exposures in Western Australia's Giralia Range and the Ediacaran Global Stratotype in the Flinders Ranges of South Australia.

Stories

Black Mountain relates the story of the evolution of molluscan groups in early shallow seas, and the rise of the cephalopods and bivalves. Black Mountain also tells the story of how geological time is correlated, as the site has been important in defining the stages of the Late Cambrian worldwide and the development of geological correlation on the Australian continent. The site provides a potential diorama of marine life in the Late Cambrian inland sea.

Chinchilla Rifle Range

Site overview

Erosional gullies at Chinchilla Rifle Range (Chinchilla Local Fauna) contain an abundant and well studied vertebrate fauna that is among the best known of all Pliocene faunas in Australia. At least 12 taxa of kangaroos, two diprotodontids, numerous birds, rodents, crocodiles, lizards, snakes and fishes have been recovered from the site. The Chinchilla site is located along the Condamine River, behind Chinchilla Rifle Range, Chinchilla, in south-east Queensland.

Scientific importance and research potential

Chinchilla local fauna is regarded as one the best Pliocene faunas in Australia. There has been extensive and high quality study of the site's faunas and stratigraphy. A detailed taxonomic list with citations leading up to 1991 was provided by Rich *et al.* (1993) with subsequent work by Hand *et al.* (1999), Wroe and Mackness (2000a, 2000b), Hutchinson and Mackness (2002), Dawson (2004) and Hocknull (2005). Small elements of the fauna (rodents, small lizards) and other rarer elements are still being discovered on site. Collection has thus far been reconnaissance with no systematic excavation, application of magnetostratigraphy or attempts at wholesale reconstruction of the environment. Definitive age controls are still required for the site.

Fossil significance

The site contains a representative fluvial assemblage of Pliocene age, with a diverse fauna. Sites of this age, extent and quality are rare. A number of taxa are endemic to the site, but some are shared with the Bluff Downs local fauna in central-eastern Queensland. The potential for palaeoenvironmental reconstruction is excellent and pertinent to modern river studies.

Taphonomy and condition of the site/fossils

Chinchilla Rifle Range is a fluvial site with well preserved and occasionally partially articulated remains in good condition. No taphonomic studies have been undertaken. The site is replete with erosional gullies and is often used as a motorcycle riding site and therefore may be under some threat.

Comparison with similar sites in Australia

The site is similar to Hamilton (Vic), Bluff Downs (Qld) and Bow (NSW). It is of equal importance to these sites and more important than Tea Tree Cave near Chillagoe (Qld), though it lacks the broad chronological and evolutionary span of sites like the older Riversleigh (Qld), Naracoorte (SA) and the Wellington Caves (NSW).

Stories

Chinchilla Rifle Range gives a detailed snapshot of both terrestrial and aquatic faunas for the Pliocene and shows the range of ecosystem elements in a riparian zone for this time. The rise of the macropods in response to aridification of the continent eight million years ago is exemplified by this and other similar sites.

Chatsworth Limestone echinoderm and trilobite localities

Site overview

The two Chatsworth Limestone echinoderm and trilobite localities are comprised of very well exposed sequences of Cambrian Limestone, in the vicinity of Lily Creek, approximately 80 kilometres south-west of Duchess, 16 kilometres east-north-east of Digby Creek and 3.5 kilometres south of Chatsworth Station in western Queensland. The fossiliferous sequences contain many trilobites, brachiopods, echinoderms and other fauna, which provide an insight into the Cambrian marine fauna of the shallow seas that flooded the Georgina Basin during the middle Late Cambrian (post-Idamean, approximately 500 Ma).

Scientific importance and research potential

The faunas and stratigraphy of both sites have been extensively studied. Classic studies were conducted by Shergold (1980) on the trilobite faunas, with the identification of 52 taxa in the Chatsworth Limestone. Several authors (Jell *et al.* 1985; Smith and Jell 1990, 1999) recorded a number of taxa of echinoderms from the Lily Creek section including large numbers of the eocrinoid *Ridersia*. Conodonts are known from the unit (Druce and Jones 1971), providing a secondary biostratigraphic framework. Taxonomic work is robust for trilobites, echinoderms and conodonts; all other work is reconnaissance in nature. The site continues to have research potential: significant other faunas are known in bulk collections from the site, and detailed middle Late Cambrian biological data would be gained from analysis and revision of the fauna.

The site is an often-visited and extremely instructive fossil site that provides evidence for the Late Cambrian diversification of trilobite groups. It is the best known Late Cambrian echinoderm locality. The rarity of such organisms makes the site important.

Fossil significance

Chatsworth Limestone is unusual for its echinoderm and other faunal abundance, and the pervasive silicification reinforces the faunal abundance. Large numbers of echinoderms of this age and type are rare worldwide, and the site provides much data regarding the evolution of the phylum. Trilobites are less uncommon in Australia but the quality and abundance of material in this sequence is extraordinary. The site provides an excellent representation of Cambrian trilobite fauna in Australia, and is an exceptional global representative of echinoderm faunas of this age. The fauna requires some investigation to interpret the palaeoenvironment fully. General information on the sedimentology and faunas is readily available.

Taphonomy and condition of the site/fossils

The fossils are preserved as silicified replaced faunal elements on bedding planes. Many trilobites are preserved as disarticulated exuviae, but some are found as complete, delicately preserved specimens. Echinoderms include complete and partial calices, complete *Cambroblastus* specimens and a number of complete edrioblastoids and edrioasteroids.

Comparison with similar sites in Australia

This is the highest quality early echinoderm site on the continent, but its fauna has been interpreted on the basis of extracted and processed material. The scientific value in the site is from removed material. It is more important than sites such as Mount Crosby, Tea Tree Cave and Chinchilla, and is equivalent to Rewan and the similarly aged Black Mountain (Qld). The site itself is less important than *in situ* 'interpretative sites' like the dinosaur track sites of Lark Quarry and Mount Morgan (Qld).

Stories

The Late Cambrian was a crucial time in the development of marine biota, and this site provides an excellent representation. Modern marine faunas include a significant diversity of echinoderms (crinoids, urchins, starfish, holothurians and other groups). These organisms all had their origins in the Cambrian at a time when radiation of all life experimented with many new body plans. In the Chatsworth Limestone some of the best representatives of these Cambrian echinoderms are preserved, along with the other well known contemporaneous fossils, the trilobites. Combined with disparate sites in the Georgina Basin it completes the story of the invasion of shallow continental seas onto the Australian craton during the Cambrian.

The site is not known to have Indigenous significance, although local chert nodules are often found in minor quarries nearby as worked tools and flakes.

Redbank Plains Formation

Site overview

The Early Palaeogene (Palaeocene–Eocene) deposits of the Redbank Plains Formation are sporadically exposed as outcrop and consist of shales with ironstone concretions, many of which contain fossils. These include fishes, crocodiles, turtles, birds, insects and plants, making it one of the few diverse early Paleogene faunas known in northern Australia. Most specimens were recovered from weathered-out nodules that occur on the surrounding slopes, much of which has undergone urban development.

The site is located along Jones and Brittains Roads, Redbank Plains, in the western suburbs of Brisbane. The Ipswich city council has reserved a small holding of the strata for a proposed conservation reserve.

Scientific importance and research potential

Some reliable study has been undertaken on the fossil fishes (Hills 1934; Kemp 1991; Li 1994), turtles (de Lapparent and Molnar 2001) and birds (Vickers-Rich and Molnar 1996). Research undertaken on insects (Tillyard 1916; Tindale 1945; Riek 1952, 1954) is in need of review. The site has minor research potential, associated with reviews of fauna.

Fossil significance

The combination of age and types of fauna make the Redbank Plains Formation an uncommon faunal occurrence. The age is unusual and poorly represented elsewhere in Australia. The site is on the edge of a major capital city with urban development encroaching upon fossiliferous outcrop. The fossils provide an excellent representation of an aquatic Early Palaeogene assemblage and the potential for palaeoenvironmental reconstruction is good, given the diversity of site.

Taphonomy and condition of the site/fossils

Material is preserved as natural casts and moulds only, with little original material preserved. Insects are small to minute and the plants are preserved as compressions, generally from nearby quarries.

Comparison with similar sites in Australia

The importance of this site is its age, as it is the oldest of the Palaeogene deposits in Australia. It contains some of the most significant fish and turtle deposits and early Cenozoic bird remains in Australia. Its suburban location means it is a more accessible site than the important Tingamarra Local Fauna site near Murgon (Qld), but this urban setting also poses threats to the fossils.

Stories

The site has modest dioramic potential. It shows the story of life 55 million years ago in the Brisbane Valley, and the animals that inhabited the cool lakes and streams. The site is important for what it shows of the broader picture of Australian faunal development during the early Cenozoic.

Mount Crosby Triassic insect locality

Site overview

The Mount Crosby Triassic insect locality is comprised of units of the Ipswich Coal Measures that are exposed in a small pipeline easement. The shales within these coal measures contain an abundant insect fauna of Middle Triassic age (approximately 225 Ma). The site is located in a cutting on the old abandoned Mount Crosby railway, a pipeline easement adjacent to the Mount Crosby road at Colleges Crossing in south-east Queensland.

Scientific importance and research potential

Insects from the site have been described by Dodds (1949); Evans (1961, 1971); Fleming (1966); Riek (1955); Tillyard (1916, and numerous papers cited in Jell 2004); Tillyard and Dunstan (1924). A review has been undertaken by Jell (2004) but substantial further review of the taxonomy of the insect groups is required. The site is widely regarded as a quality locality for insect fossils, but has no current teaching and research function.

Fossil significance

Fossil insects are sporadic throughout other sites of the Ipswich Coal Measures but are not numerous. Insects of Middle Triassic age are known from several sites in Queensland and elsewhere in Australia, but this locality contains abundant material with many site-specific taxa. Insect fossils are relatively rare, and it is unusual to have insects preserved in such numbers. The site provides an excellent representation of fossil

Figure 3.1 Exposure at Mt Crosby insect locality.
Photograph © SEWPaC

insect faunas of the Triassic, with some potential for palaeoenvironmental reconstruction.

Taphonomy and condition of the site/fossils

Very few whole animals are preserved, with most specimens found as isolated wings, elytra, and other similar elements.

Comparison with similar sites in Australia

The Mount Crosby insect locality is of similar quality and type to the Late Permian Belmont insect beds in New South Wales. However, this site is younger by about 40–20 Ma. The Permo-Triassic Dinmore Quarry contains some good insect specimens of similar age but is better known for its plant fossils.

Stories

The Triassic radiation of insect groups in the southern hemisphere is documented from a few sites in Gondwana. The Mount Crosby fauna is one such site and complements other stories.

Jack Hills Gorge and Fish Hill, Broken River Province

Site overview

The Jack Hills Gorge and Fish Hill sites are two closely located and related sites that occur in the Silurian–Devonian sequences of the Broken River Province, south-west of Greenvale, north-east Queensland (approximately 150 kilometres north-west of Charters Towers).

At Jack Hills Gorge, the Broken River cuts through near vertical exposures of the Jack Formation and the type section is 580 metres thick. The karst-forming limestone units are represented by upper and lower limestones separated by a muddy sequence. The site contains the coral gardens, trilobite localities, megalodontid bivalve beds, lingulid beds, nautiloid beds and an upper stromatoporoid coral boundstone unit. It traverses the Silurian–Devonian boundary and is an important biostratigraphic reference section.

Fish Hill is the informal name given to outcrops of allochthonous carbonates in the Burgess Marine Valley (Burgess Formation) that contain abundant fish remains in addition to a coral–stromatoporoid fauna.

Scientific importance and research potential

The existing scientific work is of good quality. The sedimentology, regional geology and stratigraphy of the sites have been reviewed by Withnall *et al.* (1988) and Jell *et al.* (1993). Conodont studies have been undertaken by Sloan *et al.* (1995) and Simpson (1999, 2000), trilobites by Holloway (1994), and corals by Munson (1979, 1987). Fish studies remain primarily unpublished.

The site has been used for teaching by the University of Queensland since the 1970s. It is regarded as a key section for the Silurian–Devonian in Australia and the coral gardens at Jack Hills Gorge are regarded as a geological monument by the Queensland Division of the Geological Society of Australia.

Both Fish Hill and the Jack Hills Gorge have potential for detailed taxonomic and biostratigraphic work. In addition, they have palaeoecological and chemo/magneto stratigraphic potential.

Fossil significance

The sequence is an excellent representation of Late Silurian to Earliest Devonian platform carbonates and is abundantly fossiliferous and well exposed. While the individual organisms found at the site are not rare, a concentration of fish remains of this age is rare. Several sequential fossil assemblages are represented, including stromatoporoid coral banks, muddy coral lagoonal systems and megalodontid banks. The potential for palaeoenvironmental reconstruction is excellent, given the abundance of data and excellent exposure.

Taphonomy and condition of the site/fossils

Some parts of the sequence show minor to medium recrystallisation. Most are in excellent preservation.

Comparison with similar sites in Australia

This is the best exposed site of its type and age in Australia. This site is more important than Late Silurian sites in the Yass Basin due to its better exposure. It is more important than the more recrystallised examples on the Chillagoe Shelf, Queensland. The site is also more important than Silurian karst limestones in New South Wales due to its better exposure, diversity and preservation. Fish fauna are represented in an unusual Early Devonian assemblage which is more concentrated than elsewhere (e.g. Taemas–Cavan–Wee Jasper NSW), although it lacks the superb three-dimensional fossils of Taemas–Cavan–Wee Jasper.

Stories

The site links strongly to the story of reef development on the north-east margin of the continent between 410–400 Ma and of the communities that lived upon the continental shelf. This also connects with the story of Early Devonian fish diversity represented at sites like Taemas–Cavan–Wee Jasper.

Jack Hill Gorge was named for Robert Logan Jack, the early north Queensland government geologist.

Riversleigh

Site overview

The Riversleigh World Heritage Area encompasses more than 200 individual vertebrate fossil localities that mostly span the Late Palaeogene to Neogene and includes a Pliocene and Pleistocene site. It is about 11 000 hectares in extent and is located 225 kilometres north-north-west of Mount Isa and 70 kilometres south-west of Gregory Downs, Queensland.

Scientific importance and research potential

A major research effort by Michael Archer and colleagues spanning over 35 years has revealed a whole suite of diverse vertebrate-yielding localities in the Riversleigh World Heritage Area (Archer *et al.* 1991, 2006). The fossil deposits span the last 25 million years but the oldest ones are in the majority and range from the latest Oligocene through to early Late Miocene in age. In total there are more than 200 localities recognised, and these have produced many new species and unusual animals. The Riversleigh sites have made the largest and one of the most important contributions to the understanding of the evolution of Australia's terrestrial vertebrates, particularly the mammals. They are predominantly vertebrate in nature, but fossil insects (Duncan *et al.* 1998) and plants have also been recovered in some localities.

Figure 3.2 Fossil outcrop at Riversleigh

Fossil significance

The Riversleigh localities have produced some of the richest mammal assemblages in the world. The unique preservation, concentration of bones and large-scale extraction operation has resulted in an outstanding collection of fossil specimens. Numerous publications and manuscripts in progress describe several hundred previously unknown species of mammals, reptiles, amphibians, birds and fish (e.g. Black and Archer 1997; Brewer *et al.* 2008; Crosby and Archer 2000; Gaffney *et al.* 1992; Hand and Archer 2005; Louys *et al.* 2009; Scanlon *et al.* 2003; Willis and Archer 1990). The cave deposits have yielded the most diverse bat assemblages known to science (e.g. Hand *et al.* 1990, 1998; Hand and Archer 2005). The Riversleigh World Heritage Area continues to have significant potential for future research.

Taphonomy and condition of the site/fossils

The main fossil-bearing sediments are included within the Carl Creek Limestone. This is a group of sediments of variable origins, composition and textures, representing a period of deposition spanning approximately 25–12 Ma (Oligocene–Miocene). The Carl Creek Limestone is found within the watershed of the Gregory River, a perennial spring-fed system. The Gregory River has carved its path through the Cambrian Thorntonia Limestone to the west of Riversleigh Homestead and Pre-Cambrian quartzites to the east. The ancestral Gregory River valley formed prior to the deposition of the Carl Creek Limestones.

The area contains a mosaic of different local limestones laid down in freshwater pools and lakes. Uplift and aerial exposure of these freshwater rocks enabled cave systems to form. Some vertebrate-yielding cave deposits are found in the Cambrian limestones of the area, but many are incised into the freshwater limestones. In addition to the lacustrine and cave deposits, there also is a Pliocene cave deposit and some Pleistocene fluviatile sediments.

The lime-saturated rainforest pools and lakes in the Oligocene–Miocene enabled almost perfect preservation of bones, skulls and teeth. After some initial experimentation with various preparation techniques, Dr Archer's research group found that the best way of extracting the fragile fossils out of the limestone was by use of diluted acetic acid. This etched the rock from the bone, leaving the latter in pristine condition.

Comparison with similar sites in Australia

The richness, diversity, preservation and completeness of Riversleigh's fossil deposits are outstanding. They are more diverse than any known Australian fossil deposit of similar age (Travouillon *et al.* 2006). Riversleigh has

Figure 3.3 Reconstruction of typical plants and animals found at Riversleigh fossil site. Painting Peter Schouten © SEWPaC

trebled the number of terrestrial mammals known from the late Palaeogene and early Neogene, and includes an equally diverse fauna of birds, amphibians and insects. The fossil deposits also show evidence of ancient and modern groups of animals existing side by side. They provide important evidence for the interpretation of the evolution and radiation of species in Gondwana. The deposits have provided new evidence to revise previously widely accepted evolutionary theory, as a number of species (including thylacines and elapid snakes) have been found to predate former postulated origin times by many millions of years. The fossil deposits have also been significant in allowing the correlation of deposits elsewhere in Australia, such as the Etadunna Formation Lake Eyre Basin localities in South Australia and Bullock Creek in the Northern Territory.

Stories

The Riversleigh fossils were first noted in 1901 by W.E. Cameron. Preliminary exploration of the area in 1963 yielded six major fossil sites of considerable scientific interest. Systematic exploration of the fossil deposits commenced in 1976 and continues today.

The completeness of the fossil record at Riversleigh not only gives us an insight into life in ancient Australian rainforests as the continent continued its northward drift, but it also helps predict the viability of individual mammal groups. Approximately 20 million years ago, the Riversleigh habitat supported five species of thylacines. Five million years later there were only two species left. Approximately eight million years ago a single species remained. As we all know too well, the lineage came to an end in Tasmania in 1936.

The dioramic potential of the Riversleigh fossils is realised in its pairing with the outstanding Naracoorte Caves National Park as the World Heritage-listed Australian Fossil Mammal Sites (Riversleigh/Naracoorte), which together illustrate key stages of evolution of Australia's mammal fauna during the Cenozoic Era (65 million years).

Homevale Station

Site overview

Homevale Station is a *Glossopteris* plant fossil site. It is located near Nebo in north Queensland and is part of the Late Permian Fort Cooper Coal Measures, in the Bowen Basin.

Scientific importance and research potential

During the Late Carboniferous and Permian periods several new plant groups evolved. These include conifers, ginkgophytes and cycads, some of which are still represented in present-day vegetation. *Glossopteris* is an early gymnosperm with features that undoubtedly favoured it and enabled it to attain a dominant status in the vegetation over more primitive plants, especially following the retreat of Early Permian glaciers. These features include secondary vascular tissue that enabled woody support for growth, more complex leaf venation and branching patterns, and more sophisticated reproductive strategies.

Initially, reproductive material from the Bowen Basin was interpreted by Gould and Delevoryas (1977) as evidence that *Glossopteris* was a member of the extinct pteridosperm group, but more recent evidence indicates that the ovules are borne on the adaxial surface of fertile structures as in early conifers. Simple pollen tubes similar to those of cycads and *Ginkgo* were observed in the ovules, a finding that has implications for pollination biology in general and in interpreting the phylogeny of seed plants (which include modern flowering plants). A recent review by Pigg and Nishida (2006) emphasised the scientific importance of the Homevale Station *Glossopteris* fossils internationally. In addition to *Glossopteris*, the Homevale Station fossils include gymnosperm trunks, ferns and fungi.

Fossil significance

Glossopteris floras of the Permian are significant because they are recognised around the world as the first definitive Gondwanan plants. Between the Silurian and Permian periods, the Australian region shifted from equatorial to near-polar latitudes and became part of Gondwana, a landmass and biogeographic complex comprising regions of modern Australia, Antarctica, India, Africa and South America. The intermediate Devonian and Carboniferous periods saw the gradual expansion and diversification of land plant floras, and included some lycophytes that attained the stature of tall trees. Relatively few detailed studies have been done on Australian material from these periods, but it is generally regarded that at least at the generic level the plants were more or less cosmopolitan. In contrast, *Glossopteris* fossils are known only from Gondwanan continents. This was one of the most powerful pieces of evidence in support of continental drift and plate tectonics (Gould and Delevoryas 1977; du Toit 1937).

Taphonomy and condition of the site/fossils

Many fossils at the Homevale site are three-dimensionally preserved (permineralised) in silicified chert, including *Glossopteris* reproductive structures and leaves (McLoughlin 1992). The remarkable preservation of the Homevale *Glossopteris* fossils was highlighted in the journal *Nature* (Nishida *et al.* 2003).

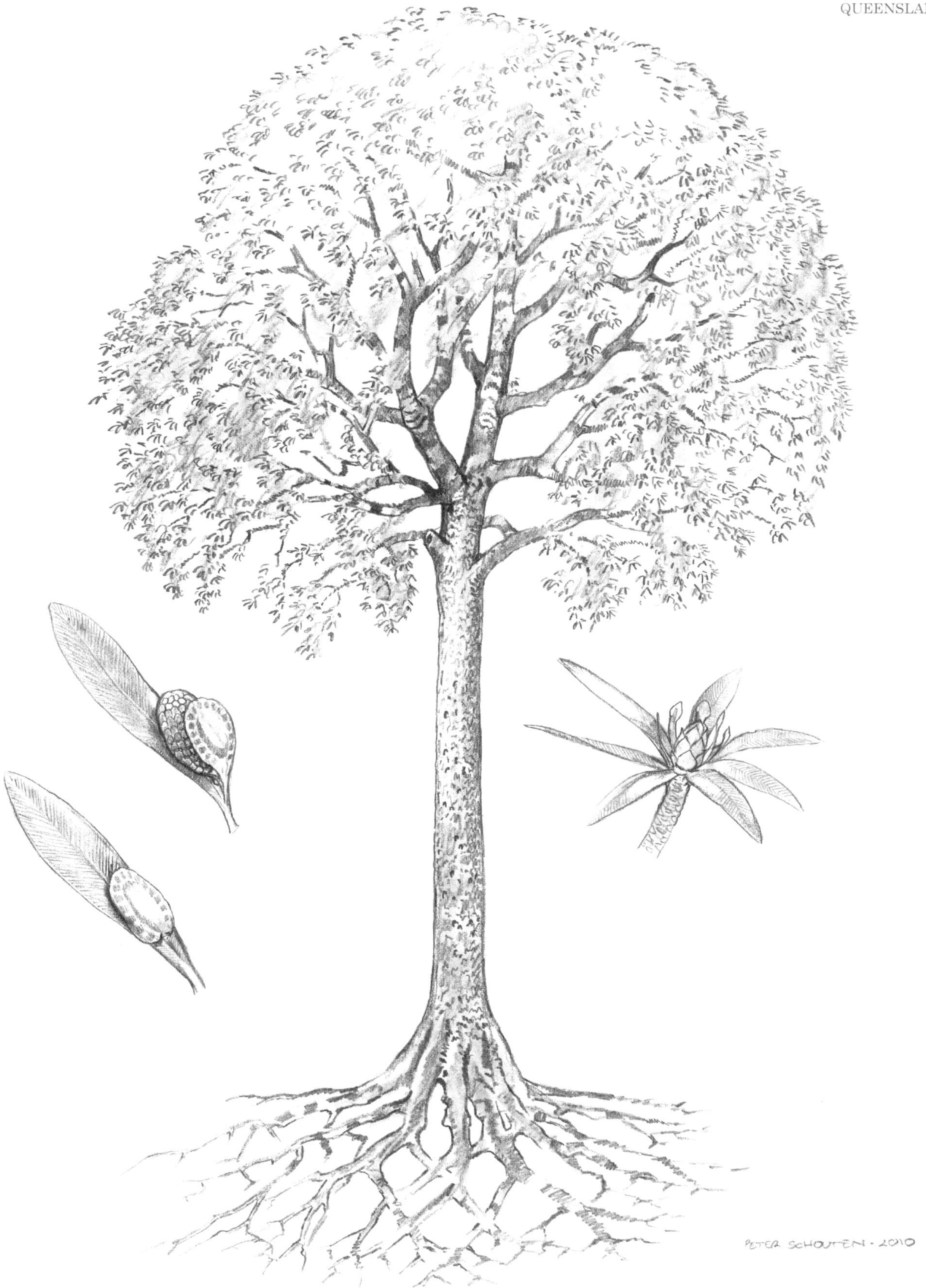

Figure 3.4 *Glossopteris linearis*. *Glossopteris* fossils are known only from Gondwanan continents, so their existence provides impor-ant evidence in support of continental drift and plate tectonics. Drawing Peter Schouten © SEWPaC

Comparison with similar sites in Australia

Similar sites include the Collie Basin near Perth (WA) (McLoughlin and Hill 1996), and notably the Cooyal locality near Mudgee (NSW). Although this latter site may host the most complete Permian impression flora in Australia, the Homevale site, now within Homevale National Park, is considered superior because of the exceptional quality of the fossil preservation.

Stories

As the first definitive southern hemisphere fossils, *Glossopteris* are important in illustrating the theory of continental drift, as described above. Historically, the deciduous *Glossopteris* vegetation has also been of great economic and social importance to Australia and other southern hemisphere countries in the form of vast coal deposits.

Dinmore Quarry

Site overview

The Dinmore quarry is an important *Dicroidium* flora site, which is part of the Late Triassic (Carnian, approximately 225 Ma) Blackstone Formation located a few kilometres east of Ipswich, south-east Queensland.

Scientific importance and research potential

The end of the Permian was marked by one of the major mass-extinction events in Earth history that saw the extinction of glossopterids and their wholesale replacement by diversifying conifers, cycadophytes, pteridosperms and ferns. The Middle and Late Triassic vegetation is often referred to as the *Dicroidium* flora after one of its most dominant plants, a member of an extinct gymnosperm group known as the corystosperms. The Dinmore Quarry assemblage is composed of a diverse array of plants including *Dicroidium* and other gymnosperm species, ferns and plants with fern-like foliage, and bryophytes.

Most of the Triassic plants described from this site are in classic papers by Walkom (1917a, 1917b) and Jones and de Jersey (1947), but revisions with supplementary modern cuticular analyses would be worthwhile. Adding to the significance of the site, a more recent (Cenozoic) fossiliferous plant exposure also occurs in the quarry. Both these and the Triassic fossils are clearly worthy of further study.

Fossil significance

The site is important because it possesses an excellent, diverse *Dicroidium* flora with more than 70 species reported (Pattemore and Rigby 2005).

Taphonomy and condition of the site/fossils

The site is given further significance because foliar cuticular preservation is apparent in the mine-dumped Blackstone Formation shale in the quarry. Cuticles can reveal much more about plant relationships, climate and environment, since a range of additional diagnostic anatomical structures such as stomata and trichomes may be preserved.

Comparison with similar sites in Australia

The best Early Triassic flora sites in Australia occur in the Sydney and Lorne basins, and the flora from the Middle Triassic Nymboida Coal Measures has recently been the subject of a series of papers (e.g. Holmes and Anderson 2005). Although the importance of the Nymboida site is well established, the Late Triassic Dinmore site is regarded as more significant: on a world scale, the plant diversity at Dinmore is comparable with the much more extensively studied benchmark Molteno Formation from southern Africa.

Stories

The Dinmore site offers a fascinating opportunity for environmental reconstruction, as many insects and other invertebrates are also preserved in the lacustrine sediments. Indeed, the Blackstone Formations in the Upper Triassic Ipswich Coal Measures were described as a principal source for Australian fossil insects by Jell (2004). (See Mount Crosby Triassic insect locality above.)

Lark Quarry Dinosaur Stampede National Monument (Winton Formation)

Site overview

The 384-hectare Lark Quarry Conservation Park records in stone the most concentrated set of dinosaur footprints known in the world: the tracks of between 170 and 200 individual dinosaurs, consisting of nearly 4000 footprints, made in 10 to 30 seconds around 98 million years ago (AHDB 2004).

During the early Late Cretaceous, a shallow and retreating inland sea covered much of what is now central Queensland. The landscape received high rainfall, and ferns, conifers, early angiosperms, cycads and ginkgos grew in abundance. Following the northern retreat of the sea, lowland delta systems developed in the greater Winton area. The resultant landscape, a vast humid plain stretching nearly 700 kilometres wide and 1200 kilometres north to south, was dominated by lake and river shoreline environments containing broad, shallow drainage channels, sand bars and plastic, muddy flats. These environments are recorded in the rocks and sediments of the 93 and 98 million year old Winton Formation (AHDB 2004, Cook 2004). In some areas the Winton Formation is up to 400 metres thick – a testament to the huge amount of sediment deposited by the inland water systems of central Queensland in the Late Cretaceous. As well as the ichnofossils of the Lark Quarry Conservation Park, the Winton Formation preserves other spectacular dinosaur, marine reptile and plant fossil sites (AHDB 2004).

In dramatic contrast to its Late Cretaceous environment, the park now lies in a spectacular setting of dissected, residual hills and mesas on the edge of the Tully Range south-west of Winton (AHDB 2004).

Scientific importance and research potential

The primary research conducted on the Lark Quarry dinosaur trackways is commonly cited as the benchmark for study into dinosaur footprints and behaviour. As the place preserves nearly all the world's dinosaur trackways made by running dinosaurs, it is also an important information source for locomotion studies and performance analysis for the types for both ornithopods and coelurosaurs ('hollow boned' two-legged dinosaurs) (AHDB 2004).

The site contains an abundance of trackways, preserved in an interpreted palaeo-landscape and behavioural context. They record the paths of a mixed herd of small, two-legged dinosaurs that ran in a single direction across a muddy shoreline area in an apparent attempt to flee a much larger carnivorous dinosaur that stood in their path. This behaviour is consistent with and has been interpreted as a dinosaur stampede event (AHDB 2004).

Sediments continued to accrue in the Middle Cretaceous and have compressed over time into the interbedded layers of sandstones and claystones of the Winton Formation. These deposits provide important palaeoecological information as well as a wealth of other dinosaur and marine reptile body fossils, which are gradually being excavated and reconstructed. The interpretation of this palaeoenvironment suggests a much wetter climate prevailed 98 million years ago in an area of Queensland that is currently arid (AHDB 2004).

The site is an international model for the *in situ* preservation of dinosaur footprints and the protection of fossil heritage (Cook 2004).

Fossil significance

Lark Quarry and Seymour Quarry are the only known fossil sites that preserve trackways made by numerous dinosaurs running in a single direction, interpreted as a dinosaur stampede (Thulborn 1990). No other known trackway site in the world indicates dinosaur stampede behaviour (Wade and Molnar 2000).

The trackways are in an area of approximately 200 square metres of exposed, almost horizontal bedding plain, roughly triangular (Thulborn and Wade 1984; Wade and Molnar 2000). The trackways at both Seymour and New Quarries – an extension of those found at Lark Quarry – are currently buried (AHDB 2004; Cook 2004; Thulborn and Wade 1984).

The trackways contain the most concentrated known set of dinosaur footprints in the world (Cook 2004). Lark Quarry and Seymour Quarry contain between 170 and 200 individual dinosaur trackways made up of nearly 4000 individual footprints (Wade and Molnar 2000). The trackways are almost entirely pointed in a single, north-easterly direction (Thulborn and Wade 1984) although there are 11 large theropod footprints comprising a single trackway that point in a south-westerly direction (Wade and Molnar 2000: p. 1).

Taphonomy and condition of the site/fossils

The tracks were preserved by burial in a fortuitous flood event soon after they were formed and are exceptionally well preserved as a result (AHDB 2004). The integrity and fine preservation of the trackways can be attributed to the characteristics of the clay-sand matrix in which they were originally formed. Fine detail such as scratch marks in the digit imprints on some *Skartopus australis* tracks as well as the presence of scrape marks across many of the *Wintonopus latomorum* tracks attest to the high level of preservation of the trackways (AHDB 2004; Thulborn and Wade 1984).

During 2001, a building was constructed over the Lark Quarry trackways to provide better security and a more stable environment, designed to be ecologically sustainable and ensure that damage to the trackways is minimised by stabilising temperature and humidity fluctuations, preventing erosion from run-off, and preventing people and wildlife from walking and standing on the trackways. The building uses the thermal mass properties of its rammed earth walls to keep environmental conditions within the building relatively stable. The fully enclosed ichnofossils are also protected from dust and harsh light (AHDB 2004; Cook 2004).

During construction, Dr Alex Cook from the Queensland Museum undertook a maintenance program on the surface of the trackways, and following the completion of the building in June 2002, the collapse of a large, rammed earth wall necessitated further work. As a result, although the surface is fragile, it is climatically and chemically stable (AHDB 2004; Cook 2004).

Comparison with similar sites in Australia

A survey of the world's largest dinosaur trackway sites published by Gillette and Lockley (1989) indicated that Lark Quarry records the highest number of preserved tracks and trackways, and among the best quality preservation for any site in the world, with more than 4000 tracks, forming more than 500 trackways. Lockley and Gillette found that the next most significant international sites were Canada's Peace River site, with around 1200 tracks and about 100 trackways; and Bolivia's Toro Toro site, with around 100 trackways (Long 2004).

The two other significant Australian dinosaur track sites are the Broome Sandstone of Dampier Peninsula (WA) and Mount Morgan (Qld); the Mount Morgan site is less diverse than both the Dampier Peninsula site and Lark Quarry. Alone of the Australian track sites, the west coast of Dampier Peninsula preserves evidence of sauropods, and ichnofossiliferous outcrops are discontinuously exposed for a length of more than 90 kilometres along that coast. The large number and variety of tracks in a range of depositional settings provides an otherwise unobtainable census of dinosaur populations and communities (Thulborn pers. comm. 2009). However, although the taxonomic diversity of the Lark Quarry site is lower than in the Broome Sandstone, its quality of preservation, its accessibility and the detailed anatomical and behavioural data enabled by the sheer number of tracks revealed led to its inscription as the first fossil site in the National Heritage List in 2004 (AHDB 2004; Cook 2004; Long 2004).

Other Australian track sites include the Ipswich Coal Measures, the Precipice Sandstone, the Razorback Beds and Walloon Coal Measures, all in Queensland. These range in age from the Late Triassic through Mid Jurassic and so are all older than both the Dampier Peninsula and Lark Quarry sites. No more than three types of tracks are preserved at any of these other sites, and the taxa represented are interpreted to be ornithipods and theropods (Thulborn pers. comm.). None of these sites approaches either the Dampier Peninsula or Lark Quarry for the numbers of prints preserved.

Stories

In the 1960s Glen Seymour, a local station manager, discovered dinosaur trackways in what became the Lark Quarry Conservation Park while fossicking for opal at the base of a small hill in an area of what was then Cork Station. This area was later named Seymour Quarry in recognition of his remarkable find (AHDB 2004). Continuing a long tradition of collaboration between academic, museum and amateur fossil hunters established in the eighteenth and nineteenth

Figure 3.5 Dinosaur stampede footprints at Lark Quarry, Winton. Photograph Jaime Rankin © SEWPaC

centuries, local people continue to work closely with palaeontologists from the Queensland Museum and the University of Queensland to find, excavate and promote Cretaceous fossils.

The fossils of the Winton Formation, and Lark Quarry in particular, have long been identified by communities in central-west Queensland for their vital role in the popular culture of the greater Winton region. The growth in dinosaur cultural and heritage experiences in the region continues with the development of fossil-based tourism projects and related events. Tourism in the Winton, Flinders, Hughenden and Richmond areas is centred on a very effectively branded 'Dinosaur Trail'. This trail includes the Dinosaur Stampede National Monument and Kronosaurus Korner <www.kronosauruskorner.com.au>. Locals continue to work with palaeontologists to ensure the protection, interpretation and promotion of the region's spectacular fossil finds.

The Dinosaur Stampede National Monument inspired stampede scenes in Steven Spielberg's blockbuster *Jurassic Park* (1993) and the BBC's award-winning series *Walking with Dinosaurs* (1999) (AHDB 2004).

Late Cretaceous (Cenomanian) Winton Formation

Site overview

These Cretaceous flora fossil sites are located in the vicinity of the township of Winton, which is also well known for its Dinosaur Trail, and most famously for the National Heritage-listed Lark Quarry Dinosaur Stampede National Monument.

Scientific importance and research potential

The Cretaceous period is of great botanical importance as it saw the rise of flowering plants worldwide. Evidence, mostly from pollen obtained from borehole sediments, shows that by the Late Cretaceous many angiosperm lineages had become established in southern Gondwana. Notably, this unique Gondwanan component included Epacridaceae, *Nothofagus* and Proteaceae, groups that are now characteristic and sometimes dominant elements of Australian and other southern hemisphere vegetation (Dettmann 1994). The Winton Formation sediments in the Eromanga Basin have great scientific importance as the only known Late Cretaceous sediments in Australia in which angiosperm leaves of any form can be found in outcrop. The geology and plant fossils of the numerous outcrops of the Winton Formation were described in detail by McLoughlin *et al.* (1995). This report also summarised previous palaeobotanical work in the region.

Fossil significance

None of the eight angiosperm leaf types recorded from the Winton Formation by McLoughlin *et al.* (1995) has any obvious resemblance to leaves of modern austral groups such as Proteaceae, perhaps reflecting pronounced heterogeneity in the vegetation (Dettmann 1994). Rather, the fossil leaves are similar to northern hemisphere forms such as *Fagus* or *Betula*. Similarly, Pole (2000) found several angiosperm cuticle types in conifer-dominated core material obtained from below the outcrops, but no evidence for Proteaceae, Myrtaceae or Lauraceae.

Permineralised fossils appear to be centred on Lovelle Downs Station approximately 50 kilometres north-west of Winton. This material includes well preserved cones of a taxodiaceous conifer, *Austrosequoia wintonensis*, as well as other conifers, ferns and angiosperms (Peters and Christophel 1978). Further work on these fossils could well be illustrative, as McLoughlin *et al.* (1995) noted: 'large collections of permineralised remains held by the Queensland Museum offer the potential for greater insights into the anatomy of plants in these austral angiosperm floras'. Other impression fossils (lacking cuticle) from elsewhere near Winton include angiosperms, ferns, *Ginkgo* and Araucariaceae cones and foliage. The latter fossils were described shortly before the discovery of the Wollemi Pine (*Wollemia*), but it has subsequently been noted how similar some of the Winton fossils are to the leafy stems and male and female cones of *Wollemia* (Chambers *et al.* 1998).

Taphonomy and condition of the site/fossils

Fossils are well preserved and include permineralised remains and impression fossils (lacking cuticle).

Comparison with similar sites in Australia

This site contains the only known Late Cretaceous sediments in Australia in which angiosperm leaves of any form can be found in outcrop.

Stories

The Winton Formation is already famous for dinosaur fossils, and tourism in the Winton region focuses on a Dinosaur Trail that includes the Dinosaur Stampede National Monument. Considerable scope exists for further publicising Late Cretaceous palaeoenvironments using the plant fossils.

South Australia

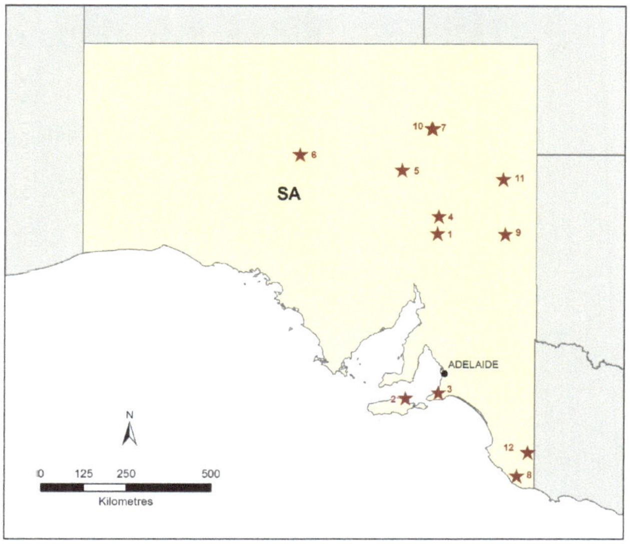

Fossil Sites ★

1 Ediacara Sites, Nilpena, Flinders Ranges
2 Emu Bay (Cape D'Estaing to Big Gully), Kangaroo Island
3 Myponga Beach, Fleurieu Peninsula
4 Ajax Mine Archaeocyathid Locality
5 Stuart, Poole and Nelly Creek Fossil Plant Sites
6 Moon Plain Cretaceous Fossil Site, Coober Pedy
7 Lake Palankarinna, Lake Eyre and Tirari Desert
8 Green Waterhole Cave
9 Lake Frome Basin, Namba Formation
10 Lake Ngapakaldi 'Leaf Locality'
11 Lake Callabonna
12 Naracoorte Caves

South Australia locality map © SEWPaC

Ediacara sites, Nilpena, Flinders Ranges

Site overview

The Ediacara site in the Flinders Ranges, South Australia, contains fossils of Ediacaran biota from the Neoproterozoic (approximately 625–550 Ma). An area of low hills includes exposure of sandstone strata bearing fossil impressions of the Ediacara biota over an area of approximately 8 square kilometres.

Scientific importance and research potential

Scientific study of the site commenced with some surface disturbance and collection of large iron-stained sandstone slabs in the years 1988 and 2000. Since 2007 selected areas have been gridded and systematically excavated, flipped and reassembled on site for detailed study of successively preserved samples of Ediacaran seafloor animal and plant communities. Some 150 square metres of excavated sandstone slabs bearing fossils have been conserved on site for current palaeoecological research and future study by qualified scientists and potentially for inspection by ecotourists.

Fossil significance

The site contains fossils of the famous Ediacara biota. These bizarre fossils include the oldest known evidence of macroscopic animal and plant communities on Earth. The site has produced some of the most spectacular Ediacara fossils ever discovered, including the largest known Ediacaran animal fossil *Dickinsonia rex*, which is currently on display in the South Australian Museum fossil gallery (e.g. Gehling 1987, 1988; Gehling and Rigby 1996; Gehling *et al.* 2005). Specimens of the earliest known ancestor of Chordates have been discovered at this site (R Jenkins pers comm.).

In 2004 a 'golden spike' was ratified in the nearby Flinders Ranges National Park, marking the global reference for the new Ediacaran Period representing the time of the origin of animals. This is the first new period division in the geologic time scale for 120 years and the first golden spike for the southern hemisphere.

Current research is directed at both palaeoenvironmental and palaeoecological reconstructions of the Ediacaran Period (e.g. Droser *et al.* 2004, 2005; Jenkins 1995; Jensen *et al.* 2005). The site is considered key to understanding our ultimate origins. As such, this site needs to be conserved and managed to prevent unauthorised, haphazard collecting and vandalism.

Taphonomy and condition of the site/fossils

The site preserves fossils at many levels and in several unique preservational styles unknown in other Ediacaran fossil localities. Recent exploration has shown that many entirely new forms of Ediacara fossils occur at many levels in the geological succession in this region. The potential of the prescribed area to produce fossil material that will recast what we know about the origins of animal life is unequalled anywhere. Furthermore, continued research at the site offers extended opportunities for palaeontologists, volunteers and students to participate in exploration and continued excavation of fossil beds.

Apart from the diversity of Ediacara fossils and modes of preservation at this site, the area is particularly valuable just because it has been so little disturbed in comparison to better known and longer visited sites such as the Ediacara Conservation Reserve.

Figure 4.1 Reconstruction of a benthic community that existed in the Flinders Ranges 580–542 million years ago. Painting Peter Schouten © SEWPaC

Figure 4.2 *Aspidella terranovica*. Image Steve Bourne © DEWHA

Comparison with similar sites in Australia

Unlike the Ediacara Conservation Reserve (Register of the National Estate ID # 5910), where Ediacara fossils were originally discovered, the Nilpena site has not been subject to 50 years of uncontrolled specimen collecting. To the contrary, recent research by geologists from the South Australian Museum and the University of California, funded by the Australian Research Council and the National Science Foundation (USA), has resulted in more than 150 square metres of excavated fossiliferous seafloor being laid out on site for scientific study. This access to complete fossil seafloor surfaces has revealed unprecedented new information about the oldest known communities of marine animals (Droser *et al.* 2006).

The Ediacara Reserve, approximately 15 kilometres north-north-west, consists of the same geological formations and is a source of Ediacara fossils similar to those at Nilpena. However, as it was a mining area from 1868 to 1914 and with no direct control over access, the reserve has been subject to almost unrestricted collecting by both permitted and non-permitted scientists and amateur and commercial fossil collectors. Consequently the reserve is unsuitable for detailed palaeoecological field research involving whole-surface investigation.

The beds at Nilpena involve much larger slabs, and fossils of many different kinds occur at many more levels than at the Ediacara Conservation Reserve. The diversity of fossils at the Nilpena site already greatly exceeds that of the Reserve.

The Ediacara Member of the Neoproterozoic Age can be traced throughout the central Flinders Ranges. The 1994 Register of Field Sites for the Ediacara fauna under the provenance of the Australian Heritage Commission indicates the extent of outcrops. Most, but not all, occur within the Flinders Ranges National Park, but none offers the availability of study from outcrop like the Nilpena site. The orientation of bedding and the topography at this regolith-covered site almost coincide, yielding significant bedding surfaces and access to fossils.

The potential for research at the Nilpena site is greater than other Ediacara biota sites because it offers better accessibility and establishment of new assemblages, and because it does not have a history of looting and extensive collecting. These other sites are situated in the north–south trending Heysen Range, which extends from Wilpena Pound through the Flinders Ranges National Park and north beyond Parachilna Gorge.

The first site in the Heysen Range is at Bath Tub Gorge, registered as PL94.04 of the National Estate Program Register of Field Sites 1994. Differences in the preservation of the two most common fossils in the Ediacara Member, *Phyllozoon* and *Dickinsonia costata*, indicate that they were made from contrasting structural materials, and the taphonomy suggests different means of transportation to the site of burial. Bath Tub Gorge is also important for the examples of feeding traces, sponge and new body fossil occurrences. The site has been looted, and is both accessible and outside the national park. There is also more limited exposure of fossiliferous material, and access to the site is more difficult than access to Nilpena.

Crisp Gorge (Dead Goat Soak), which is further north than Bath Tub Gorge along the Heysen Range, is registered as Field Site PL94.05 under National Estate Program 1994. The site has provided slabs of rock that were reconstructed as a jigsaw of a single bedding plane containing some 200 fossil specimens over 7 square metres. This now forms the Ripple Bed display in the Ediacara fossil gallery of the South Australian Museum. The 150 *Dickinsonia costata* specimens displayed on this one bed make it a unique Ediacara fossil exhibit. Problems with access to Crisp Gorge rate it as less important than Nilpena, where access and the potential for future research are hard to overstate.

Brachina and Bunyeroo Gorge sites, registered as Field Sites PL94.03 and PL94.03, 02 09, respectively, are on the south end of the Heysen Range within the Flinders Ranges National Park boundaries. They have been significant in early research on Ediacara fossils (Jenkins and Gehling 1978; Wade 1972a, 1972b). Both sites exhibit significant numbers of fossils in place. However, they have also suffered from significant unauthorised collection despite their position within the national park. The Brachina Gorge Geological Trail includes an outcrop of Ediacara fossils and is the only Ediacaran site flagged for public awareness and education.

Another comparable Ediacara fossil site, in the eastern end of the Chace Range (south of Wilpena Pound), is registered as PL94.05 and 94.15 of the National Estate Register of Field Sites 1994. Research at this site extended the range of the echinoderm phylum into the Ediacaran Period, with a resultant relevant evolutionary impact. Likewise, the first Ediacaran sponge, *Palaeophragmodictya*, was described from this site. Comparisons with other Ediacaran and Early Cambrian trace fossil and body fossil impressions with radial symmetry suggest that another fossil at this site, *Inaria*, was a non-burrowing precursor to actinian (anemone) infauna – again an exhibition of an evolutionary process. This Chace Range site, which is restricted to a band about 5 by 40 metres, has been thoroughly studied and collected for description of its fossil material. It offers less potential for further research than many other Ediacaran fossil sites in the Flinders Ranges.

Stories

Due to the excellent preservation of both fossils and depositional environments, and the diversity of fossils, the dioramic potential of the site is high.

Emu Bay (Cape d'Estaing to Big Gully), Kangaroo Island

Site overview

This site, located at Emu Bay (Cape d'Estaing to Big Gully) on Kangaroo Island, contains Early Cambrian Burgess Shale-type fossil deposits.

Scientific importance and research potential

Emu Bay is one of three major Cambrian *Konservat-Lagerstätten*, along with the World Heritage-listed Burgess Shale in Canada and Maotianshan Shale (at Chengjiang) in China. These sites are unique in that they preserve soft tissue anatomy as impressions in the rock. The sites are of fundamental importance in evolutionary studies as they record the sudden appearance in the fossil records of many of the extant phyla and give a valuable insight into perhaps the most important radiation among multicellular animals. The Emu Bay Shale has been firmly established in the scientific community as one of the most important of these sites.

University theses (Dinnick 1985; Nedin 1995a) have provided the most extensive research on these sites. Papers by eminent palaeontologists have further revealed the uniqueness and rarity of preservation and biota at the site for the age of the deposit (e.g. Briggs and Nedin 1997; Conway-Morris and Jenkins 1985; Glaessner 1979; McHenry and Yates 1993; Nedin 1995b, 1999; Paterson and Jago 2006). The site has significant research potential and might provide answers to questions including:

- Do species from the Emu Bay Shale fauna represent phylogenetic intermediates that are transitional between their relatives in earlier (e.g. Chengjiang) and later (e.g. Burgess Shale) faunas?
- Are there common patterns in the faunal composition in all Burgess Shale-type faunas?

- Do any of the species from the Emu Bay Shale push back the origins of forms that otherwise first appear in the later Burgess Shale fauna?

Fossil significance

The fossils from Emu Bay are from Early Cambrian rocks (approximately 542–488 Ma), which are immediately younger than those with Ediacara fossils. The preservation at the Big Gully location is exceptional and the term *Konservat-Lagerstätten* has been extended to this biota. This type of soft tissue preservation is extremely rare and enables soft-bodied organisms to be recorded as fossils. The site contains the enigmatic *Vetustovermis*, *Naraoia*, *Xandrella*, *Palaeoscolex* and *Anomalocaris* on which international research is currently being conducted in similar Cambrian fossil deposits. Affinities have been established between some Burgess biota and those at Big Gully, e.g. *Myscolex/Opabinia* species and the predator *Anomalocaris*. *Vetustovermis* has been identified and studied from Kunming (South China). The arthropod *Naraoia* from the Emu Bay Shale has recently been linked to the Ediacara fossil *Parvancorina* and to an arthropod *Skania* from Chengjiang (Eastern Yunnan Province) (Lin *et al*. 2006).

Originally described as numerically abundant, but low in diversity, the fossil assemblage at Big Gully has continued to reveal new aspects of the environment and habits of Cambrian animals, such as evidence of predation and healing of wounds, early eyes, and transitional stages in arthropod evolution (Conway-Morris and Jenkins 1985; Nedin 1999). Current research aims to investigate this and other Ediacaran–Cambrian sites, with plans to assess marine palaeocommunities during the Cambrian explosion and to examine what environmental factors influence faunal composition of Burgess Shale-type deposits by systematic study of the Emu Bay fossil beds.

Taphonomy and condition of the site/fossils

The Emu Bay Shale was deposited in relatively shallow water, roughly comparable in depth to that of the slightly older Maotianshan Shale but shallower than the deep-water Burgess Shale. Anoxic bottom conditions suppressed bacterial decomposition long enough for sediment to cover the body of organisms completely and preserve an imprint of the outline of soft parts. The assemblage is preserved as iron-stained calcite, commonly with a core of fluoroapatite. As such there are spectacular specimens but with fragile and easily destroyed surfaces. The specimens are in fracturing shale that does not part along the planes of sedimentation, so many specimens can be lost through inexpert removal. Cliff face and beach zones of exposure add to a delicate balance of preservation.

Comparison with similar sites in Australia

Vying for contention with the Emu Bay Shale outcrops on Kangaroo Island are two significant sites of comparable age, but removed from the Emu Bay site by approximately 700 kilometres: the Castle Rock and Angorichina localities.

In comparison with Emu Bay Shale sites, the Castle Rock site in the Flinders Ranges has been quoted as of extreme relevance as it also addresses arthropod ancestry into the Ediacaran. At Castle Rock there is the first evidence that Ediacaran biota survived into the Cambrian Period, which has not been observed anywhere else in Australia so far. In that sense the Cambrian fossils of Castle Rock are of great importance to collectors and specimens would be highly sought.

The Early Cambrian exposures at Angorichina Station also are significant for the quality of preservation of assemblages of trilobites, molluscs, bivalve arthropods, brachiopods and 'small shelly fossils'.

However, as a fossil *Lagerstätten* the exceptional preservation of the biota of the Big Gully site of the Emu Bay Shale has been recognised internationally and contributes a disproportionate amount of palaeobiological information on the lower Cambrian of Australia. The Emu Bay sites are Early Cambrian (mid–late Botomian) and thus represent a period transitional between other well known sites elsewhere, such as Canada's Burgess Shale (Middle Cambrian) and the Chengjiang of China (Early Cambrian, Atdabanian).

The Emu Bay Shale offers more potential for research, and it also offers opportunities for teaching, tourism and community involvement. A significant threat to the site is its attractiveness to collectors, because of the quality of the preserved material.

Stories

Research into the association of these fossils with the Burgess and Chengjiang Cambrian faunas, and also the Castle Rock and Angorichina sites, has a marked potential for revealing evolutionary aspects to early life-forms. The transition to the Emu Bay faunas from the Ediacara faunas of the Flinders Ranges has high dioramic potential.

Myponga Beach, Fleurieu Peninsula

Site overview

Myponga Beach on South Australia's Fleurieu Peninsula preserves an Early Cambrian trace fossil site. It is a valuable teaching site that is close to Adelaide, and is in Cambrian sediment undeformed by the Delamerian Orogeny (the mountain-building episodes that formed the original mountains that eroded into the Adelaide Hills and Flinders Ranges).

Scientific importance and research potential

Cambrian sequences have been noted at Myponga Beach since 1927. Research by Gravestock and colleagues (e.g. Alexander and Gravestock 1990; Debrenne and Gravestock 1990; Gravestock *et al.* 2001) has documented the value of the stratigraphic sequences readily viewed there. The fossil record and geological sequence provide a view of the early Pre-Cambrian/Cambrian history in Australia, particularly by providing evidence of the building of the continental margin of the Gawler Craton in the early phase of the agglutination of the rest of Australia to the cratons of the west and south-west.

Fossil significance

The site provides an extensive palaeontological record of the Early Cambrian in Australia (Atdabanian, Botomian and Toyonian, approximately 526–510 Ma), especially in the form of trace fossils and small invertebrates such as molluscs, archaeocyathae, and acritarchs. The trails of early life at this site consist of prolific horizontal trails and burrows, with less common vertical burrows. The bedding has not been destroyed by this bioturbation and thus much is revealed of the sedimentary history. Many of these trails have not been described, but their variety, mode of preservation (and beauty) are preserved. In some of the five fossil facies, fossil sponges are recorded between archaeocyatha and encrusted with the calcimicrobes that bind a bioherm of fossil life together.

Geopetal remnants that indicate the orientation of the sedimentary succession are common. Bioclastic packstone is composed of fragments of archaeocyatha, hyoliths, chancellorides, sponge spicules, gastropods and tommotiids, and rare trilobite fragments. Palaeoenvironmental evidence suggests the environment was unlike that of similar age rocks in the Flinders Ranges.

Taphonomy and condition of the site/fossils

The fossils are found in loose surface slabs called 'float' that are easily examined but are not removed. The coastal access poses some dangers to visitors.

Comparison with similar sites in Australia

A comparable site is Mount Scott Range in the Uratanna Formation, which is the type section for the earliest Cambrian strata in South Australia. However, at Mount Scott Range the trace fossil exposures are not easily discernible, unlike those at Myponga Beach. In terms of relative significance, the potential for research and teaching scientific values of Myponga are far greater.

Stories

Some historical significance is attached to the Sellick Hill area in general and in the strata of the Myponga Beach site, for it was in the Fork Tree Limestone in 1897 that Sir Tannant William Edgeworth David, along with Sir Walter Howchin, discovered the Archaeocyatha that provided the first proof of Cambrian rocks in the Mt Lofty Ranges. Myponga has been a teaching site since the beginning of the twentieth century.

Ajax Mine archaeocyathid locality

Site overview

The Ajax Mine (Puttapa/Beltana) site is an Early Cambrian marine invertebrate site containing Archaeocyatha ('ancient cups', vase-shaped sponge-like life-forms). It is located 50 kilometres north of Parachilna and 6 kilometres south of Leigh Creek in eastern South Australia.

Scientific importance and research potential

The Ajax Mine archaeocyathid locality has been extensively researched by Bedford and Bedford (1934, 1936, 1937, 1939). In 1969 it was revised by Debrenne (1969) and compared to the other South Australian exposures containing early Cambrian taxa by Gravestock *et al.* (2001). This site, along with the limestones at the Mount Scott Hookapunna Well locality and Wirrealpa Mine, are some of the most important for the study of early Cambrian Archaeocyatha in the world (Gravestock 1984; Thomson *et al.* 1976).

Fossil significance

This was the site of a reef composed of Archaeocyatha, bound together by calcimicrobes to form a bioherm 530 Ma, and providing evidence of the Early Cambrian environment of the area. The site also includes Archaeocyatha taxa that are index fossils for this time period, due to their distinct temporal distribution that ended 15 million years later.

Taphonomy and condition of the site/fossils

The fossils outcrop as a rounded hill 600 metres north of an open cut willemite (zinc silicate) and haematite mine. Preservation has been termed 'classic' to reflect the colours and contrasts in the silicified limestone. Bioherms of Archaeocyatha form solid reefs of outcrop. The archaeocyathid skeletons have been replaced by silica with little loss of detail. Specimens are etched at the rock surface but are too fragile to be preserved intact.

Comparison with similar sites in Australia

Archaeocyatha are regarded as index fossils for the Early Cambrian. Ajax Mine is significant as the type locality of many late Early Cambrian archaeocyathids. These specimens helped to delineate the earliest sponge-like organisms of the early Middle Cambrian elsewhere in Australia (Kruse 1990) and their subsequent disappearance from the geological record.

The outcrops of Ajax/Wilkawillina limestone are of Atdabanian–Botomian age (approximately 526–522 Ma) and are scientifically important compared to international fossil faunal assemblages of the same age. On a generic basis, the oldest faunal assemblages are correlated with horizons of the Altai-Sayan region of the former USSR. The silicified specimens at Ajax Mine are older than the other Early Cambrian reef sites in South Australia (namely the limestone reefs at Wirrealpa Mine, Flinders Ranges, and Horse Gully, Yorke Peninsula). Neither of the latter exhibits the abundance or diversity of the Ajax limestone.

The 73 species identified at the comparable sites of Mount Scott, Hookapunna Well and Wilkawillina Gorge are not known outside of Australia, thus increasing the rareness of these Early Cambrian 'sponges'.

Other than an outcrop in Brachina Gorge in Flinders Ranges National Park, Ajax Mine is the most accessible of the South Australian Cambrian Archaeocyatha sites for teaching. The Mount Scott and Wilkawillina sites, despite their diverse fauna, are significant but do not offer the opportunity of accessibility and appeal.

Stories

Ajax Mine is historically important as a site of Early Cambrian outcrop that has been studied since 1907. As an accessible teaching site it is an ideal locality for both education and tourism visits.

Stuart, Poole and Nelly Creek fossil plant sites

Site overview

The Stuart, Poole, and Nelly Creek palaeochannels at Lake Eyre South, South Australia, are Palaeogene to early Neogene plant fossil sites.

Scientific importance and research potential

Current research on the Stuart, Poole and Nelly Creek sites is healthy and the sites have been the subject of several journal publications over the last four decades (e.g. Alley 1989; Alley *et al.* 1996; Alley and Pledge 2000; Christophel *et al.* 1992; Conran *et al.* 2003; Greenwood *et al.* 1990; Hill and Christophel 2001). The sites continue to have potential for contributing to understanding the floral history of central Australia during the late Palaeogene to early Miocene.

At Nelly Creek, only a small area has been examined within the streambed. A further 5 kilometres of creek bed is expected to contain more fossil deposits. Some of the Gondwanan flora preserved in South Australia has not yet been completely studied, and there is potential for Nelly Creek to contribute to this understanding.

Palaeochannels, such as those found at Stuart, Poole and Nelly Creek, are landforms with a long evolutionary history. As such, they provide evidence of past climates. The physical qualities, preservation and plant diversity of the sites at each location have not been diminished by heavy access over the years by collectors and scientists.

Fossil significance

The leaf fossils in the central Australian palaeochannels are remarkable for their abundance, diversity, preservational detail and beauty (Alley and Pledge 2000). The fossils and their host rock are late Palaeogene or early Neogene in age. The flora is interpreted as having formed a marginal monsoonal forest with tropical elements. The climate was warm, with distinct dry and wet seasons typical of monsoonal climate today.

Hill and Chistophel (2001) established that a *Dacrydium* specimen from Nelly Creek took the total number of fossil species to six. This increased the known diversity of this macrofossil, extended its range out of south-eastern Australia, and its age from early Oligocene to Middle Eocene. The specimens are the best preserved examples discovered to that date. They have been important in furthering understanding of the genus and of conifer history in general. The site greatly expands the knowledge of Eocene central Australian floras.

Poole Creek contains mummified remains of eucalyptus leaves and fruits preserved in a stratigraphically dated sequence of two floras: the Eyre Formation and the Etadunna Formation (Palaeocene/Eocene and Oligocene/Miocene respectively).

The Stuart Creek Palaeochannel has leaf fossils in the Willalinchina sandstone of middle Palaeogene to early Neogene. At least 117 leaf forms, 40 fruit, seven seed types and two major kinds of wood have been identified in the floras so far (Alley and Pledge 2000).

Taphonomy and condition of the site/fossils

The extent of deposit at Nelly Creek is small and the material is particularly fragile. However, the level of preservation is extremely high, conserving leaves as mummified objects right down to the cellular level in some instances. The exposures exist as clay lenses in hard rock of the Eyre Formation and when extracted the leaves curl off the face of the bedding planes as the saturated sediments dry in the sun.

Stuart Creek has a preserved flora in silicified sandstone, younger than Poole Creek. The silcretes at Poole Creek are as spectacular as Nelly Creek and the younger deposits at Stuart Creek. Known as the silcrete floras, they are preserved as quartz-impregnated rock (silcrete), but Poole Creek also has plant material (spores, pollen and leaves) mummified as at Nelly Creek. Both preservational styles are in close proximity.

Comparison with similar sites in Australia

Nelly Creek is taxonomically distinct and physiognomically has more sclerophyllous specimens than other south-eastern floras, including the Poole and Stuart sites. Other south-eastern floras of comparable age include Nerriga (NSW); Anglesea (Vic); and Golden Grove and Maslin Bay (SA). Anglesea, Golden Grove and Maslin Bay show more tropical and higher moisture regime features, thus certain Gondwanan families are shared with Nelly Creek (Myrtaceae, Proteaceae, Casuarinaceae, Podocarpaceae).

The sites of Poole, Nelly and Stuart Creeks, as a group, have been well studied. They retain the potential for further research, especially as they have the potential to reveal so much about climate in south-eastern Australia during the late Palaeogene to early Neogene. They are significant as a group and as separate sites.

Stories

The sites together have high dioramic potential, and with the other well studied south-east Australian flora sites promise to aid in the elucidation of late-Gondwanan and post-Gondwanan environments and climates of Australia.

Other stakeholders

Marree Arabana Peoples Committee Inc.; North East Lands Aboriginal Corporation; Kuyani Aboriginal Corporation; Native Title Claims: Dieri SC97/4, The Arabunna Peoples SC 98/2, Kuyani-Wilyaru SC06/2.

Figure 4.3 Silcrete flora slab, 34 centimetres across. Eocene, Willalinchina Sandstone, equivalent to Eyre Formation, Stuart Creek Palaeochannel, South Australia. Image © The Museum Board of South Australia 2010

Moon Plain Cretaceous fossil site, Coober Pedy

Site overview

The Moon Plain site is located 20 kilometres north of Coober Pedy and contains Mesozoic marine fossils recovered from the Bulldog Shale.

Scientific importance and research potential

The invertebrate biota of the Moon Plain is reasonably well studied. Reports from existing research on the site have appeared in international journals (*Molluscan Research*, *Alcheringa*, *Cretaceous Research*), reflecting the robustness of the work. Marine vertebrates, particularly reptiles, are the subjects of current research (e.g. Kear 2003, 2004, 2005, 2006, 2007). However, the area continues to have considerable research potential for both vertebrate and invertebrate fauna, particularly through the study of the palaeoenvironment and palaeoecology of the deposits.

Fossil significance

The Bulldog Shale of the Early Cretaceous has produced a large number of new taxa. The fauna from the Moon Plain is unique as a record of a cold, Mesozoic, high-latitude marine site, and contains many endemic taxa for that time. More than anything else, this area has considerable potential for palaeoenvironmental reconstruction: it is considered to be a key to understanding the total faunal record of the Bulldog Shale. Within the early Bulldog Shale, fossiliferous Early Devonian boulders have been found (Flint *et al.* 1980). These alone are not of great significance from a fossil taxa viewpoint, but the manner of their transport is. The ice-rafted transport of the binding matrix helps researchers interpret the environment of the Cretaceous.

Taphonomy and condition of the site/fossils

The taphonomy includes coquinoid layers (storm accumulations) and *in situ* benthic faunas. Vertebrates are typically articulated and preservation is frequently excellent.

Comparison with similar sites in Australia

The Moon Plain with its Bulldog Shale (Eromanga Basin) has a lateral equivalent in the Wallumbilla Formation localities in New South Wales (White Cliffs, Lightning Ridge). However, the Moon Plain site preserves material at the surface rather than underground, and in great volume. The Bulldog Shale exhibits a similar macro-invertebrate fauna to that of the Wallumbilla, as well as palaeoclimatic indicators such as ice-rafted erratic boulders, glendonites and growth-banded fossil wood. The latter all provide evidence of cold climates in southern Australia during the Early Cretaceous (Frakes *et al.* 1995).

Figure 4.4 Platypterygius cf. longmani. Ichthyosaur vertebrae, 30 centimetres across. Aptian, Bulldog Shale, Moon Plain, South Australia. Image © The Museum Board of South Australia 2010

Fossil finds in the Bulldog Shale have shown that as many as five families of plesiosaurs may be present (Kear 2005), and the formation shares this dominant faunal element in Late Cretaceous deposits with the northern hemisphere, Antarctica, and White Cliffs in New South Wales. The faunal composition of this Cretaceous high latitude deposit contrasts markedly with the low latitude faunas of the Toolebuc Formation in north-eastern Queensland, and is considered to be reflecting palaeoenvironmental/climatic differences that are critical to our understanding of that period.

Stories

The surface preservation at the site lends itself to easy interpretation by non-specialists, and the preservational indicators and taxa provide evidence of evolutionary adaptation to cold climates over time.

Comparable terrestrial Cretaceous vertebrate sites include Queensland's Winton dinosaur trail and the ichnofossils of the Broome Sandstone in Western Australia. The polar dinosaur sites of southern Victoria also offer opportunity for reconstructing demographic and climate differences across Cretaceous Australia.

The Indigenous significance of Moon Plain is unknown, but Aboriginal stone tools have been noticed in the area generally. The site is in close proximity to opal mines and by association would have some historical significance.

Lake Palankarinna, Lake Eyre and Tirari Desert

Site overview

These Oligocene–Pleistocene vertebrate fossil sites within the Lake Eyre Basin and Tirari Desert region include those at Lake Palankarinna, Lake Eyre and the northern lakes of the Tirari: Lakes Kanunka, Pitikanta and Ngapakaldi, as well as Cooper Creek and Warburton River site localities.

For the purpose of this catalogue, the region is considered as two geographic areas: the Tirari Desert region and the Lake Eyre Basin. This has been done to convey the significance of the region as a whole and the interrelationships of the sedimentary record more effectively, given that the sediments are from Oligocene to Pleistocene ages at the same sites.

Scientific importance and research potential

Tirari Desert region

The Pliocene Tirari Formation, first recognised at Lake Palankarinna and northwards, contains faunal groups dominated by extinct species and genera not found in Quaternary assemblages. These sites differ from the older Late Palaeogene and Early Neogene faunas of the Etadunna Formation (Lake Palankarinna, Ngapakaldi, Kanunka and Pitikanta) in that their taxa are more closely related to groups of animals that are alive today

Figure 4.5 Lake Palankarinna. Photograph N. Pledge © SEWPaC

or are recently extinct. The oldest Quaternary deposits of the Tirari, Kutjitara and Katipiri Formations (the latter two are the ancestral rivers of the Cooper and Warburton drainages) are dominated by aquatic vertebrates and water birds, while terrestrial mammals are rare.

Deposition of the late Pleistocene Tirari Desert formations (Kutjitara and Kitipiri) occurred during early phases of the last two glacial cycles (Tedford and Wells 1990). The vertebrate taxa recovered from these sites reflect the last forms of many fossil vertebrates, particularly the large species, and are associated with warm/wet interglacials and cold/dry glacials of the Pleistocene. The area is thus of great significance to the palaeoenvironmental record of central Australia.

Etadunna Formation

Since the 1950s, research has been conducted on the fossil sites of the Etadunna Formation, which are Late Palaeogene to Early Neogene in age (Stirton 1955). Expeditions were mounted regularly: six by the renowned University of California Berkeley palaeontologist Ruben Stirton and his students (until 1962), and annual visits by the South Australian Museum since 1976. The first marsupial fossil, a koala, was found in 1953 at Lake Palankarinna. Lake Palankarinna exhibits a sequence of 'layer cake' stratigraphy enabling determination of the relationship of one fauna to another. Mammal index taxa and magnetostratigraphic data show potential for developing the Etadunna Formation as a reference standard for land mammal correlations within Australia and the relationship of these to land mammal zonations beyond the Australian continent (Woodburne *et al.* 1994).

Fossil significance

Tirari Desert region

The Pliocene–Pleistocene deposits of the Tirari Desert contain terrestrial reptiles, giant goannas (*Megalania*) and pythons, which were some of the largest contemporary carnivores (Tedford *et al.* 1992). Palaeontologists Richard Tedford and Roderick Wells suggested in 1990 that the Tasmanian tiger and marsupial lion must be counted among their prey. Early finds included a crocodile snout and scutes (dermal armour), alligator coprolites, and pelican tarsals.

Fossils recovered since 1963 include late Pliocene to early Pleistocene flamingoes, pelicans, masked owls, falcons, dromornithids, *Sthenurus*, *Protemnodon* (giant kangaroos), and diprotodontids (giant graviportal herbivores), fish (catfish and perch), turtles, and many other taxa, particularly water birds. In essence, the aquatic lower vertebrates and water birds dominate the fauna while terrestrial mammals are rare. The rich fauna of aquatic vertebrates includes many forms that live in the modern Lake Eyre region.

Lake Eyre Basin

The Lake Eyre Basin received Geological Monument status in 1977 after being listed as a fossil reserve under the Crown Lands Act in 1954. The sites contain unusual fossil specimens such as pelicans, flamingoes, ducks, crocodiles, lungfish, and numerous extinct marsupials. Each locality contains a different fauna; some are not exceptional but all are representative of taxa from this late Palaeogene to Neogene timeframe (Wells and Callen 1986). Lake Palankarinna has the richest Oligocene–Miocene avifauna in Australia.

Woodburne *et al.* (1994) divided the fossils of Lake Palankarinna (Etadunna Formation) that are both rare and diverse into four mammalian fossil zones, based on the area's stratigraphy, which is visible (unlike other locations of this age), and accessible. These zones have been used to establish relationships of relative age between similar sites in Australia through the magnetostratigraphic dating of the Etadunna Formation. Lake Kanunka, which has been securely dated as late Pliocene by palaeomagnetic dating, contains evidence of the earliest rodents to appear in Australia.

Taphonomy and condition of the site/fossils

Preservation is generally good, subject to weathering and floods. Specimens would not be easily carried from the region.

Comparison with similar sites in Australia

Tirari Desert region

The faunal exposures in the Tirari Desert region, including the Katipiri and Kutjitara formations, can be correlated with later Pleistocene sequences in the Lake Frome Basin at Lake Callabonna. This would indicate that little taxonomic change has occurred in the large marsupial assemblages during the later Pleistocene. Lake Victoria, central Murray Basin, New South Wales, contains the same taxa as the Katipiri and Kutjitara formations in the Lake Eyre Basin and shows general similarities to the Victoria Cave assemblage in south-eastern South Australia. Thus major faunal changes in the later Pleistocene, as exhibited in the Tirari Desert, are associated with the later part of the last glacial event.

Lake Eyre Basin

The Etadunna Formation has been magnetostratigraphically dated to the late Palaeogene–early Neogene, which has enabled a relative age to be

Figure 4.6 Lake Eyre from Hermit Hill. Drawing George French Angas © National Library of Australia

placed on a number of Cenozoic vertebrate deposits in Australia that share taxa, e.g. the Riversleigh sites in north-west Queensland (Woodburne *et al.* 1994). The deposits of the Lake Eyre Basin reveal a type of fauna similar to that of Riversleigh but of course nothing of the quality or quantity. The base of the mammal-bearing sequence in the Etadunna and the nearby Namba (Lake Frome) Formations pre-date the Riversleigh succession (Woodburne *et al.* 1994).

Lake Palankarinna area is the Rosetta Stone for stratigraphic correlations for the Tertiary (RT Wells pers. comm. 2006). The combined mammal and palaeomagnetic data show promise of developing the Etadunna Formation (Lakes Palankarinna, Pitikanta, Ngapakaldi and Kanunka) as a reference standard for land mammal correlations within Australia, and for the relation of these to land mammal zonations beyond Australia (Woodburne *et al.* 1994). The vertebrate fossils of the Etadunna Formation have aided in reconstruction of palaeoenvironments and understanding of the Australian faunas. Furthermore a comparison of the Lake Eyre Basin sites with the other major Australian terrestrial vertebrate sites of Riversleigh, Queensland, and Bullock Creek, Northern Territory, is relevant on a taphonomic basis. Limestones at the latter sites preserve specimens in excellent detail but as individual bones. At Lake Eyre Basin, the bones may be crushed but the skeletons are articulated. The two types of preservation provide complementary data on the animals of the Tertiary. For that reason again, the Lake Eyre Basin sites are very significant.

Stories

The Lake Eyre basin sites enable the reconstruction of Neogene to Quaternary Australian mammal faunas, supplementing information from other important Neogene and Pleistocene sites like Naracoorte (SA), Riversleigh (Qld), Bullock Creek, and Alcoota (NT).

Their discovery and exploration is tied to the rejuvenation of Australian vertebrate palaeontology during the 1950s and the search for 'Tertiary Australia' by Berkeley palaeontologist Ruben Stirton and his colleagues and students.

Green Waterhole Cave

Site overview

Green Waterhole Cave, also known as Fossil Cave, is located in the Mount Gambier district in south-east South Australia. It contains a Pleistocene bird deposit as well as some mammal fossils.

Scientific importance and research potential

Research has shown that the accumulation of bird material in this drowned cave relates to flocking in the cave at time of drought, where water in the cave attracted the birds. The fossil animals collected reflect local population size and flocking habits rather than a bias caused by predators. Diving continues at the time of this report to collect further fossil material of fine preservation (T Worthy pers. comm.).

Fossil significance

This drowned cave has provided both the type and only occurrence for two named fossil birds and continues to produce a significant unique fauna. Sixteen species of birds have been identified from the site (to 1985); all but three are referable to modern species. Three new species have been identified that, along with one extant taxon, provide examples of Pleistocene gigantism. Of the three new species, the accipitrid (diurnal bird of prey) may be larger than any extant member of the family; the coucal (cuckoo) is unusual for its enormous size, which is a third larger than its closest Australian relative and in an advanced state of flightlessness; the passerine (song bird) is also unique for its adaptations to terrestriality and large size. As with the mammalian fauna at the site, the avian fauna of both extinct and extant species is of late Quaternary age, which is considered unusual.

Taphonomy and condition of the site/fossils

Fauna is from sediments below water at a depth of 15 metres. The quality of preservation (RT Wells pers. comm.) is excellent. The taphonomy at the site is unique in Australia (Baird 1985, 1991), as accumulation of the avian assemblages is thought to be due to the cave acting as a natural water trap.

Comparison with similar sites in Australia

The full potential of Green Waterhole Cave has not yet been established, as diving is difficult (RT Wells pers. comm.). However, the bird fauna diversity at the site is very good. A comparison between the Victoria Fossil Cave, Naracoorte World Heritage Area (SA) and Green Waterhole (the sites from the region that have the largest fossil bird assemblages) reveals different composition of species reflecting different modes of accumulation (Reed and Bourne 2000). Further research on the bird faunas may reveal other taphonomic biases of interest.

Stories

The interpretation of the cave as a drought refuge inserts the site into the canonical Australian climate story of Pleistocene drying and climate variability.

Figure 4.7 *Simosthenurus occidentalis*. Short-faced kangaroo skull, 19 centimetres across. Pleistocene, Green Waterhole Cave, South Australia. Image © The Museum Board of South Australia 2010

Lake Frome Basin, Namba Formation

Site overview

The Early Miocene vertebrate sites from Lake Frome are located at Frome Downs south-east of Lake Frome, at Lake Frome itself and at Lakes Namba, Pinpa, Tinko and Tarkarooloo to the south.

Scientific importance and research potential

These sites have been studied extensively since the 1970s by international and Australian scientists and were the location of discovery of several important vertebrate fossils, particularly the first Cenozoic fossil platypus (Woodburne and Tedford 1975) and the first river dolphins (Fordyce 1983).

The site offers potential for palaeoenvironmental reconstruction. Pollen and spores from the Namba Formation are dominated by swamp plant species and very rare grass pollen. This shows that grasslands had not yet developed in central Australia by the early Neogene, an interpretation supported by the dominance of browsing rather than grazing mammals in the vertebrate faunas.

There is ongoing taxonomic and systematic study and revision of the vertebrate fauna from these sites, and there has been small-scale recovery of new material over the last 20 years, with potential for continued excavation.

Fossil significance

A number of extremely significant vertebrate taxa have been recorded from the Lake Frome Basin. In the case of many mammals these are the oldest representatives of living families, e.g. koalas and wombats, and is the type locality for many previously undescribed extinct reptile and mammal taxa (e.g. Burke *et al.* 1983; Fordyce 1983; Woodburne and Tedford 1975). It also includes the first and only fossil material of Rhabdosteidae dolphins discovered in Australia. This group of dolphins was considered largely of northern hemisphere distribution, but are found in late Oligocene/Miocene of Argentina and New Zealand. Their discovery has aided in establishing the age of the Lake Frome deposits.

Comparison with similar sites in Australia

The late Oligocene faunal assemblages of the Namba Formation have been biocorrelated with faunas from the dated Etadunna Formation of the Lake Eyre Basin and are considered to be slightly older than the Riversleigh sites in Queensland, with some overlap (Travouillon *et al.* 2006; Woodburne *et al.* 1994). The quality of preservation and abundance of fossil material found in the Lake Frome Basin is also comparable to the Lake Eyre Basin, but both are lesser in comparison to the Riversleigh sites. Current knowledge of past faunas is predominantly based on Pleistocene vertebrate fauna. Therefore, the faunal remains found in this area have international, as well as national, scientific importance.

Stories

The area is significant biogeographically in both botanical and faunal studies. It is particularly significant in the understanding of the evolution and dispersal of the flora and fauna of Central Australia with increasing aridity in the post-Miocene period. The area provides significant evidence for the reconstruction of Late Oligocene vegetation in central Australia, particularly in reconstructing the subtropical riparian rainforests of that time.

Lake Ngapakaldi Leaf Locality

Site overview

The Lake Ngapakaldi Leaf Locality is a vertebrate and plant fossil site located on the eastern shore of Lake Ngapakaldi, South Australia. It contains a rich assemblage of Early Miocene marsupials, birds, reptiles and plants (the Kutjamarpu local fauna), accumulated in a fluviatile channel deposit.

Scientific importance and research potential

The first published scientific report on Lake Ngapakaldi site was by Ruben Stirton, Richard Tedford and Michael Woodburne (1967a). A team led by Ruben A. Stirton (University of California, Berkeley) was investigating the taxonomy, biostratigraphy, geochronology and evolutionary history of Australia's vertebrate fauna. The discovery and interpretation of the Kutjamarpu local fauna at Lake Ngapakaldi underpinned their greatest achievement, which was to produce a land mammal (marsupial) biochronological framework for the continent. The work of the Stirton team established marsupial palaeontology as a serious academic discipline in Australia (Tedford 1991).

Although Stirton and colleagues had already laid a foundation for an Australian land mammal biochronological framework before the discovery of the Leaf Locality, its discovery (Stirton *et al.* 1967a) was the keystone linking the older known sites (especially the Late Oligocene Etadunna Formation) and significantly later ones (especially Late Miocene Alcoota and all younger, i.e. Pliocene and Pleistocene, sites). The description of the primitive zygomaturine *Neohelos tirarensis* (Stirton 1967) in turn led to the description of an intercalary lineage of zygomaturines useful in correlation (Stirton *et al.* 1967b). This was central to the first articulation of a continental land mammal biochronological scheme (Stirton *et al.* 1968).

Subsequent work has testified to the inherent robustness of the achievements of Stirton and his students, and has resulted in elaborations and refinements rather than any major revision of their insights (Megirian 1994; Murray *et al.* 2000; Woodburne *et al.* 1985, 1994). The site has potential for further significant research. More collecting needs to be done, and not all the material already collected has been fully analysed.

Fossil significance

This is a unique assemblage with some rare species. The Leaf Locality has produced the biochronological index fossils for the Early Miocene of the Australian land mammal biochronological scheme (Woodburne *et al.* 1985). The Leaf Locality is one of the few Cenozoic deposits containing both plant and animal remains, allowing the local environment to be reconstructed without extrapolation from distant, correlated sites.

Taphonomy and condition of the site/fossils

The site contains both floral and faunal fossils. This includes fragmentary skeletal remains in silty, sandy and gravelly fluvial channel deposits, and leaf impressions in clay sediments. Bones are often heavily fractured by post-depositional expansion and contraction of the host sediments. The extraction of fossils from the site is very time-consuming and requires technical skill.

Although the Leaf Locality has potential to produce more material, yields are low. Other sites of the same age (Kangaroo Well, NT, and Riversleigh, Qld) have produced some of the same species, in some instances with more complete representation. However, the Kutjamarpu local fauna will always remain the reference assemblage for the interpretation of these sites.

Comparison with similar sites in Australia

The Leaf Locality assemblage (Kutjamarpu local fauna) is the land mammal biochronological marker, or reference standard, for the Early Miocene. Strata of approximately similar age containing comparable faunal assemblages are present in the Ulta Limestone and Kangaroo Well local fauna, Northern Territory, and the freshwater limestone of the Riversleigh World Heritage Area, Queensland. These other occurrences are interpreted with reference to the Leaf Locality assemblage.

Stories

Specimens and data from the site could be used to reconstruct the flora and fauna of an Early Miocene channel environment in the Lake Eyre basin. Australia's fauna underwent a marked turnover from assemblages typical of the Oligocene (dominated by, for example, wynyardiids and ilariids) to assemblages characteristic of the Miocene (dominated by zygomaturine diprotodontids). The Kutjamarpu local fauna contains some of the last representatives of typically 'Oligocene' lineages and the earliest representatives of typically 'Miocene' lineages. As such it is a key transitional fauna.

Pioneering investigations of the site by R.A Stirton, T.H. Tedford and M.O. Woodburne under the intellectual leadership of Stirton are of enormous historical significance in the development of an Australian marsupial biochronological scheme. Stirton's work in the Lake Eyre Basin began in the 1950s with the support and encouragement of eminent figures of Australian Earth science history including Sir Douglas Mawson, C.W. Bonython, F.W. Whitehouse, R.C. Sprigg, M.F. Glaessner, and N.B. Tindale (Tedford 1991).

Material from the Leaf Locality is exhibited in the South Australian Museum. The Kutjamarpu local fauna is referred to in the Vickers-Rich and Rich (1993b) account of evolution of Australia's fauna. The Leaf Locality is also the type section of the Wipajiri Formation (Geoscience Australia stratigraphic number 27617).

Figure 4.8 Dromornithid sp. Avian tibiotarsus (legbone), 46 centimetres long. Mid Miocene, Wipajiri Formation, Lake Ngapakaldi Leaf Locality, South Australia. Image © The Museum Board of South Australia 2010

Lake Callabonna

Site overview

Lake Callabonna is a Pleistocene vertebrate fossil site located in north-east South Australia.

Scientific importance and research potential

The area (once called Lake Mulligan) is a unique accumulation of Pleistocene lacustrine deposits. Articulated specimens of these Quaternary fossils have been discovered and recovered consistently. The last large-scale expedition to the area was in 1983, but there is potential for further research.

The Lake Callabonna deposit provides a significant insight into the environment and fauna of Australia during the Pleistocene. The Callabonna fauna includes species of *Sthenurus*, *Diprotodon*, other large marsupials, and large flightless birds. The presence of such complete material enables a wider array of studies to be undertaken (e.g. detailed functional anatomy, intraspecific variation, and population dynamics). The site also provides support for the hypothesis that *Diprotodon*, the largest marsupial to have lived, survived into the last glacial cycle (Tedford 1984).

Fossil significance

Skeletons are articulated and fossils representing nine extinct mammal species (e.g. giant wombats and kangaroos) and two extinct bird species (e.g. *Genyornis*) have been recovered to date. The accumulation of remains, such as of the genus of extinct giant short-faced kangaroos *Sthenurus*, has provided an opportunity for comparative study of individual animals of three closely allied species: *Sthenurus stirlingi*, *S. tindalei* and *S. andersoni* (Wells and Tedford 1995).

The stratigraphic record indicates dry climate of yearly or greater cyclicity but with enough rainfall to support a woodland vegetation, as shown by plant remains such as wood, twigs, *Callitris* cones and pollen. The preservation of some mammalian soft tissue (impressions of skin and hair) allows aspects of the external morphology to be described.

Furthermore, finds are unique in that complete sets of articulated foot bones have been found along with vegetable matter from the stomach area of a *Diprotodon* skeleton, and the partial skeleton of a pouched young *Diprotodon* was found in the pelvic region of another specimen.

Taphonomy and condition of the site/fossils

The fauna is dominated by heavy mammals and large flightless birds, which were trapped and bogged in Lake Callabonna by sticky clay sediments. They underwent very little transportation or disruption after death, as shown by the large number of closely packed, articulated skeletons. The articulated remains and impressions of skin tissue are preserved because of the mode of entrapment and the chemical composition of the sediments, which is very rare. Footprint trackways are also preserved in the laminated clays and travertine lenses in the sediments, and reveal much about the method of locomotion of these mammals and birds.

Comparison with similar sites in Australia

Sites in South Australia that are comparable to Lake Callabonna include the Pleistocene sites at Burra (Baldina Creek), mid north Flinders Ranges, and Black Creek Swamp, Kangaroo Island.

At Baldina Creek, articulated or closely associated specimens indicate low levels of transport and rapid burial following death. This provides a direct association and correlation between the fauna and any climate proxy record contained within the sediments of the deposit, and thus is suitable for multiple dating techniques (Grun *et al.* 2008; Wells pers. comm.). While the site is significant for these reasons, the area has reached the end of extensive research potential and is well protected through the Burra Conservation Management Plans and regional interests.

Black Creek Swamp site has provided extensive fossil horizons. Work at the site continues and a report of island dwarfism of giant marsupials is in press. From a geochronological perspective, this site is one of several 45 000-year-old megafauna sites that are scattered all over Australia and is therefore of less significance than Callabonna. Lying within the Flinders Chase National Park, the Black Creek Swamp site is well protected and not under direct threat.

Lake Callabonna is significant and unique in Australia because of the large number of articulated skeletons. These are some of the best preserved large Pleistocene marsupials and giant flightless birds in Australia.

Lake Callabonna has extraordinary historical significance as well. Its discovery in 1893 led to a vertebrate fossil bonanza, and unprecedented interest in vertebrate palaeontology in the state. It also featured in Ruben Stirton's 1953 expedition, which first uncovered the wealth of the Lake Eyre Basin vertebrate sites (Douglas 2010).

Stories

The Callabonna Lake area is historically important, as the finds in the late nineteenth century were reflective of the developing vertebrate palaeontology scene in Australia. The evocation of dying megafauna in a drying lake has fostered ongoing popular fascination with the site as a place of mystery. The cycles of

Figure 4.9 Exposed skeletons of diprotodons at Lake Callabonna. Images © National Library of Australia

drought and replenishment echo the modern interior lake system.

Naracoorte Caves

Site overview

The 26 caves of the 307-hectare Naracoorte Caves World Heritage area are located in a flat coastal plain interrupted by a succession of low fossil coastal dune ridges, about 10 kilometres south-south-east of the town of Naracoorte in South Australia's Coonawarra region. These stranded shorelines, formed by a series of marine transgressions and regressions, are parallel to the present coast. Geologically, the region is dominated by sediment-covered karst, exposed in outcrops and cliffs. The caves occur in a ridge of Oligocene to Miocene Gambier limestone capped by the Naracoorte East Dune. In the Late Pleistocene the caves were open to the surface, which led to the accumulation of sediments and bones that preserves a rich record of past faunas and environment. The best known of these accumulations are those in Victoria Fossil Cave, discovered by cavers in October 1969 (UNEP-WCMC 1994/2006).

Scientific importance and research potential

One of the richest vertebrate fossil localities in Australia, the Naracoorte Caves fossils illustrate faunal change spanning several ice ages, demonstrating the impacts of climate change on Australia's vertebrates from at least 500 000 years before the present. The oldest dated fossil-bearing strata are 528 000 years from Cathedral Cave (Prideaux *et al.* 2007). Research at the Naracoorte Caves sites has begun to document snapshots of Pleistocene environments in south-east Australia, including details of climate and vegetation associated with the fauna (for example, Desmarchelier *et al.* 2000; Fraser and Wells 2006; Forbes *et al.* 2007). The caves are predominantly formed from Oligo-Miocene karstic limestone and Pleistocene sediments. The Naracoorte fossils span the probable time of the arrival of humans in Australia, and may illuminate modern debates about the cause or causes of the extinction of the Pleistocene megafauna (Grun *et al.* 2001; Moriarty *et al.* 2000; Prideaux *et al.* 2007; UNEP-WCMC 1994/2006).

Fossil significance

Specimens of more than 100 vertebrate species have been discovered. These range from tiny frogs to *Zygomaturus trilobus* and *Palorchestes azael*, both weighing around 500 kg. The fossils occur in deposits spanning several glacial and interglacial periods. Fauna found in the caves are a mix of species that became extinct during the Pleistocene and species that persisted until historical times. Approximately 20% appear to have become extinct during the Pleistocene, 5% totally extinct following European settlement of Australia, 28% locally extinct but surviving elsewhere in Australia, and 47% of species represented in pitfalls are still extant in the area today (Reed pers. comm. 2011).

From the 3–4-metre-deep fossil bed, in a series of deposits, tens of thousands of specimens have been recovered. The fossil material includes complete post-cranial remains, many of them still partially articulated, and skulls so well preserved that even the most delicate bones are still intact. This enables both reconstruction and detailed anatomical descriptions and functional morphology of crania and post-cranial skeletal elements. The fossil remains steadily accumulated in the karst over thousands of years as animals fell through pit traps to perish in the cave below, until sediments seeping into the cave completely blocked the entrances to the chambers (Forbes and Bestland 2006, 2007; Reed 2008; Reed and Gillieson 2003; UNEP-WCMC 1994/2006).

Taphonomy and condition of the site

Many specimens are exceptionally preserved, including articulated specimens and a high proportion of cranial remains. Fossils include Late Pleistocene megafauna and many modern species, including locally or recently extinct animals like the Tasmanian devil, thylacine and others. An outstanding feature of the deposits is the detailed ecological information preserved in the caves and environs (Bourne and Reed 2009; Darrenougue *et al.* 2009; Desmarchelier *et al.* 2000; Forbes *et al.* 2007; Fraser and Wells 2006; Moriarty *et al.* 2000; Prideaux *et al.* 2007; Reed and Gillieson 2003).

Comparison with similar sites in Australia

The Pleistocene vertebrate deposits of Victoria Fossil Cave and the other Naracoorte caves are considered to be Australia's largest and best preserved, both in volume and faunal diversity (Reed and Bourne 2000; Wells *et al.* 1984). The many sites within the caves span the late Holocene right through to around 530 000 years ago (Middle Pleistocene) (Desmarchelier *et al.* 2000; Forbes and Bestland 2006 and 2007; Fraser and Wells 2006; Grun *et al.* 2001; Pate *et al.* 2006; Reed and Bourne 2000, 2009). The Naracoorte fossil locality is one of the richest in the world. Together with Riversleigh in north-west Queensland, it forms the Australian Fossil Mammal Sites World Heritage property, representing key stages in the evolution of Australia's mammal fauna (Bourne and Reed 2009; Reed and Bourne 2000, 2009).

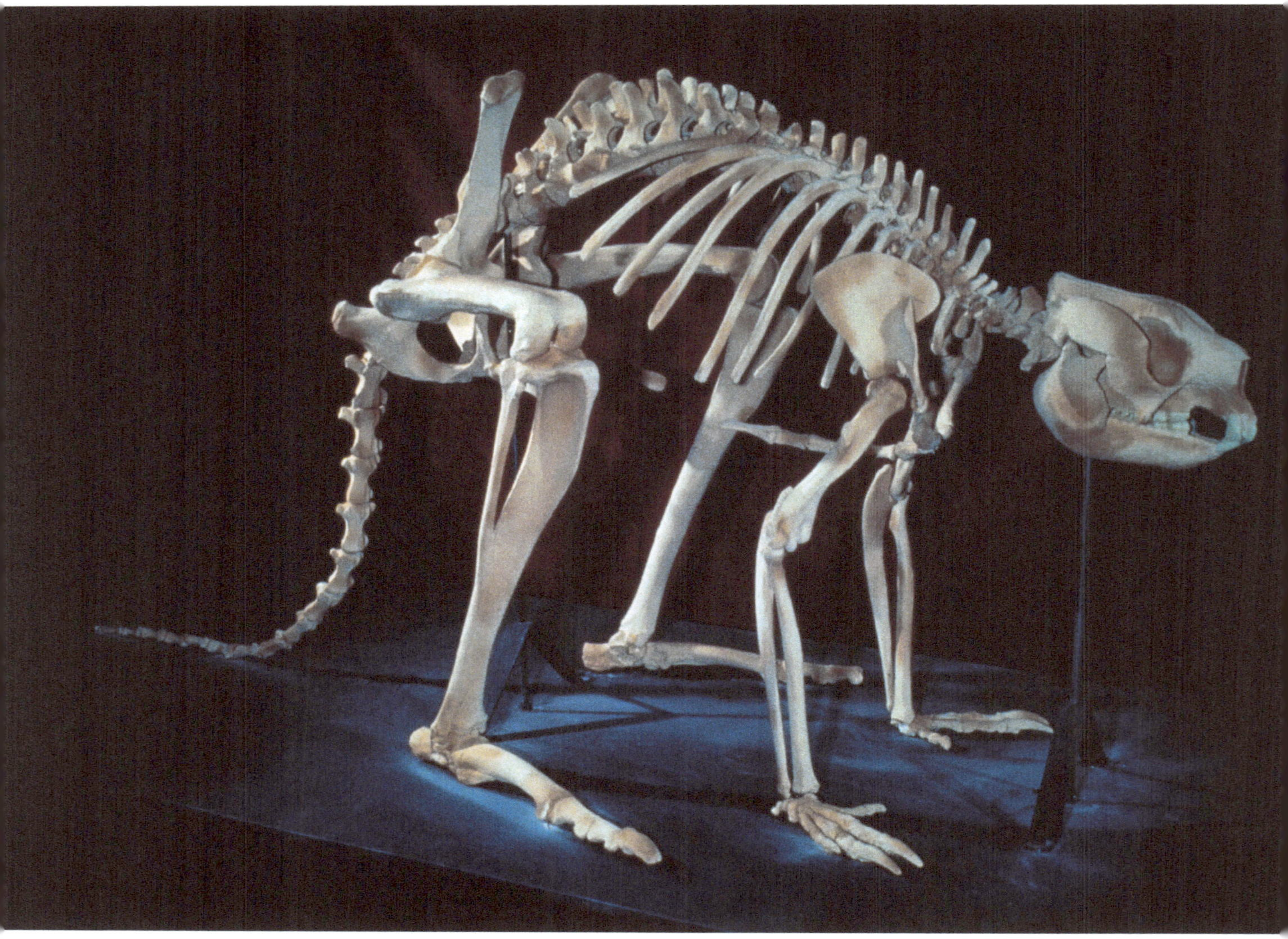

Figure 4.10 Sthenurine kangaroo, Naracoorte. Image © DSEWPaC

Stories

The dioramic potential of the Naracoorte Caves National Park fossils is realised in its pairing with the outstanding Riversleigh as the World Heritage-listed Australian Fossil Mammal Sites (Riversleigh/Naracoorte), which together illustrate key stages of evolution of Australia's mammal fauna during the Cenozoic Era (65 million years) (Bourne and Reed 2009; Grun *et al.* 2001; Reed and Gillieson 2003; UNEP-WCMC 1994/2006).

The Naracoorte fossils contribute to recent debates about the cause or causes of the extinction of the Late Pleistocene megafauna, and the possible role of human beings (Grun *et al.* 2001; Pate *et al.* 2002; Pate *et al.* 2006; Prideaux *et al.* 2007). Some of the caves provide a modern analogue for the fossil deposits, as animals continue to fall into pit traps dissolved into the karst limestone. The palaeoenvironmental details that the cave deposits preserve allow palaeontologists to recover and present a meticulous reconstruction of Quaternary and Late Neogene terrestrial ecologies of south eastern Australia through time (Darrenougue *et al.* 2009; Desmarchelier *et al.* 2000; Fraser and Wells 2006; Moriarty *et al.* 2000; Reed 2006, 2008; Reed and Gillieson 2003).

Tasmania

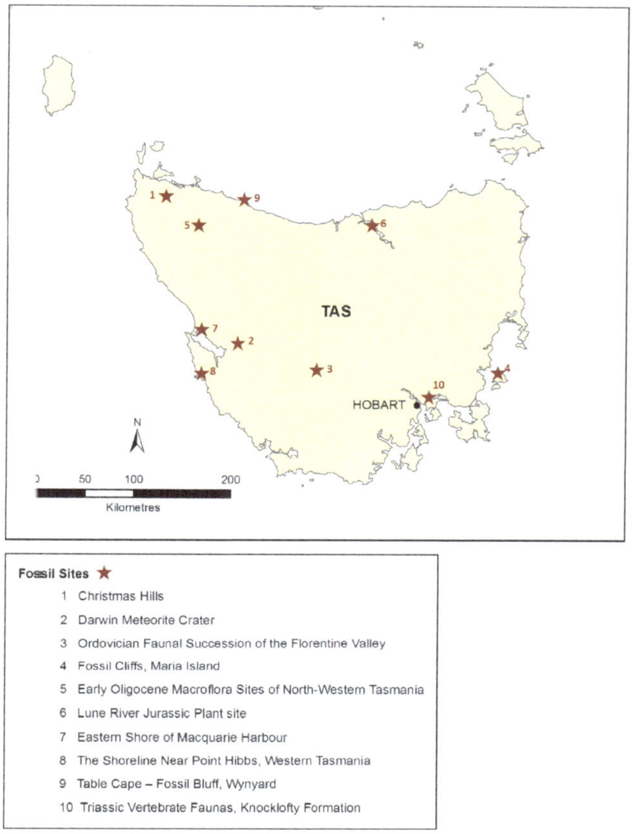

Tasmania locality map © SEWPaC

Christmas Hills

Site overview

Late Middle Cambrian invertebrate fossils were discovered in laminated siltstone at Christmas Hills, north-west Tasmania by Gulline (1959). Banks (1962) referred to a 1959 personal communication from Armin Öpik that suggested a *Lejopyge laevigata* II Zone age.

Scientific importance and research potential

J.B. Jago and J.H. Buckley (1971) noted that there are two faunas at Christmas Hills with an abrupt faunal change between the lower and upper faunas, despite their similar ages. Jago (1973) included the lower fauna in his agnostoid trilobite faunal assemblage 3 and the upper fauna in assemblage 2, which is regarded as the more offshore of the two faunas.

Jago and Daily (1974) and Jago (1976) described the agnostoids, as then known, from both the upper and lower faunas. The first polymerid trilobite described was the wuaniid *Tasmacephalus platypus* from the lower fauna (Bentley and Jago 2004). Quilty (1971) described the dendroid graptolites from the upper fauna, and Jago and Anderson (2004) reported a bivalved arthropod from the same fauna. The geological setting of the fossil locality is described in Brown (1989).

Jago and Bentley (2007) published a paper on the remaining undescribed polymerid trilobites from the lower fauna in honour of John Shergold. This significant paper establishes two new genera and includes an analysis of the composition of the fauna. With over 1700 specimens available this is one of the best records anywhere of a Cambrian fauna; it will certainly be the best in Australia. A future paper will deal with the upper fauna and compare the different results from the faunal analyses of the lower and upper faunas. The upper fauna at Christmas Hills also includes rare late Middle Cambrian dendroid graptolites.

Fossil significance

This is a significant site in that there are two faunas of very similar late Middle Cambrian age that are poorly documented in Australia. Although the faunas are of similar age they differ in taxonomic composition, reflecting differing palaeoenvironmental conditions. Both are offshore faunas, and so provide a contrast to the shallower water faunas of similar age found in northern Australia.

Jago (1973) reported the presence of three agnostoid-based trilobite assemblages in the late Middle and early Late Cambrian fossil assemblages of Tasmania. The Christmas Hills fossil faunas are the best representatives of agnostoid faunal assemblages 2 and 3. This work is significant because intercontinental Middle and Late Cambrian correlations are based largely on agnostoid trilobites.

Taphonomy and condition of the fossils

In common with almost all Tasmanian Cambrian faunas, the Christmas Hills fossils are tectonically distorted, although this fauna is one of the least distorted. The fossils are very abundant and all specimens described occur as internal and external moulds in weathered siltstone; there is no original shell material.

Comparison with similar sites in Australia

There are no comparable sites in Australia containing two juxtaposed Middle Cambrian faunas of very similar age that have very little in common. As noted above, the difference is due to different palaeoenvironmental conditions. This matter is the subject of ongoing research. There are similar faunas near St Valentine's Peak and Native Track Tier in Tasmania, but these are not as richly fossiliferous.

Darwin Meteorite Crater

Site overview

The Darwin Meteorite Crater was formed through a meteorite impact approximately 816 000 years ago. This resulted in the melting of thousands of cubic metres of rock, and it sprayed fragments of this melted rock in the form of 'Darwin glass' over an area of about 400 square kilometres (Corbett and Howard 2006). The lower part of the crater filled with debris, but the upper 60 metres of the crater became a deep lake that slowly infilled over thousands of years with clay and sand, preserving a record of the local and regional vegetation. The 60 metres of lake sediments contain abundant pollen and plant remains washed in from surrounding hills, thereby preserving a relatively complete record of vegetation and therefore of climate changes over the last 800 000 years (Corbett and Howard 2006).

The Darwin Crater is located 26 kilometres south-south-east of Queenstown, Tasmania, and 2 kilometres east of the Franklin Track.

Scientific importance and research potential

This site is unique and recognised by Australian and international scientists for its contribution to the study of impact craters. Furthermore it is the most important microfossil site in Tasmania, as it provides significant information about the Late Pleistocene and Holocene in Tasmania, and one the longest continuous pollen records in Australia (Colhoun and Van de Geer 1988, 1998; Heusser and Van de Geer 1994). It also provides data for interpreting when the first Tasmanian Aboriginals arrived and occupied the region, as well as the glacial history of Tasmania (Jackson 1999a).

A number of studies have been conducted on the site (e.g. Fudali and Ford 1979), culminating in the PhD thesis of Howard (2004). Colhoun and Van de Geer (1988, 1998) focused on the microfossil component of the lake sediments. Further research on the 60-metre lake sediment profile could prove significant.

Fossil significance

The fossils found at the site are typical, indicative pollen grains from the later Quaternary, illustrated by the pollen profile from the Darwin Crater (see Colhoun and van de Geer 1988; Colhoun 1989). The site is also the only known location for the Late Pleistocene fossils of three groups of plants now extinct in Tasmania (the rainforest shrubs/trees *Symplocos*, an unknown daisy *Tubulifloridites pleistocenicus* and another unknown flowering plant *Stephanocolpites oblatus*) (Macphail *et al.* 1993). The assemblage provides a good representation of a late Quaternary palaeoflora scenario in general (see Jackson 1999b).

The Darwin Meteorite Crater is unusual in that it provides significant fine-scaled information about the late Pleistocene and Holocene in Tasmania, and is the longest continuous pollen record in Australia (see Balmer *et al.* 2004). This gives the site excellent potential for palaeoenvironmental reconstruction.

Taphonomy and condition of the site/fossils

The fossils are mostly very well preserved.

Comparison with similar sites in Australia

The Darwin Meteorite Crater provides significant fine-scaled information about the late Pleistocene and Holocene in Tasmania, and is the longest continuous pollen record in Australia (see Balmer *et al.* 2004). Another Australian site of comparable significance is Lake George (NSW); however, although the Lake George record is longer, it has not been as well dated, it has large gaps of unknown duration in the record, and the fossil pollen is significantly less well preserved.

Stories

The site has wide dioramic potential, for both the meteorite impact and the pollen profile. Key stories are the impact of the meteorite, the formation of the crater, and the subsequent formation of the lake sediments. Also the diorama could show how the analysis of pollen allows us to interpret the changing climate during the Pleistocene and provides evidence for the arrival of Aborigines in south-western Tasmania.

The main evidence provided on evolution by this site is related to plant extinctions. The history of plant extinctions during the Quaternary is better preserved in Tasmania than in any other part of the southern hemisphere (Macphail *et al.* 1993; Jordan 1997b). The Darwin Crater provides significant evidence for understanding the timing of these extinctions.

The site is also of historic importance. Tasmanian Aboriginals discovered the Darwin Glass more than 25 000 years ago (Corbett and Howard 2006). The site is significant to scientific history, with the development of the pollen profiles from the lake sediments.

Ordovician faunal succession of the Florentine Valley

Site overview

A thick, essentially continuous, sedimentary succession spanning the Ordovician Period (488–443 Ma) is found in the southern Florentine Valley in southern Tasmania. The sequence, some 2500 metres thick, is richly fossiliferous at many horizons, and has been the subject of numerous taxonomic and biostratigraphic studies into the contained invertebrate shelly fossils, corals, conodonts and graptolites. It is an index succession for the Ordovician biostratigraphy of shallow-water marine faunas in Australia, contrasting with (and supplementing) the deep-water graptolite succession of the Ordovician of Victoria. The site embraces numerous localities over an area of approximately 200 square kilometres. The succession is predominantly limestone (known as the Gordon Subgroup), with siltstone at the base (Florentine Valley Formation) and sandstone at the top (Arndell Sandstone). The Arndell Sandstone ranges up into the earliest Silurian (approximately 443 Ma).

Scientific importance and research potential

The Tasmanian Ordovician faunal succession (largely based on that of the Florentine Valley) forms a significant part of syntheses of Australian biostratigraphy such as the *Australian Phanerozoic Timescale* by Young and Laurie (1996).

The site has been extensively studied. An early phase of reconnaissance palaeontological work is reflected in the papers of Kobayashi (1940), Brown (1948), Banks (1957), Banks and Johnson (1957) and Thomas (1960). Corbett and Banks (1974) described the geological setting, gave an overview of the palaeontology of many of the localities, and foreshadowed the potential for further systematic work. Stait and Laurie (1980) documented seven consecutive assemblages based on brachiopods and trilobites in the Florentine Valley Formation. Banks and Burrett (1980) set up a preliminary biostratigraphic zonation incorporating the seven assemblages of Stait and Laurie (1980) and including a further fourteen assemblage-zones in the Gordon Group and Arndell Sandstone, thereby spanning the whole of the Ordovician and the earliest Silurian.

PhD theses covering the systematic taxonomy and biostratigraphy of particular groups, with a significant focus on the Florentine Valley, include Burrett (1978) on conodonts, Laurie (1982) on brachiopods, and Stait (1981) on nautiloids. Laurie (1980) described the brachiopods of the Florentine Valley Formation and Jell and Stait (1985) the trilobites. Laurie (1991) erected a biostratigraphic scheme based on twenty brachiopod assemblages spanning the Ordovician and earliest Silurian, based largely on the Florentine Valley sections. Details of the nautiloid faunas were published by Stait (1988). Webby (1991) described the stromatoporoids. Baillie *et al.* (1978), Sheehan and Baillie (1981) and Banks (1988) dealt with aspects of the palaeontology and correlation of the late Ordovician to earliest Silurian, Arndell Sandstone. Rickards and Stait (1984) and Rickards *et al.* (1991) dealt with the systematics and taphonomy of the graptolite *Psigraptus jacksoni* from the Florentine Valley Formation.

Research on the site has declined in the last 15 years but much potential remains for further systematic taxonomic and biostratigraphic work in several major groups, including corals, molluscs and trilobites; and integration of the biostratigraphic work into a formal zonal scheme.

Fossil significance

The Tasmanian Ordovician shelly fauna has a high degree of endemicity. Many species and genera in the Florentine Valley succession are known only from this

Figure 5.1 The (Upper) Florentine Valley, Tasmania. Image © Catherine Rogers

area or from one or two other places in Tasmania. Examples of endemic genera include the brachiopods *Tasmanorthis*, *Lepidomena* and *Maydenella* (Laurie 1991). However, the main significance of the site lies in its stratigraphic continuity, structural simplicity, richness in fossils, and the duration of time (approximately 45 Ma) represented by the succession.

The Florentine Valley faunal succession, together with sedimentological studies, has been very significant in palaeoenvironmental interpretation of Tasmania during the Ordovician. For most of the Ordovician the site was part of an isolated, subtropical carbonate platform (e.g. Banks and Baillie 1989).

Taphonomy and condition of the site/fossils

Preservation of fossils at the site is variable and often excellent. *Psigraptus jacksoni* from the Gordon Road locality is the first graptolite in which preserved zooids have been found (Rickards and Stait 1984). Preservation of trilobites and brachiopods as moulds in the mudstones of the Florentine Valley Formation is generally very good. At many horizons within the limestones of the Gordon Group, the shelly fossils are silicified, allowing extraction of three-dimensionally preserved fossils by acid treatment.

The state of the exposures is variable. Some of the road cuttings referred to above have deteriorated by weathering in the 30–50 years since road construction. The natural vegetation is dense temperate rainforest and most of the limestone exposures (e.g. localities 4 and 5) were originally cleared and made accessible by forestry operations. Dense regrowth has tended to obscure these exposures over time, but continuing forestry operations in the area generally provide similar exposures along strike.

Comparison with similar sites in Australia

The only comparable Ordovician section in Tasmania is in the Mole Creek area, but there the fossiliferous succession covers a significantly shorter time interval (Banks and Burrett 1980). The Florentine Valley site is unusual in the Australian context in the apparent continuity and duration of the biostratigraphic succession. Being a predominantly shallow-water shelly succession, it is difficult to correlate directly with the index Victorian graptolite zonal scheme, which is comparably complete. Shallow-water Ordovician successions are found in western New South Wales but are less complete than in the Florentine Valley. Geographical affinities of the Florentine Valley faunas are generally with east Asia, North America and South America, and the Florentine Valley succession offers the possibility of improved correlation between these areas (Banks and Burrett 1980).

Stories

Because of the long time interval represented in this section, evolutionary development of many lineages is well displayed. For example, tabulate corals underwent their initial radiation during the Middle to Late Ordovician and increase markedly in diversity through the upper part of the Florentine Valley succession.

Fossil Cliffs, Maria Island

Site overview

An outcrop and adjacent former quarry site at the Fossil Cliffs near Darlington on Maria Island form an exceptional site for viewing and studying an extraordinarily rich assemblage of Early Permian marine invertebrate fossils. These lived in a shallow sea, cold enough to support floating ice.

The site was known as early as 1836, when George Frankland, Surveyor-General of Van Diemen's Land, published a paper about it in London (Frankland 1836). Only two years later, Lady Jane Franklin, wife of the Governor Sir John Franklin, noted petrified shells from Darlington. As early as March 1825, convicts were sent to Maria Island and a settlement established at Darlington. A new phase of development of the island began in 1884, leading to the development of a cement works using limestone from a quarry at the Fossil Cliffs. The cement works closed in 1930.

The Fossil Cliffs are one of the features that led to the declaration of Maria Island as a national park in 1972. Of key significance at this site are the assemblage of fossils in the Basal Beds of the Early division of the Parmeener Supergroup (Early Permian), especially that in the Darlington Limestone. The rocky shore platform exposes the Basal Beds, 7–8 metres of limestone and siltstone including a unit rich in dropstones (called the 'Lower Erratic Zone' by Johnston in 1888). The dropstones include some of local derivation, but others have travelled a considerable distance. Some are oriented with their long axis vertical (e.g. an angular granite clast, which disrupts and crumples the bedding, and a thick bilaminar colony of *Stenopora*, probably *S. johnstoni*). Fossils are generally abundant and include spiriferid brachiopods, pectinoid molluscs, and thick-shelled clams (*Eurydesma* spp.). Stenoporid bryozoa predominate, and some of these are preserved in their life position. Gastropods, conulariids and crinoid fragments are also present. The beds are almost horizontal.

Following the Basal Beds is the Darlington Limestone, just over 17 metres thick, and consisting mainly of *Eurydesma*-rich calcirudites, with some spiriferid calcirudites and, more rarely, laminated bryozoal siltstone. This formation is extraordinarily

Figure 5.2 Fossil Cliffs, Maria Island. Image © National Library of Australia

fossiliferous. Over 20 species of marine invertebrates, such as foraminifera, bryozoa, brachiopods, bivalves, gastropods and crinoids, have been recorded (e.g. Runnegar 1970; Reid 2003). However, the main feature is the abundance of well preserved specimens with their original shells. No one has attempted to count the shells but there must be many thousands of complete valves of *Eurydesma*, possibly more than a million, exposed to form a spectacular view.

Scientific importance and research potential

Johnston (1887) described some of the shells. Many, if not most, of the fossil species in the Darlington Limestone and the underlying beds have been described and illustrated, albeit in some cases from places other than Maria Island. Accessibility of the site and clarity of the exposures within the site allow spatial relationships between strata and between fossils within strata to be established readily. Isotopic studies by C.P. Rao show that water temperatures were low (<5°C), consistent with transport of the dropstones in floating ice (Rao and Green 1982). More detailed sedimentological studies of the several types of sediments might well reveal more information on the nature of the cyclic sedimentation (e.g. temperature, current velocities and depth variations). The abundance of fossils and the clarity of their spatial relationships should allow palaeoecological parameters to be determined.

At least four ecological assemblages can be postulated, based on existing knowledge: *Eurydesma* calcirudite, spiriferid calcirudite, stenoporid-rich assemblages (mainly in arenite layers), and siltstone layers rich in fenestrate bryozoa. Quantitative studies of these assemblages have not yet been made.

Eurydesma-rich sequences are known elsewhere in Tasmania but these mostly lack the excellent outcrop shown at the Fossil Cliffs. Similar sequences are known in New South Wales, Queensland, Western Australia, India and Pakistan, South Africa and South America (a Gondwanan distribution) so that detailed studies at the Fossil Cliffs might well have wide application in reconstruction of past environments in many places.

Clarke and Baillie (1984: fig. 4, p. 21) depict the palaeogeography of Maria Island as an archipelago of islands in a sea floored by shelly limestone in some places and by arkosic sandstone and boulder, cobble and pebble conglomerate in others. In what ways did the faunas of a sub-polar archipelago in the Early Permian differ from those of sub-polar regions today?

Johnston (1887) identified some of the fossils from the site and figured (1888: pls. XVII, XVIII) some of them, and noted (1888: pp. 123–124) that 'nowhere throughout the island [Tasmania] is a finer section of the Lower Marine Beds exposed.' Montgomery (1891) was concerned with the extent of the 'hydraulic' limestone and its nature, and Johnston (1902) gave further details. The next significant works appeared in 1957; Everard (1957: fig. 53) provided a geological map and Banks (1957: pp. 57–59) presented a detailed stratigraphic section of the Darlington Limestone.

Fossil significance

Eurydesma, the most obvious fossil component of the limestone, is found in other parts of Tasmania, many places in New South Wales and Queensland, Western Australia, the Indian subcontinent, South Africa, and South America (see Runnegar 1979: figs 2, 3). *Eurydesma* has long been a prominent Gondwanan fossil and is associated at the Fossil Cliffs with foraminifera, a conulariid, fenestellid and stenoporid bryozoans, *Trigonotreta*, other spiriferid and terebratulid brachiopods, gastropods, other bivalves including pectinids, and crinoid stems. Runnegar (1979) reviewed the evidence for *Eurydesma* to have occupied a rigorous sub-littoral environment in cold water, and much of the evidence for this reconstruction can be seen at the Fossil Cliffs.

In and beneath the Darlington Limestone at the Fossil Cliffs, the fossils noted above occur in abundance, especially *Eurydesma* spp. Fossils characteristic of the interval, referred to as Invertebrate Zone 2 (see Clarke and Banks 1975), occur widely in Tasmania and are used to infer contemporaneous deposition of the rocks containing them. The same principle is used to correlate rock units with a similar suite of fossils throughout eastern Australia and, with less certainty, elsewhere. Because of the excellence of the exposure and the number and variety of the fossils at the Fossil Cliffs, the section there provides a very useful reference point for biocorrelation.

The Fossil Cliffs should be preserved, not for the rareness of the fossil species but for the unusual abundance of life in a clearly visible and readily understandable example of the proclivity of life to bloom under hostile conditions. As an added bonus to palaeontological studies, the well exposed cycles of deposition provide an excellent opportunity to study the interaction of processes leading to different types of sediment with the various elements of the marine fauna. The potential for palaeoenvironmental reconstruction has been demonstrated by Runnegar (1979) and by Clarke and Baillie (1984: pp. 21–22).

Taphonomy and condition of the site/fossils

The site consists of a shore platform washed by storm waves and on which exposure of bedrock is virtually complete. The platform varies up to about 10 metres in width and is backed by steep cliffs cut into the Darlington Limestone and higher beds, and is affected by waves and sea spray during stormy weather. Outcrop in these cliffs is virtually 100%. There are further and slightly higher outcrops in a cliff-top quarry, possibly 100 metres long by 30 metres wide, and this also exposes *Eurydesma*-bearing beds.

Most of the *Eurydesma* is preserved as conjoined or single valves composed of original shell material. Runnegar (1979) noted that most of the shells are articulated and in life orientation. Some of the shells are pierced by tubular holes, possibly caused by boring by sponges present in Permian seas. Because of wave, swash and spray action, the exposures are such that the casual observer might be excused for thinking that they are looking at beds of living shells.

Comparison with similar sites in Australia

The assemblage of fossil species at the Fossil Cliffs provides scope for comparative studies with sites in New South Wales, Queensland, Western Australia, the Indian subcontinent, South Africa, and South America. Rocks at these places (some listed by Runnegar 1979) record oceanographic conditions, climatic conditions and somewhat variable faunal conditions that could cast light on the limiting factors controlling the occurrence and abundance of *Eurydesma* and the associated fauna.

Stories

The site is of significance in the history of European exploration and colonisation of Australia. Maria Island was discovered by Tasman in 1642. It was circumnavigated by Péron in 1802, but he missed or misinterpreted the shells. Settlement on Maria Island began in 1825, when Governor Arthur established a penal settlement, named Darlington after Governor Darling of New South Wales; this was abandoned in 1832. Cement works were first proposed for Maria Island in 1887 with cement produced there from 1924 to 1930. Maria Island has a long history of protection and conservation. The first ranger was appointed in 1968; in 1971 it was listed as an animal sanctuary; and in 1972 as a national park. Although Tasmanian Aborigines are known to have lived on or visited Maria Island, no specific references to the Fossil Cliffs are known. This history is dealt with by Pretyman (1970) and Weidenhofer (1976).

The site also tells the story of key geological processes. It shows the deposition of sand and mud, and the folding and intrusion of granite (mountain chain erosion). It provides evidence of the Late Palaeozoic Ice Age (Late Carboniferous to Early Permian), and the retreat of the ice. Maria Island was an archipelago in a sea with floating ice, and was the site of deposition of sediments and proliferation of clams and other animals (Crowell and Frakes 1971).

Early Oligocene macroflora sites of north-western Tasmania

Site overview

There are three significant plant macrofossil sites in north-western Tasmania: Little Rapid River, Lea River, and Cethana.

Location 1: Little Rapid River

This site is a small, well circumscribed area in far north-western Tasmania. It consists of exposures of organic-rich, soft sedimentary rocks in two cuttings, in a small, basalt-capped rise. The fossiliferous sediments are presumably continuous between the exposures and may extend some distance under the basalt. The site is listed on the National Estate.

Location 2: Lea River

There are two main exposures at Lea River: the main river cutting and in a nearby gravel pit. The river cutting site was described in the 1995 report by R.S. Hill (1995b). The gravel pit exposure is a shallow exposure of mudstones and sandstones made by a small creek running through the gravel, which is probably of Quaternary age.

Location 3: Cethana

Located in a cutting on the uphill side of a road above Cethana Dam.

Scientific importance and research potential

The combination of the three contrasting and complementary sites of similar age in a small geographic area makes for a richer and more complex overview of the Early Oligocene (approximately 34–30 Ma) vegetation. There is almost no overlap in species between the sites, and this allows for inferences of much higher diversity than could be gleaned from any single site. The ecological contrasts between the three sites can be used to make inferences about ecological patterns.

These sites represent a key phase in the major transition from a conifer-dominated Gondwanan flora to the modern sclerophyll-dominated flora. The Early Oligocene was characterised by the onset of cooling climates (starting in the Middle Eocene, but most emphatically at the Eocene–Oligocene boundary, 34 Ma), but was before the onset of aridity in the Oligocene–Miocene.

Little Rapid River

The site has been studied intensely. Fossils from the site provide the main, or at least significant, components of 35 papers published in international scientific journals or books (see below). Little Rapid River provides a detailed record of the local vegetation at that time, showing a major response of Australia's vegetation to changing climate. For further details, see Hill (1995a) and Jordan and Hill (1996).

Little Rapid River has a remarkable diversity of conifers. All three major families of southern hemisphere conifers are represented: 14 genera and 26 described species. No other fossil or modern vegetation approaches this generic or specific diversity – there are only about 35 genera and 200 species in the southern hemisphere in total. The site also has a high diversity of some angiosperms, with 18 recorded species. Research on the site is not complete, and the number of recorded species is likely to increase.

Little Rapid River was the first site in Australasia where macrofossils (both leaves and fruit) of the South American group of *Nothofagus* (subgenus *Nothofagus*) and the tropical group of *Nothofagus* (subgenus *Brassospora*) were found. These macrofossils have had significant implication for understanding the implications of the break-up of Gondwana. It is also the only known locality for 27 species: *Agathis tasmanica*, *Araucaria fimbriatus*, *Austrosequoia tasmanica*, *Dacrycarpus mucronatus*, *D. linifolius*, *D. arcuatus*, *D. cupressiformis*, *D. linearis*, *Dacrydium tasmanicum*, *D. aciculare*, *Dicksonia dissecta*, *Epacriphyllum macphailii*, *E. mesibovii*, *Gymnostoma tasmanianum*, *Lagarostrobos marginatus*, *Libocedrus mesibovii*, *L. jacksonii*, *Nothofagus peduncularis*, *N. cooksonii*, *N. mucronata*, *N. serrata*, *N. smithtonensis*, *Phyllocladus lobatus*, *Richeaphyllum jonesii*, *Smithtonia jonesii*, *Sigmaphyllum tasmanensis* and *Willungia oppositifolia*.

The site is the type locality for 29 species of fossil plants (all those listed in the previous paragraph and also *Nothofagus bulbosa* and *N. lobata*).

Lea River

The Lea River site is rich in mummified leaves and fruit. A general description of the flora was given in the 1995 report by R.S. Hill (1995b). In particular it records a mixture of predecessors of Australia's cold-climate montane floras and warm-climate floras. It is the only known locality of four species: *Euproteaciphyllum papillosum*, *Fitzroya tasmanensis*, *Orites scleromorpha*, and *O. milliganoides*. It is also an important research site, with at least seven scientific papers published on it, and further studies ongoing.

Cethana

Cethana is rich in compressions of leaves and fruit. The flora at this site was described in the 1995 report by R.S. Hill. It is complementary to the Little Rapid

River and Lea River sites in that it contains a strong representation of sclerophyllous plants, which are much rarer than those plants found at the other sites.

This site is the richest known for the important Gondwanan family Proteaceae, with some 16 identified species. It is also rich in other angiosperms and conifers. It is the only known locality for 21 fossil species: *Acsmithia grandiflora*, *Banksieaeformis dentatus*, *Banksieaephyllum orientalis*, *B. linearis*, *Euproteaciphyllum lomatioides*, *E. tridacnoides*, *E. gevuininoides*, *E. cethanicum*, *E. linearis*, *E. rugulatum*, *E. attenuatum*, *E. ornamentalis*, *E. integrifolium*, *E. microphyllum*, *Lomatia xeromorpha*, *Macrozamia australis*, *Nothofagus cethanica*, *Tmesipteris tasmanica*, *Schizomeria tasmaniensis*, *Weinmanniaphyllum bernardii*. It is also a research site, with 22 published papers and ongoing research.

Taphonomy and condition of the site/fossils

Little Rapid River

The fossils are in good condition, with three-dimensional organic preservation with some compression, and many anatomic features preserved. Since they occur in moderately coarse-grained riverine sedimentary rocks, they presumably represent accumulation of plant litter transported by a river and deposited in slower water. The fossil-bearing sediments are oxidising slowly but are not showing strong signs of erosion.

Lea River

The Lea River site contains a dense assemblage of fossils preserved in poorly lithified organic-rich fluvial sandstones. The fossils are well preserved, with three-dimensional organic preservation and little sign of compression. The main site is a large, actively eroding river cutting, from which large blocks appear to be falling off at frequent intervals.

Cethana

The Cethana site contains compressions and impressions of leaves, flowers and fruit preserved in lacustrine siltstones. The fossils show good detail of fine and coarse venation, and many of them have well preserved cuticles showing microscopic detail that is essential for good identifications. The site has been heavily collected over the years because of the attractive fossils and good access. As a result much of the high quality exposed material has been removed.

Comparisons with similar sites in Australia

Little Rapid River is the most intensely studied plant macrofossil site of any age in Australia. The other two sites complement the Little Rapid River site, with Lea River adding a cool-climate assemblage and Cethana adding a sclerophyll component to the story told by Little Rapid River.

There are a number of sites elsewhere in Australia that show the transition from warm-wet Eocene climates to dry Late Neogene climates. They are either older (e.g. Maslin Bay, Anglesea) or younger (Latrobe Valley Coal, Berwick Quarry). However, these sites are less well studied, and certainly have not yielded such detailed information on past floras as Little Rapid River.

Stories

The key stories associated with these sites are to do with the dramatic changes in our vegetation over the last 50 million years, with the massive decline in rainforest and increases in the sclerophyllous vegetation often dominated by eucalypts. Associated with this is the fact that Tasmania (along with places like the wet tropics of North Queensland and the rainforests of northern New South Wales and southern Queensland, New Zealand and New Caledonia) remains a refuge for some of the ancient kinds of plants and animals. It also shows the ancient links between the landmasses of the southern hemisphere and even the northern hemisphere (*Austrosequoia* is a close relative of the redwood, *Sequoia*).

Lune River Jurassic plant site

Site name, type and location

The Lune River fossil plant site is situated about 100 kilometres south of Hobart on Tasmania's south-east coast. The Lune River site preserves a wealth of Jurassic plants, including numerous conifers and ferns, occurring in sandstone-mudstone beds of up to 1 metre thick, dated to the Toarcian Stage, late Early Jurassic (189–176.5 Ma) (see Bromfield 2004; Hughes *et al.* 2007).

Scientific importance and research potential

This site is unique and scientists around the world recognise its well preserved Jurassic plant specimens and its large fossil trees. The site comprises a suite of rocks confirming Tasmania's close links to Antarctica prior to the rifting of Gondwana (Bromfield 2004).

The work of the American palaeobotanist William Tidwell and co-workers makes it clear that the site is a world centre of diversity for tree fern fossils, and there are records of ferns and modern families including Cyatheaceae, Dicksoniaceae, Matoniaceae and Osmundaceae. Some of the petrified stems of Osmundaceae are of great significance because they enable an understanding of evolutionary development in one of the oldest known extant fern families (Tidwell and Ash 1994). A number of studies have made significant and lasting contributions on the fossil floras (Tidwell 1987; Tidwell and Jones 1987; Tidwell *et al.* 1987, 1991; Tidwell 1991; Tidwell and Skogg 1992). Bromfield (2004) conducted a palaeoenvironmental reconstruction using plant fossils and stratigraphic information.

Fossil wood anatomy can be used as a climate indicator to help interpret the palaeoenvironmental conditions that existed in this area during the Jurassic. New taxa have been recognised from the site and await formal description, including the first silicified cycadeoid reproductive structures known from Australia and a cycadeoid or cycad stem. The full extent of the fossil plant localities and associated basalts is to be determined. The taphonomy of the site requires further study.

The 2005 Forestry Tasmania annual report referred to a more recent collection of Lune River material containing the remains of an araucarian tree and the reptile *Tasmaniosaurus*, a representative of a primitive group from which the dinosaurs are thought to have evolved.

Fossil significance

This site is unique for its well preserved silicified (Jurassic) plant specimens and large fossil wood specimens. The site comprises a suite of rocks confirming Tasmania's close links to Antarctica prior to the rifting of Gondwana (Bromfield 2004). The available evidence is not complete but it seems likely that some trees are in growth position. It is the principal site of Jurassic-aged flora in Tasmania.

The site provides a good representation of Early Jurassic gymnosperm and pteridophyte (early vascular plant) floras. The leaf remains provide links to other Early Jurassic floras in Australia, for example Durikai in south-eastern Queensland. The silicified ferns at Lune River are known only from this site. The plants found at Lune River provide valuable climate data due to the range of species present. They indicate that the environment during the Jurassic was humid, seasonal and temperate (Bromfield 2004). Lune River offers a good potential for palaeoenvironmental reconstruction.

Taphonomy and condition of the site/fossils

The fossils from Lune River are thought to outrank other Jurassic sites because of their amazing three-dimensional preservation. Spectacular coloured photographs of transverse sections of Lune River fossils published in White (1991) are testament to the quality and significance of the site, as they reveal detail down to the level of individual cells. The fossilised wood is particularly well preserved (Bromfield 2004). More detailed research on the taphonomy of the site is needed to determine whether the tree trunks are in life position as a standing forest.

Comparison with similar sites in Australia

Silicified ferns are described from sites in south-eastern Queensland and elsewhere. However, the diversity of ferns and the presence of silicified cycadeoids (extinct group of cycad-like seed plants) is unique to Australia.

Stories

The Lune River site is of global importance in terms of climatic reconstruction and Jurassic plant studies. A key story is the evolution of gymnosperm and pteridophyte plant species in the early Jurassic. The ferns provide some of the earliest fossil evidence of the family Cyatheaceae, are an excellent representation of the Osmundaceae, and also rare fossil evidence of the Matoniaceae in Australia. The site has excellent dioramic potential within the realms of climatic reconstruction.

The site is a well known fossicking site. Amateur collectors have helped drive much of the research on the site by providing specimens to professional palaeontologists for study. Silicified plants have been described from the site for over 40 years and it has been known to amateur collectors for a much longer time.

Eastern shore of Macquarie Harbour

Site overview

The northern portion of the eastern shore of Macquarie Harbour contains a series of significant plant macrofossil locations. The key area extends from the mouth of the King River to the Strahan township on the west coast of Tasmania. The known fossil localities in this area occur within 200 metres of the shore of Macquarie Harbour. Within this broader area, several specific localities contain important fossils. These are of two distinct age ranges:

1. Early Pleistocene (2.2–0.8 Ma), primarily based on palynostratigraphy (see Jordan 1997b; Macphail *et al.* 1994).
2. Early Eocene, based on palynostratigraphy but supported by marine microfossils (see Macphail *et al.* 1994).

Three key localities within the area are at Regatta Point, at a road cutting on Lowana Road near Lettes Bay, and in a cutting on a side road off Lowana Road just north of Blackwood Creek (the Lowana site). There are a number of additional sites in the region.

Regatta Point site

This site contains fluvial gravels, probably of Middle Pleistocene age, overlying Early Eocene sediments. These were exposed in vertical cuttings during construction of the Mt Lyell Company wharf early in the twentieth century. The basal part of these gravels contains large angular mudstone clasts that contain fossils. These have been transported some unknown, but probably small, distance from their original site of deposition. The underlying Early Eocene mudstones and sandstones, which extend much further, also contain rich macrofossil floras.

Lowana site

A recently located site, this contains the best preserved representation of the Early Eocene fossil assemblages. It occurs in a road cutting on a private road running uphill off Lowana Road.

Road cutting site

Well preserved Early Eocene macrofossils also occur in a new road cutting at Lowana Road south of the junction of Lowana Road and Lettes Bay Road, immediately above and north of the sand quarry.

Other sites

Fossils of Early Eocene age occur at a number of sites in the area defined above, most notably in the railway cutting (the former path of Lowana Road, and now the ABT Railway line) beside the sand quarry just south of the road cutting site. This site was discussed by Pole (1998a) and Pole and Macphail (1996). There are relatively large exposures of the Early Eocene fossil-bearing sediment, which are weathering at present but remain in reasonably good condition. It is highly likely that further exploration will reveal many other sites with significant macrofossils in the Early Tertiary sediments of Macquarie Harbour. This is a large area of Tertiary sediment and is relatively poorly explored by palaeobotanists.

Scientific importance and research potential

The site contains rich Early Eocene macroflora and the only diverse Early Pleistocene macroflora in Australia. It is part of a suite that provides evidence of the climatic and vegetation history of Australia and the southern continents during the Cenozoic. The Macquarie Harbour locality contains fossils that record the transition from flora of 55–2 million years ago to modern vegetation; it records this time in far more detail than any other site in Australia (Jordan and Hill 1991). The early Pleistocene sediments show the transition between floras of the Palaeogene and Neogene (often dominated by diverse rainforest) and the modern vegetation (dominated by sclerophylls with species-poor rainforests).

The site is also significant for past climate interpretations: the Early Eocene included a 'greenhouse' interval linked to high global atmospheric carbon dioxide levels. This is reflected in the generally large size of the Macquarie Harbour fossil leaves and the fact that fossils of plants such as mangroves and cycads have been recovered (Carpenter *et al.* 1994; Pole and Macphail 1996). The Early Eocene sediments provide the most detailed record of the last period of true greenhouse climates in Australia. Research is ongoing, with the Lowana site being a focus of current study.

Fossil significance

The Early Pleistocene sediments contain the only diverse Early Pleistocene macrofossil flora in Australia (with at least 44 distinct macrofossil taxa and more than 70 pollen and spore types). It is also one of the richest floras of this age in the world. The Early Eocene sediments contain the richest Early Eocene macroflora in Australia, although this is not completely researched.

The site is the only known locality for 18 species of fossil plants. Eleven of these (*Acacia bulbosa*, *Banksia strahanensis*, *Callitris strahanensis*, *Dacrycarpus carpenteri*, *Laurophyllum australum*, *Myrtaceaephyllum pleistocenicum*, *Nothofagus pachyphylla*, *Orites truncata*, *Oxylobium pungens*, *Quintinia tasmanensis*, *Rubus nebuloides* and *Telopea strahanensis*) are from the Early Pleistocene sediments, and seven species (*Acmopyle glabra*, *Araucaria readiae*, *Araucarioides linearis*, *A. sinuosa*, *Eucryphia microstoma*, *Nypa australis*, *Pterostoma hirsutus*) are from the Early Eocene sediments. The site is the type locality for all 19 species listed above. It is a site of research, with some 23 scientific papers published and research continuing.

Taphonomy and condition of the site/fossils

The Early Pleistocene Regatta Point site contains well preserved mummifications and compressions of leaves, flowers, fruit and pollen. Macro- and microfossils of both modern Tasmanian species and extinct species are found together in the Early Pleistocene sediments. Different mudstone blocks contain different floras, and presumably were derived from different strata of the parent sediment. Three different major floras have been recognised. The RPU2 flora is derived from a single clast, which contained three clearly defined, densely fossiliferous layers. The composition of these three layers was essentially the same. The RPU1 flora is a composite of several clasts. The RPA flora is from a single small sample with a distinctly subalpine flavour.

The Early Eocene sediments contain fossil leaf fragments and fruit of the mangrove palm *Nypa*, leaf fragments of the cycad *Bowenia*, *Gymnostoma* (Casuarinaceae) twigs, and a range of conifers (*Araucaria readiae*, *Araucarioides linearis*, Podocarpaceae, Cupressaceae) and dicotyledons (e.g. *Eucryphia*).

The Early Pleistocene site is extremely restricted in area (about 15 metres by 15 metres), and the fossils are restricted to occasional clasts within these sediments. The site was last visited in March 1997. Exposures of the fossil-bearing clasts have weathered seriously in the last few years, and no good material is now exposed. However, considering the presence of fossiliferous clasts in all of the exposures at this site, it is almost certain that high quality material remains incorporated in the matrix at no great depth. It is vulnerable to collection, especially since it is within a couple of kilometres of a major tourist centre along a well-used road. This site has very significant conservation value, and its preservation deserves high priority. The site is not under immediate threat from development, but the West Coast Council is lobbying for development proposals for the nearby Mt Lyell Company wharf site.

Comparisons with similar sites in Australia

There are a number of other Early Eocene sites in Australia (Brooker Highway near Hobart and Buckland, north of Hobart and several sites in the Tamar Valley (Tas), and Mt Hotham (Vic)). However, none of these has such good preservation and diverse ecology. No other Early Pleistocene site with macrofossils has been studied in detail anywhere in Australia, although one significant site occurs in central Victoria at Stony Creek Basin. The latter site is currently under study.

Stories

This area has considerable potential for tourist and educational interpretation. Strahan is already a centre for ecotourism with considerable efforts having been made to communicate evolutionary, ecological and historical aspects of the area to the general public. The fossil sites have the potential to contribute significantly to this.

One key story is the dramatically different environment of this region 50 million years ago. Even though Tasmania was near the Antarctic Circle at this time, the region was an estuary with mangroves, palm trees, conifers with leaves 20–30 centimetres long, cycads, and tropical rainforest trees. This was a time when Tasmania was a peninsula on the south-eastern corner of Australia, and Strahan itself was only a few hundred kilometres from the Antarctic mainland. The fossils in this area are therefore the plants that grew immediately after Australia broke free of Antarctica. The Pleistocene story shows how extinctions have helped to mould our current vegetation.

To the best of our knowledge, this site does not have Indigenous significance. The site has minor historical significance in that it was an early area of palaeobotanical interest in Tasmania. R.M. Johnston recognised the significance of the area and described a number of fossils from it in the late nineteenth century.

The Shoreline near Point Hibbs, Western Tasmania

Site overview

Point Hibbs is a promontory on the coast of western Tasmania just over 46 kilometres south of the mouth of Macquarie Harbour. The site consists of a coastal strip about 1.25 kilometres wide from the low-water mark. Within the area there are fine coastal exposures (shore platforms and cliffs) backed by fairly gently sloping but heath-covered surfaces intersected by stream valleys. The coastal outcrops include Ordovician, Devonian and Permian.

Ordovician:

- Sandstone, granule conglomerate with minor siltstone, bioturbated in places and with some fossils.
- Ordovician limestone, dolomite, calcareous mudstone and minor quartz sandstone with some fossils (Gordon Subgroup Limestone).

Devonian:

- Fossiliferous marine limestone interbedded with calcareous mudstone (the Point Hibbs Formation). Coral heads up to 0.5 metres in diameter are a feature of outcrops of this unit on shore platforms.
- Non-fossiliferous red-bed sequence of siltstone and fine-grained sandstone with interbedded coarse-grained sandstone and conglomerate (Red Reef Cliff Siltstones).
- Well bedded marine quartz sandstone with minor siltstone and conglomerate, with some fossils (Whitehorses Beach Sandstone).

Permian:

- A sequence containing poorly sorted, polymict conglomerate, some of it possibly tillitic, sandstones, siltstones, some with glendonites and limestones. Some horizons contain abundant marine fossils. The sequence has been affected by Devonian thrust faulting from the east-south-east, and by major meridional faulting down to the west subsequent to the dolerite intrusions (possibly Late Cretaceous or Early Cenozoic).

The many and varied fossils in the Devonian Point Hibbs Formation are better exposed and younger than most Devonian fossils elsewhere in Tasmania, but the fossils in the Ordovician and the Permian rocks are also of considerable interest from several points of view.

Scientific importance and research potential

Limestone was first reported from Point Hibbs by Charles Gould (1862, 1866). He regarded it as equivalent to the limestones on the Gordon River, which he considered Silurian in age. It is now known that most of the limestones on the Gordon River are Ordovician in age, but an Early Devonian limestone is also recognised and correlated with the Point Hibbs Limestone (Gee *et al.* 1969).

Hills (1914) noted the limestone but the first study of fossils from the area was Hill (1942a), who recognised the Devonian age of the Point Hibbs Limestone on the basis of a study of corals. Banks visited the area with the late R.J. Ford early in 1958 and collected specimens from limestones within the nominated area. Later examination of these specimens revealed that some limestones contained Ordovician conodonts and the Ordovician coral *Tetradium*, whereas others contained Early Devonian corals and conodonts. Initially the Ordovician conodonts and coral were taken as 'remanié' but later two separate limestones were recognised.

Brachiopods were described from the Point Hibbs Limestone by Flood (1974). Conodonts from the Ordovician limestone were studied by Burrett in 1978. Burrett (1984) also noted and interpreted the difference in the Colour Alteration Index of conodonts in both limestones. Microvertebrates were described by Burrow *et al.* in 1998, and chitinozoa by Winchester-Seato and Carey in 2001. The microvertebrates are the oldest fossil vertebrates at present known in Tasmania.

During a visit to Tasmania in 1964–5, Professor J.E. Conkin collected samples from the Permian from which he obtained and described foraminifera (Conkin and Conkin 1993). In all, the fossils from the Point Hibbs area have been an important resource for research. However, the fossils in the Ordovician rocks (other than the Point Hibbs Limestone), Devonian units and the Permian sequence have barely been touched. Further studies of the Ordovician, Devonian and Permian rocks and fossils should allow better appreciation of the palaeoecology and palaeogeography of the area than is presently established.

Taphonomy and condition of the site

The site is relatively untouched because it is not easily accessible. Because the rocks of interest in the area are subject to marine and shoreline processes they should provide fresh and clean exposures for examination for decades into the future. Most of the fossils in the limestones (Ordovician, Devonian and Permian) are well preserved with little, if any, deformation. Some fossils in the sandstone and siltstone units are preserved as internal or external moulds.

Fossil significance

The fossils have been useful in attributing ages to the beds in which they were found. They have provided links with other outcrops in Tasmania in the Ordovician; New South Wales, Queensland, China and North America in the Devonian; and Tasmania, New South Wales, Queensland and Western Australia and to some extent elsewhere in Australia in the Permian. Observations on the conodonts in the Ordovician and Devonian limestones have provided some idea of the original thickness of the Ordovician to Devonian in the area. The age determinations based on the fossils have contributed significantly to the determination of the complex geological history of the area.

Comparison with similar sites in Australia

Fossils in the Ordovician rocks provide links with other parts of Tasmania and with North America that are yet to be fully explored. Fossils in the Devonian rocks show links with New South Wales, Queensland, China and North America, and in the Permian with other parts of Tasmania, New South Wales, Queensland and Western Australia.

Stories

There is an interesting and instructive story about the age of the Point Hibbs Limestone outlined above. No Indigenous significance is known to the author, but archaeological remains have been reported from the area.

Table Cape – Fossil Bluff, Wynyard

Site overview

The Early Miocene Table Cape site is a highly prolific source of invertebrates and unique vertebrate fossils (mainly marine) in southern Australia. Miocene marine sediments, approximately 23–22 million years old, occur along a 10-kilometre stretch of the north coast of Tasmania. They are best exposed at Fossil Bluff between the township of Wynyard and Table Cape, a volcanic prominence and well known landmark.

The site at Fossil Bluff consists of a cliff exposure

that is several hundred metres in length and approximately 30 metres high. There are two horizontal geological formations; a lower Freestone Cove Sandstone approximately 1–1.5 metres thick, and the overlying Fossil Bluff Sandstone. The two formations comprise the Table Cape Group. The Miocene outcrop sits on a magnificent exposure of the older Permian glacial unit, the Wynyard Tillite and is overlain by basalt, which protects the cliff section from erosion. The outcrop of the Wynyard Tillite on the extensive intertidal platform provides additional value to the site.

Scientific importance and research potential

As noted below, it is some time since the very diverse fauna of Fossil Bluff has been subject to comprehensive revision. At present, much of the description has been taxonomic and not enough attention has been paid to the palaeoenvironmental aspect, nor to the role of the species present as ancestors of the modern fauna of southern Australia. To illustrate the latter, it is clear that the Miocene bivalve mollusc *Eucrassatella oblonga* is the ancestor of the modern *E. kingicola*. The same is true for many other fossil species that led to well known modern forms. It would be relatively simple to perform a detailed taxonomic study to document the rate at which evolution had occurred and compare these results with those to be derived from amino-acid racemisation studies to help quantify the latter technique. There is need not only for taxonomic revision of all elements of the biota but also to extend the results of that study into a broader context.

At the time of deposition, Bass Strait had just opened to allow the passage of marine conditions from the west across southern Victoria and northern Tasmania. Before that time, very little water flowed between southern Australia and Antarctica because Australia had not moved far enough north to allow such circulation. Thus there is scope for comparing coeval faunas from opposite sites of the Australian–Antarctic Gulf as the gulf widened.

Fossil significance

The site is an extremely rich source of abundant and extremely diverse fossils, very dominantly marine.

Figure 5.3 Fossil Bluff, Wynyard. Photograph Stephen Spurling © National Library of Australia

By 1896, Pritchard could note that Fossil Bluff had produced mammals (one species), fishes (three species), cephalopods (one species), gastropods (153 species), bivalves (65 species), brachiopods (17 species), bryozoans (nine species), echinoderms (three species), corals (19 species) and foraminifera (10 species), but this is a gross underestimate as later work would show. It is now known that this list can be expanded greatly and much needs to be done to revise the list and its taxonomy.

The key vertebrate species that deserve redescription include:

- *Prosqualodon davidis* Flynn, a virtually complete representative of an order of extinct toothed whales (Flynn 1948). Much of the original is lost but good casts remain in the Tasmanian Museum and Art Gallery and the known remaining bones are held in the vault at the School of Earth Sciences, University of Tasmania.
  Other articulated whale vertebrae, isolated vertebrae and various fragments are recovered from time to time and indicate that the locality will yield further vertebrates over time.
- *Wynyardia bassiana* Spencer, a marsupial species that is related to wombats. The animal may have died onshore (or drowned in a river) and drifted out to sea to become buried in the shallow marine environment. The specimen has a deformed jaw and had something wrong with one of its hind legs, but the fossil is fairly complete and is housed in the Tasmanian Museum and Art Gallery.
- Fish: A flathead (*Platycephalus*) skull and the teeth of several species of shark are well known.

The very diverse invertebrate fauna was described in many papers, principally in the late nineteenth and early twentieth century, and a record is being consolidated at present. Darragh (1970) noted that the type specimens of over 30 species of bivalves and over 135 species of gastropods were described from Table Cape material. The diverse and abundant foraminiferid fauna has been the subject of sporadic small studies over many years and the then new genus *Sherbornina* was described by Chapman (1922). It has become an important genus in southern Australian stratigraphy. Quilty, in a series of papers (Quilty 1972, 1974, 1977, 1982), described and illustrated the foraminiferids and included the new genus *Crenulostomina*. The entire invertebrate fauna is ripe for re-analysis.

Taphonomy and condition of the site/fossils

Preservation of fossils at the site is excellent. They are extremely abundant and highly accessible. Bivalves are commonly articulated, and in one instance separating the shells revealed the sand-filled digestive tract and traces of the attachment of the soft parts to the shell. The potential for reconstructing the living animal is high in some cases. In the lower unit, tidal energy has caused breakage and mixing of faunas, but the environment of deposition is easy to envisage and reconstruct. It is comparable to conditions within a few metres of the modern site. In the upper, thicker unit, most of the fauna is in a condition that shows it has moved very little since death and many burrowing forms are in life position, thus allowing the biota in the environment at the time to be reconstructed in some detail.

The entire section seems to reflect an advancing sea initially with shallow, partly intertidal, high energy conditions, grading with time to a quieter, embayment environment. Water depth was never great, perhaps 20 metres at the deepest. Whether the transgression was caused by rising sea level or subsiding land cannot be determined from this outcrop, but a widespread incursion at this time suggests a global sea level rise.

Comparison with similar sites in Australia

Examination of the literature on the taxonomy of southern Australian invertebrates makes it clear that there are other sites of highly diverse faunas but most differ considerably in age, environment of deposition, or accessibility.

Stories

A significant component of the invertebrate fauna was described by a few authors, including: R.M. Johnston, who wrote the first comprehensive account of the geology of Tasmania in a classic work in 1888; the Reverend Julian Tenison Woods who, among many other things, was mentor to Mother Mary MacKillop; Professor Ralph Tate, University of Adelaide; and Professor T.T. Flynn, father of actor Errol Flynn.

One value of the site is educational. The potential for further educational development is strong, particularly with issues such as evolutionary change through time and the ancestry of modern invertebrate faunas.

Triassic vertebrate faunas, Knocklofty Formation

Site overview

A number of localities in south-eastern Tasmania have yielded Early Triassic vertebrates. These localities are treated as an assemblage of interrelated sites on the basis that they are of similar age and faunas. The vertebrate fauna is dominated by temnospondyl amphibians and a variety of fishes, including lungfish and ray-finned fish. The collection of Early Triassic vertebrate localities occurs in the Knocklofty Formation of south-eastern Tasmania (see Cosgriff 1974). The localities vary from quarry sites to sites close to sea level, some of which are within tidal range. According to Cosgriff (1974), the most complete exposures are in brick-clay quarries, particularly the Crisp and Gunn Quarry, West Hobart. These localities have yielded a range of vertebrate taxa collected mainly in the 1960s, although vertebrate remains have been known for close to 150 years (Banks *et al.* 1978).

The Tasmanian Knocklofty Formation is part of the Upper Parmeener Supergroup, Unit 2 quartz sandstone sequence and associated rocks, and consists mainly of well sorted, commonly cross-bedded and glistening quartz sandstone, feldspathic quartz sandstone and coloured lutite (Forsyth 1989).

The Knocklofty Formation is a non-marine unit as it contains no known marine fossils, but it does contain lycopods, equisitales, ferns and seed ferns, worm castings and burrows, conchostraceans and possibly malacostracans and insects, fish, amphibians and a proterosuchid reptile (Camp and Banks 1978; Cosgriff 1974; Dziewa 1980). It was deposited in the south-eastern part of a terrestrial basin that occupied much of Tasmania during the Early Triassic.

The Knocklofty section reveals that the Knocklofty Formation was deposited by water action, often under floodplain conditions, and consists of pond and lake deposits. This Formation is composed of three members (an upper sandstone member 50–55 metres thick; a middle member of extremely varied lithology 40–55 metres thick; and a lower sandstone member at least 90 metres thick); the upper and lower sandstone members mainly consist of massive sandstone layers, with occasional beds of siltstone and quartz and clay pebble conglomerates. The middle layer mainly comprises siltstones and shales (Cosgriff 1974). Most of the abovementioned fossil localities seem to correlate with either the middle or lower members of the Mount Knocklofty section, but closer correlation is prevented by limited exposure, often widely separated due to heavy vegetation, and also by disruption to exposures caused by block faulting (Cosgriff 1974).

Scientific importance and research potential

The site has been moderately well studied. A number of the localities have proved to be nationally and internationally important in terms of taxa and correlations (Banks *et al.* 1978). The Old Beach locality has proved to be one of the richest beds of vertebrate material of this age in the world (Banks *et al.* 1978), with the new Kooyna site appearing to offer this sort of potential. Camp and Banks (1978), Cosgriff (1974) and Dziewa (1980) in particular have made significant contributions to the study of the vertebrate remains. The amphibian species, and to a lesser extent the fishes, have been researched in detail by Cosgriff (1974) and Dziewa (1980) respectively. However, both groups require taxonomic revision in the light of new research (Bender and Hancox 2004; Borsuk-Bialynicka *et al.* 1999; Damiani 1999).

The Griesbachian Regional Stage, or earliest Triassic, is a time of particular interest for a number of reasons. The geology of this Stage is not well known and has relatively few exposures. The faunal elements are regarded as the survivors of the greatest disaster the biosphere has experienced, living immediately after the massive extinction of the Paleozoic (Erwin 1993). There is very little agreement on the conditions that caused the extinction, and this makes any new research invaluable. The Triassic Period and particularly the Early Triassic Epoch were interesting times, with the changeover from the archaic faunas of the Permian to those more typically associated with the Triassic and the later Mesozoic (Borsuk-Bialynicka *et al.* 1999).

Early Triassic fishes are rare in most Australian deposits, thus making this series of sites with its fish faunas important (Cosgriff 1974; Dziewa 1980; Turner 1982). Although fish remains are also relatively rare in the Knocklofty Formation, they provide important evolutionary links between Australia and other countries that were formerly joined as the southern continent Gondwana. Fishes, as opposed to land animals, can also provide information on ancient waterways. In particular, the lungfish *Ptychoceratodus phillipsi* belongs to a lineage of lungfishes and is a possible ancestor to the modern Australian lungfish (see Bender and Hancox 2004). The other fish species also form important links within the Gondwanan context and provide information on the evolution of lineages that led up to the evolution of the first mammals.

Fossil fishes in particular are good indicators of palaeoenvironments and have also been useful for global biostratigraphic and correlative purposes (see Bender and Hancox 2004); thus more detailed study of the Tasmanian material could prove worthwhile in this regard.

The Amphibia are the most abundant vertebrate group in the south-east Tasmanian Early Triassic vertebrate assemblage. These temnospondylid amphibians underwent a radiation in the Early Triassic, with Australia preserving the fullest record of the Gondwanan Triassic amphibians (see Anderson 1999). The presence of temnospondylid and lydekkerinid amphibians in the Knocklofty Formation and deposits of similar age in mainland Australia,

Figure 5.4 Reconstruction from the Knocklofty Sandstone Formation, depicting plants and animals that lived in the early Triassic. Painting Peter Schouten © SEWPaC

South Africa, Antarctica, India, Madagascar, Russia, North America and Greenland provides a strong basis for global correlative research between the Australian assemblages and fossiliferous outcrops of similar age globally (Banks *et al.* 1978; Neveling 2004). The reptile *Tasmaniosaurus* from the Crisp and Gunn Quarry site is of great evolutionary significance in that it belongs to the primitive animal group Thecodontia from which dinosaurs evolved (Banks *et al.* 1978).

Continuous production of new outcrop by erosion and the discovery of new localities such as the Kooyna site indicate that the unearthing of new material will continue. Planned future research includes development of the new Kooyna site, which promises to yield more vertebrate fossils than any of the other sites have to date, including the rich Old Beach site. Most of the other sites have not been investigated recently for new remains, thus providing promise for future collecting.

Fossil significance

The sites offer very good potential for palaeoenvironmental reconstruction. Fossil fishes in particular are good indicators of palaeoenvironments and have also been useful for global biostratigraphic and correlative purposes (Long 1993b). The Knocklofty Formation sites are unusual because of the range of vertebrate taxa represented (reptile, fish and amphibian). Several taxa have a global distribution but are rare enough to be important indicators for the time period, particularly the proterosuchian reptile *Tasmaniosaurus* and most of the fish taxa.

Taphonomy and condition of the site/fossils

Many of the fossils in the conglomerates are not articulated and are often only moderately well preserved. However, fossils within the finer grained sandstone and siltstone horizons, such as Parkers Beach, are articulated and more concentrated, and are often in good to excellent condition.

Comparison with similar sites in Australia

The Knocklofty Formation appears to be contemporaneous with the Arcadia Formation (Qld) (Northwood 1999), which is regarded as older than the Early Triassic Blina Shale (WA) and possibly Griesbachian Regional Stage (251–250 Ma) (Northwood 1999; Damiani 1999). Knocklofty contains a good assemblage of Early Triassic amphibians and fish, comparable to well represented sites in Australia including the Arcadia Formation sites.

Stories

Owing to the collective nature of the sites, these localities not only provide excellent potential for dioramic insight into pre-dinosaur and pre-mammal vertebrate faunas of Australia and Gondwana but also have the potential to provide environmental reconstruction (see Banks *et al.* 1978).

Australia preserves an important Triassic record which contributes to the Gondwanan picture at this time, with the Tasmanian Triassic a vital component. The Early Triassic time period is one of the most important in the history of the Earth, because it occurred immediately after the end-Permian global extinction. This was the greatest of the big five extinctions and set the scene for the most explosive of all terrestrial radiations in the mid and late Triassic, leading to the rise of the mammals and the radiation of the dinosaurs (Erwin 1993; Borsuk-Bialynicka *et al.* 1999).

Fossils have been known from this series of localities for close on 150 years. The history of the site has included national and international collaboration, and has made a significant contribution to the global story of vertebrate evolution within the early amphibian, reptile and fish groups. The site should continue to be the subject of collaboration among scientists statewide, nationally and internationally. It has tourism potential for Tasmania, particularly the south-east region.

Victoria

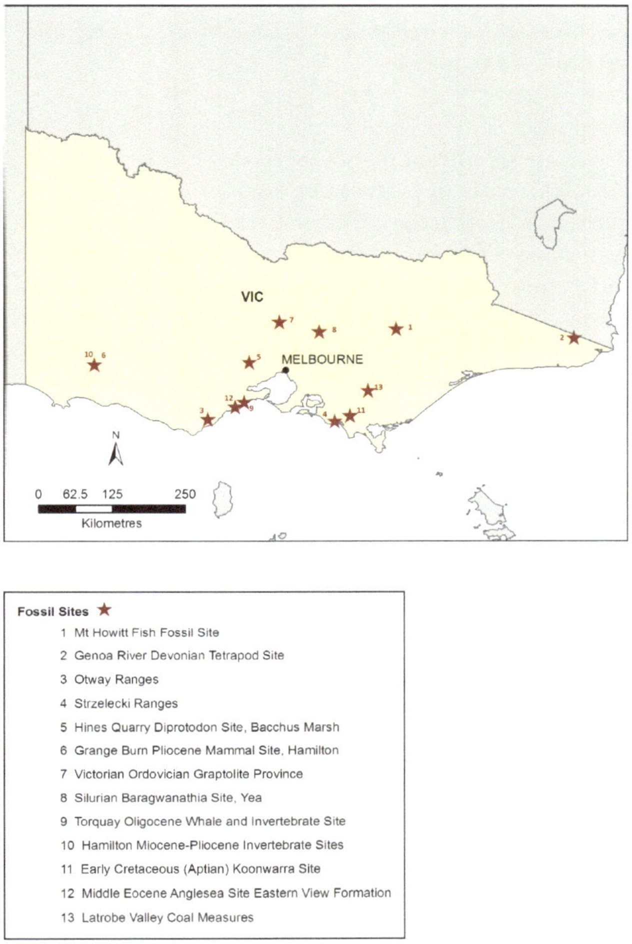

Victoria locality map © SEWPaC

Mt Howitt fish fossil site

Site overview

The Mt Howitt fish fossil site is located in central Victoria. Fossils occur in shale outcrops along the old vehicle track next to the meeting of the north and south branches of the Howqua River, about 4 kilometres due west of Mt Howitt.

Scientific importance and research potential

The site has been excavated over one 6-week field season in 1971–1972 by a Monash University team led by Professor James W. Warren, and by smaller collecting parties from the Australian Museum (Dr Alex Ritchie). The geology of the site was first mapped by Mark Marsden and then written up in the 1986 version of *Geology of Victoria*. Subsequent work has clarified the geological status of the fossil site as lying within the Bindaree Formation, of late Givetian age (Middle Devonian, Young 1999). The original excavations identified two subsites within the area: a main quarry site and a snig track site formed by a backhoe cut up the hill, about 60 metres up (east) from the quarry site, next to the meeting of the Howqua River branches.

The fossil remains from the Mt Howitt site include well preserved fishes and fragments of plants. The 1971–72 excavation recovered around 1000 specimens of fossil fishes. It is the only site of Middle Devonian age to preserve complete whole fishes in all stages of growth, making this the most significant fossil vertebrate site of its age in Australia. To date, some 15

Figure 6.1 Outcrop of the Mt Howitt fossil beds. Image © John A Long

species of fishes have been described, some uniquely placed in their own orders and families.

Much of the published work on the Mt Howitt fish fossils has been heavily cited and is still widely used in reference to all recent work on the interrelationships of basal placoderm, acanthodian and osteichthyan fishes. For example, recent work focusing on ray-finned fish evolution by Gardiner, Schaeffer and Masserie (2005) cites Howqualepis in a pivotal position on the cladogram, as does work by Cloutier and Arratia (2004). Similar work on acanthodian evolution always cites the Mt Howitt forms (e.g. Hanke *et al.* 2001; Hanke and Wilson 2004). Papers reviewing sarcopterygian phylogeny always mention of *Marsdenichthys*, *Gavinia*, and the Mt Howitt lungfishes (e.g. Schultze 2001; Johanson 2004). Work on *Bothriolepis* in recent papers has cited the Mt Howitt species (e.g. Thomson and Thomas 2001; Arsenault *et al.* 2004), as have all recent papers on phyllolepids (e.g. Ritchie 2005; Young 2005; Young and Long 2005). The robustness of the scientific work on the Mt Howitt site is based on the fact that it is long-standing and still widely cited as of 2006.

Although the formal taxonomy of most of the fishes has been done, some forms are poorly known and require better examples before they can be fully described (e.g. *Gavinia*, *Marsdenichthys*, *Beelarongia*, and an unidentified, possibly new form of osteolepid). In addition, new studies on the growth and allometric change in some of these fishes needs to be done, utilising the fact that most of the fishes preserved in large numbers from the site show all stages of growth from larvae to adults.

Fossil significance

Mt Howitt has a good representation of the fossil fish fauna of its age as it contains a diverse assemblage of forms. Such diversity is seen in only one other Middle Devonian site in the southern hemisphere, the Aztec Siltstone of Antarctica.

The fish at Mt Howitt are generally common once the right layers are excavated. Common species include *Bothriolepis* spp.; *Austrophyllolepis*, *Groenlandaspis*, *Howqualepis*, *Barwickia*, *Howidipterus* and *Howittacanthus*. The remaining taxa are known from only two or three examples, except for *Gavinia* and *Beelarongia*, which are known from one specimen each.

The Mt Howitt site is unusual in several aspects. First, it is a well preserved lake deposit preserving fishes and plants within a dark-coloured shale, but it is entirely devoid of invertebrate fossils (which are normally abundant in lake deposits). Coprolites (fossil fecal matter) are found occasionally, testifying to the unusually quiet conditions of preservation. Second, the fishes are preserved in all stages of growth, indicating that periodic and regular mass kill events took place. There is good potential for reconstructing the palaeoenvironment because we have a diverse assemblage of fish fossils coupled with numerous plant remains, thus giving an idea of the nature of early land vegetation at the time.

Taphonomy and condition of the site/fossils

The most significant feature of the taphonomy of the Mt Howitt site is the complete preservation of the fishes in all stages of growth, indicating quiet lacustrine conditions of deposition with regular mass kill events. In some cases, very delicate bones not normally preserved in fishes of this age are identified from impressions in the mudstone, for example otoliths or ear stones in both the palaeoniscoid *Howqualepis* and the placoderm *Austrophyllolepis*, and also remains of the tail in the placoderms. The bone is not well preserved in the shale due to surficial weathering, so the remains are best studied by negative preparation: immersion in a weak hydrochloric acid solution to dissolve away any weathered bone, then casting the surface with latex to give a clear impression of the bones and their structures.

Comparison with similar sites in Australia

The Canowindra fish fossil site (NSW) is Late Devonian in age, which is slightly younger than the Mt Howitt site. The scientific significance of the Canowindra site lies in its preservation of large tristichopterids, which are end members of their clade (Johanson 2004) and in the fact that the earliest complete rhizodont *Gooloogongia* occurs here. Mt Howitt is clearly a much more significant site scientifically because:

- it has a larger diversity of fish species (14 named species, another undescribed)
- many more unique or endemic generic taxa are present
- some monotypic new families are represented here and nowhere else (e.g. Howittacanthidae, Culmacanthidae)
- fishes in all stages of growth are present.

Furthermore, the site is located in a picturesque setting where visitors can see the kind of intramontane environment that the Devonian fishes lived in.

Stories

Mt Howitt is very amenable to reconstructing the ancient diorama of the Devonian lake at the time Victoria's first forest was being established on land. The site also helps tell the story of the origins of the first fishes, especially our distant ancestors

Figure 6.2 A fossilised skull of the primitive lobe-finned fish *Marsdenichthys*. Image © John A Long

international scientific importance and led the expedition to recover the fossil slabs by helicopter. These specimens are now housed by Museum Victoria and are on public display in the Evolution Gallery. The robustness of the scientific work on the Genoa River site is based on the fact that it is longstanding and still widely cited as of 2006.

the osteolepiforms, from which tetrapods and all subsequent land animals evolved. It also helps show the origins of air-breathing in early lungfishes, and the evolution of new species by hox gene changes in growth and development (heterochrony).

The site sits within a historic national park framework, so stories of the high mountain country, the early pioneers and local Indigenous stories could all be woven into the same tourism attraction.

Genoa River Devonian tetrapod site

Site overview

The site is located at Genoa River in eastern Victoria, where Red Sandstone and shale outcrops along the margins of the river.

Scientific importance and research potential

The site was first studied in 1971 by Monash University scientist Norman Wakefield, who discovered the fossil tetrapod trackways and contacted Professor James Warren. Warren identified them as being rare fossils of

The Genoa River tetrapod trackways are among the oldest dated trackways of tetrapods in the world, and are frequently referenced in the literature dealing with early tetrapods (e.g. Clack 2002). The outcrops along the banks of the Genoa River have huge potential for further exploration to find the oldest tetrapod remains in Australia and even the southern hemisphere. If a well preserved tetrapod skull were found here it would be ground-breaking scientific news.

Fossil significance

The fish fossils at Genoa River are abundant as isolated plates or parts, and unknown as articulated or complete specimens. However, the tetrapod trackways are preserved as a set or complete trackway (Warren and Wakefield 1972). This is akin in ichnotaxonomy to having an articulated fossil, as the gait and size of the track-maker can be fully studied. The trackways are indeed very rare, but in times of low river level it would be possible to search for others. Genoa River has a fair representation of the fossil fish fauna of its age, but not a diverse assemblage like Mt Howitt.

Figure 6.3 Genoa River. Image Robert Bruce © National Library of Australia

There is good potential for reconstructing the palaeoenvironment because we have a small assemblage of fish fossils coupled with plant remains, thus giving an idea of the nature of early vegetation on land at the time. The genera found here are also similar to those at the Canowindra fish fossil site (NSW).

Taphonomy and condition of the site/fossils

The most significant features of the taphonomy of the Genoa River site are the incomplete preservation (fragmentary and isolated nature) of the fish remains, coupled with beautiful preservation of trackways of the tetrapods. The bone is well preserved in the red mudstone facies, although exposed blocks in the creek generally show surficial weathering of bone with impressions commonly preserved. The remains are best studied by negative preparation: immersion in a weak hydrochloric acid solution to dissolve away any weathered bone, then casting the surface with latex to give a clear impression of the bones and their structures. The tetrapod trackways are also studied by casting and low light photography, so replicas could easily be made for display at or near the site should the need arise.

Comparison with similar sites in Australia

The Genoa River site is the only Australian and southern hemisphere site to show Devonian tetrapod trackways, and among the oldest dated tetrapod trackways in the world. It could be compared favourably with the Canowindra fish fossil site (NSW) as it has similar species of fish (*Bothriolepis*, *Remigolepis* and a large tristichopterid), but not anywhere near as perfectly preserved as at Canowindra, nor as many species. The Genoa River site is located in a picturesque setting where visitors can see the kind of fluviatile environment that the Devonian fishes lived in.

Stories

The site is very amenable to reconstructing the ancient diorama of the Devonian lake at the time of Victoria's first forest being established on land. The genera found here are also similar to those at the Canowindra fish fossil site (NSW), so there are reconstructions and models for some of the fish already made. A model of an early tetrapod could be made easily using the East Greenland models of *Acanthostega* made by Jenny Clack's team at Cambridge University. The site has links to the origins of the first amphibians and the first land animals, from which all subsequent land animals evolved. The site sits within a historic national park framework, so stories of East Gippsland, the early pioneers and local Indigenous stories could all be woven into the same tourism attraction.

Otway and Strzelecki Ranges

Site overview

The Otway and Strzelecki Ranges sites, which are from two locations, are comprised of arkosic sandstone, claystone, and mudstone outcrops exposed on the shore platforms flanking the Otway and Strzelecki ranges. The Otway Ranges Coastal Cretaceous site extends all the way from Lorne to Moonlight Head on the western Victorian coastline. The Strzelecki Ranges Coastal Cretaceous site extends from Inverloch through to San Remo on the eastern Victorian coastline.

Scientific importance and research potential

Systematic study of the Early Cretaceous Victorian coastal platform deposits began in 1978 and continues to the present. Approximately 40 person years of effort have been put into collecting the two principal sites, Dinosaur Cove in the Otway Ranges and Flat Rocks near Inverloch. Fossils are collected by breaking up the indurated rock manually and examining each surface thus exposed. Adding to the significance of these sites, in addition to those factors which make fossil land vertebrates of that age generally interesting, is that they were located in a polar environment (Vickers-Rich *et al.* 1999).

Dinosaur Cove (Otway Ranges)

Dinosaur Cove is located 12 kilometres north-west of Cape Otway (Rich and Vickers-Rich 2004). Fossil vertebrates were first discovered at Dinosaur Cove in 1980 by Michael Archer and Timothy Flannery. Systematic excavation began in 1984 and continued through to 1994. Dinosaur Cove is Early Albian in age (Early Cretaceous) and the productive unit is in the Eumeralla Formation of the Otway Group (Rich and Vickers-Rich 2004; Wagstaff and McEwen-Mason 1989). There were three sites where excavations were carried out at Dinosaur Cove. From west to east these are Dinosaur Cove West, Dinosaur Cove East, and Slippery Rock. Research on the material collected from Dinosaur Cove is current and ongoing.

The fossils from Dinosaur Cove constitute one of the very few known polar dinosaur assemblages from either hemisphere and one of the most diverse (Rich *et al.* 2002b). Work ceased at Dinosaur Cove at the end of the 1994 season. As almost all the known fossiliferous rock there had been excavated and the low scientific return in excavating the remainder did not justify further effort. At that final stage, there was only one productive site and when work was terminated there, the fossiliferous layer was 5 metres below sea level. To work it, strenuous effort was required because of flooding by the sea.

Study of the material collected between 1984 and 1994 continues. The most recent addition to knowledge of the fossil vertebrates from Dinosaur Cove was the publication at the end of 2005 of a new monotreme, *Kryoryctes cadburyi* Pridmore 2005. Unique insights about hypsilophodontid dinosaur physiology and visual acuity has been gained from specimens collected from Dinosaur Cove (Rich and Rich 1989; Chinsamy *et al.* 1998).

Flat Rocks (Strzelecki Ranges)

Flat Rocks is located on the shore platform 5 kilometres west of Inverloch. The first fossil vertebrates were found at the Flat Rocks site in 1991 by Lesley Kool and Michael Cleeland. Based on fossil pollen, these rocks are regarded as Early Aptian in age (Rich *et al.* 1999), and thus older than Dinosaur Cove.

The mammals from the Flat Rocks site are at the centre of a controversy about the most fundamental aspects of mammalian phylogeny. Do they represent placentals more advanced than contemporaneous ones from the northern hemisphere (Rich *et al.* 2001b)? Or are they the first known representatives of a radiation of Gondwana mammals that mimic placentals and have never been recognised before (Luo *et al.* 2002)?

The first excavation at the Flat Rocks site took place in the summer of 1992. Subsequently there has been an excavation at the Flat Rocks site every summer since in 1994. The diversity of mammals represented continues to increase, albeit at a slow rate, and the plan is to continue the summer excavations for the foreseeable future. The assemblages from both the Dinosaur Cove and Flat Rocks sites have been extensively monographed.

Figure 6.4 Reconstruction of Dinosaur Cove at Cape Otway from the Cretaceous period. Painting Peter Schouten © SEWPaC

Fossil significance

As far as can be determined the tetrapod fossils from the Early Cretaceous of Victoria are representative of the polar dinosaurs that lived in south-eastern Australia between 106 and 115 million years ago. As there are no other contemporaneous fossil tetrapod sites in south-eastern Australia, it is impossible to be sure of this.

Evidence of dinosaurs and Mesozoic mammals is rare in Australia. Polar dinosaurs are internationally rare. Palaeoenvironmental reconstructions based on the site have been published in both popular (Vickers-Rich and Rich 1993a) and technical (Vickers-Rich *et al.* 1999) publications.

Recovery of fossil vertebrates at Dinosaur Cove required the construction of a mine from scratch. Elsewhere, dinosaurs have been recovered from existing mines. The 'Dinosaur Mine' at Dinosaur Cove may well be unique, in that the mine was created specifically to recover dinosaurs.

Taphonomy and condition of the site/fossils

Vickers-Rich *et al.* (1999) and Rich and Vickers-Rich (2000) deal in detail with the taphonomy of Dinosaur Cove and Flat Rocks.

Comparison with similar sites in Australia

There are no other Cretaceous polar dinosaur sites in Australia. It is the polar aspect of these two sites that makes them so important (Rich *et al.* 2002b).

Stories

The dioramic potential of Victoria's polar dinosaur sites is realised in the following sources:

- Part 5 of the BBC program *Walking with Dinosaurs* was based on Victoria's polar dinosaurs.
- Rich (1995), which gives the viewpoint of a child about a dinosaur excavation.
- Illustrations by Peter Trusler in Vickers-Rich and Rich (1993b) and Rich and Vickers-Rich (2000).

A key element of the story associated with the site is of the polar habitats as refuges where species persist after becoming extinct elsewhere. This is coupled with concept that in a polar habitat, species arise and spread to elsewhere. The site has also been important for the development of scientific concepts relating to the origin of mammals, e.g. Luo *et al.* (2002) and Rich *et al.* (2002a, 2005).

A history of dinosaur excavation in Australia including Victoria is given in Rich and Vickers-Rich (2003).

Hines Quarry Diprotodon site, Bacchus Marsh

Site overview

The Hines Quarry megafauna site is located 9 kilometres south-west of Bacchus Marsh.

Scientific importance and research potential

The skulls of *Diprotodon* (the largest species of marsupial ever alive) that have been recovered from Bacchus Marsh are of better quality than those from anywhere else. Illustrations of the skulls were made in the late 1970s by Peter Trusler and will be published in a collaborative effort between the artist Peter Trusler and the palaeontologists Patricia Vickers-Rich and Thomas H. Rich. The objective of the book will be to convey to the reader how an artist and two scientists work together to produce scientific illustrations. Dr Tim Senden of the Australian National University (ANU) will carry out a structural analysis of the skulls of *Diprotodon* using an elaborate computer program that has recently become available.

A preliminary account of the Pleistocene Bacchus Marsh locality was given in Rich (1976). A more thorough treatment, which is the principal study of the taphonomy of the site, is found in Long and Mackness (1994).

Fossil significance

The Bacchus Marsh assemblage is typical of the Pleistocene in Australia. What makes the site so unusual is the exquisite preservation of the *Diprotodon* skulls. Because the cranial roof is only a few millimetres thick, all specimens of *Diprotodon* found elsewhere are invariably crushed. Those from Bacchus Marsh are preserved in three dimensions. Skulls of *Diprotodon* of the quality of those found from Bacchus Marsh are known from nowhere else.

Taphonomy and condition of the site/fossils

The Bacchus Marsh assemblage appears to be the result of mass dying of a mob of female *Diprotodon* at the end of a prolonged drought. The evidence for this is that the fifteen individuals of *Diprotodon* known from the site are virtually identical in their stage of growth. All of them have their most posterior molar in the process of erupting, which indicates that they were young adults at the prime age for reproducing. It is this age group among red kangaroos that are the last to succumb to a drought.

Comparison with similar sites in Australia

There are other Diprotodon sites in Australia with greater historical significance, megafauna species

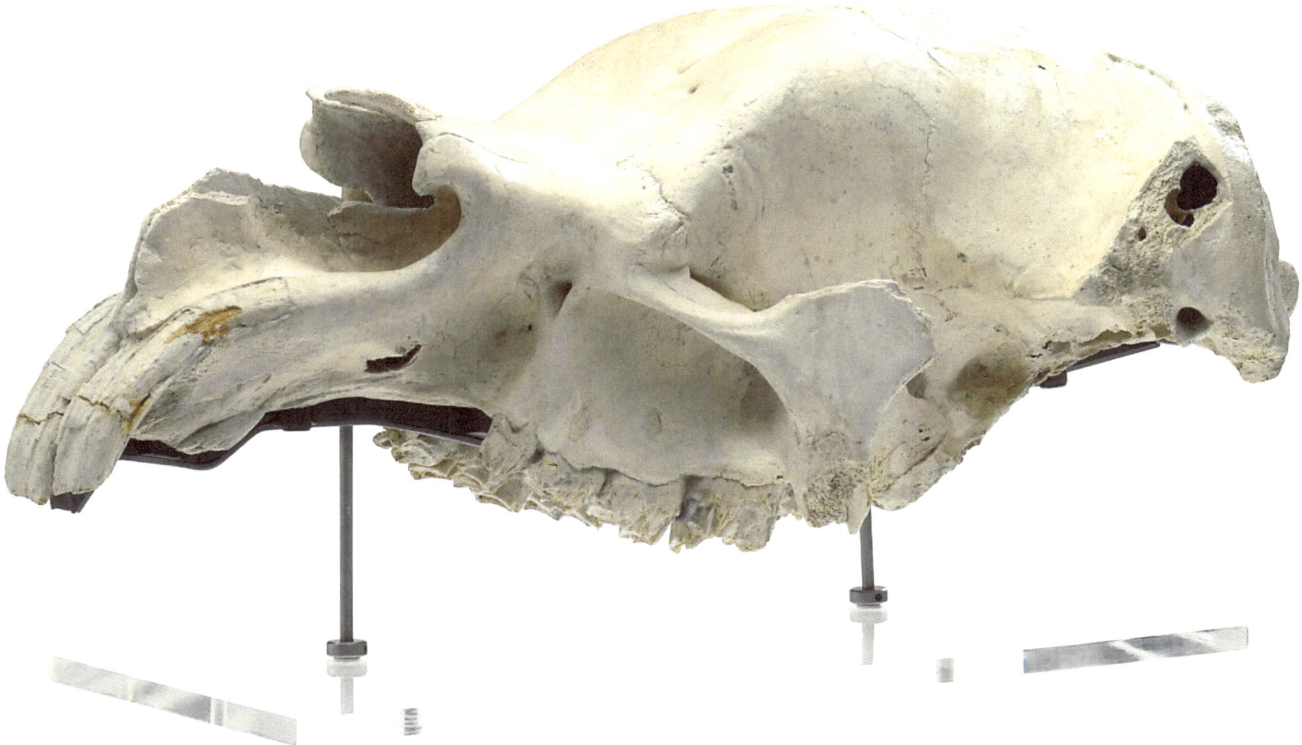

Figure 6.5 Diprotodon (marsupial) skull, Pleistocene (1.8 million–10 000 years old), Bacchus Marsh, Victoria. Image © Museum of Victoria

diversity or quantity of specimens, like Lake Callabonna (SA) and the Wellington Caves (NSW), but the Bacchus Marsh site is the most important one for the recovery of skulls of *Diprotodon* because of the quality of the specimens and their undeformed nature.

Stories

Restoring the *Diprotodon* and giant kangaroos found at Bacchus Marsh would be dramatic. The accumulation of *Diprotodon* material at Bacchus Marsh appears to be the result of a mob of females dying *en masse* at the end of a drought. The evidence for this integrated with a reconstruction of the death assemblage would make an interesting palaeoenvironmental reconstruction. The site was discovered in 1973 by a local schoolgirl, Kerry Hine.

Grange Burn Pliocene mammal site, Hamilton

Site overview

The Hamilton local fauna is a Pliocene terrestrial mammal site. It is located 7 kilometres west of Hamilton, Victoria, on the south side of the Grange Burn.

Scientific importance and research potential

The Hamilton local fauna was discovered by Edmund Gill, then Curator of Palaeontology at the National Museum of Victoria, Melbourne. In the company of Ken Bell, Gill found a single tooth, which became known as the Grange Burn Cuscus (Stirton 1957). The first systematic collecting there was carried out by William D. Turnbull of the Field Museum, Chicago, and Ernest L. Lundelius of the University of Texas, Austin. Their results were reported in a monograph (Turnbull and Lundelius 1970) and a separate paper (Ride 1993). Beginning in 1977, Thomas H. Rich of the National Museum of Victoria continued excavation of the site. Although that year had meagre results, the following two seasons yielded over 80% of the fossil mammals known from the Hamilton local fauna. At the end of the 1979 field season, the yield of fossils abruptly declined for no apparent reason. The following year Robert MacDonald led a field party to the site, but little was found and 1980 was the last year the site was worked. The results of the 1977–79 fieldwork were summarised in a series of papers: Flannery *et al.* (1987, 1992), Rich (1986), Turnbull *et al.* (1987a, 1987b, 1987c). The Hamilton local fauna is continually referred to in the mammalian palaeontological literature because of both its biostratigraphic and palaeoecological significance.

An unnamed lithic unit is overlying the marine Grange Burn Formation and underlying a basalt. The unnamed lithic unit, which produced the fossil mammals, is about 1.3 metres thick and is interpreted as a fossil soil by Gill (1957). Both units bracketing the productive lens have been dated as Pliocene. Turnbull *et al.* (1965) published a potassium-argon date of 4.35 ± 0.1 Ma on the overlying basalt. Utilising the constants for the potassium-argon technique recommended by Steiger and Jager (1978), McDougall (pers. comm. 1980) has recalculated the date as 4.46 ± 0.1 Ma. Ludbrook (1973) regarded the molluscan fauna of the underlying Grange Burn Formation as indicative of the Kalimnan Stage.

Because the age of the Hamilton local fauna is tightly constrained, it is an important tie point in the framework used to establish the chronology of Australia's Cenozoic terrestrial mammals. There are only two other robust tie points on which to construct a biochronology of Australian Cenozoic mammals in all of Australia: Beaumaris (Vic) and Risdon (Tas). The site will therefore always be of importance, as long as there is a program of collecting and analysing such fossils elsewhere in Australia.

Fossil significance

Many of the fossil mammals in the Hamilton local fauna belong to genera and even species that are still living today. What makes them rare is that they can be dated very precisely. The Hamilton local fauna is one of the most diverse Cenozoic terrestrial mammal assemblages known from Australia. Coupled with its refined dating, this makes it an ideal tie point for the correlation of Australian terrestrial mammalian assemblages on this continent.

In the conclusions of the first major study of the Hamilton local fauna, Turnbull and Lundelius (1970) note the similarity of the species composition of the Pliocene site with the mammals in a modern Australian rainforest. With six times as much material as was available to Turnbull and Lundelius for their 1970 study, a reanalysis of the Hamilton local fauna for the purposes of palaeoenvironmental reconstruction could profitably be carried out.

Taphonomy and condition of the site/fossils

When the animals that make up the Hamilton local fauna were alive, the area was close to a marine shoreline, perhaps no more than a few hundred metres away. Shortly after their remains were buried, the area was covered by a basalt flow.

Today there is no obvious place to continue excavating at the Hamilton site. The last season spent there (1980) resulted in the recovery of an extremely meagre collection of specimens. In addition to this, the Grange Burn is choked with water plants owing to the run off from fertilisers. Therefore, it is no longer possible to use the underwater sieving technique to recover the fossils as was previously done, without first undertaking a major task of clearing this vegetation.

Comparison with similar sites in Australia

The Hamilton local fauna is similar to the penecontemporaneous Bow local fauna (NSW). However, because the Bow local fauna cannot be dated as precisely as the Hamilton local fauna, the latter is more significant. Other terrestrial mammal faunas in Australia that are comparable include the Pliocene Bluff Downs local fauna (Qld).

Stories

The mammalian assemblage from Hamilton is diverse enough (and for the most part indicative of a rainforest fauna) that a diorama could be made to represent it that could be contrasted with a modern rainforest biota. Two aspects of the fauna lend themselves to an evolutionary story: first, many of the taxa are related to modern taxa; and second, several taxa represent completely extinct taxa.

Victorian Ordovician graptolite province

Site overview

The Victorian Ordovician graptolite province comprises a series of invertebrate fossil sites located in central and eastern Victoria.

Scientific importance and research potential

The Ordovician graptolite faunas of Victoria have been extensively studied since the middle part of the nineteenth century. The basis of the presently accepted subdivision of the sequence into 31 zones was established between 1892 and 1938 (Harris and Thomas 1938a), and since then the taxonomy and zonal subdivision have been further refined. Scientific research on the Province has been carried out by leading graptolite workers, including T.S. Hall, W.J. Harris, R.A. Keble, Sir F. McCoy and D.E Thomas in the late nineteenth and first half of the twentieth centuries (e.g. Thomas 1960), and more recently by W.B.N. Berry, R.A. Cooper, R.B. Rickards and A.H.M. Vandenberg (e.g. Vandenburg 2003). The work is regarded as of high quality.

The Victorian sequences are the most important scientifically in Australasia, and the graptolite zonal subdivision based on them is the standard for the

Pacific Province. The zonation has been applied in New Zealand and parts of North America, and for world-wide correlation of Ordovician sequences. Although the graptolite sequence is well known there is scope for further research to refine and update the taxonomy.

Fossil significance

The Ordovician graptolite fauna of Victoria is one of the richest and most complete in the world, and this is the reason for its importance in the correlation of strata and in understanding graptolite evolution. The Victorian graptolite assemblages are the basis for the zonal subdivision of the Ordovician in Australasia. However, the potential for palaeoenvironmental reconstruction is probably limited. It is generally accepted that the graptolite-bearing sequences were deposited in a deep marine environment beyond the edge of the continental shelf.

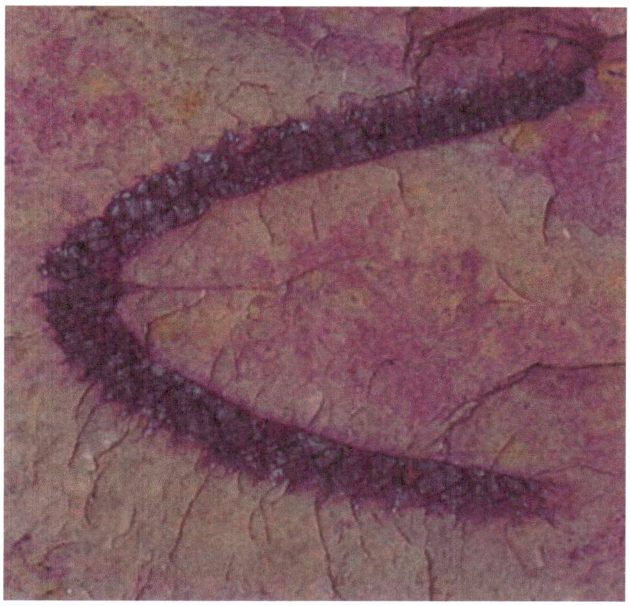

Figure 6.6 *Isograptus victoriae* fossil from Ordovician period. Image Invertebrate Palaeontolgy Museum Victoria © Museum Victoria

Taphonomy and condition of the site

The Ordovician graptolite province of Victoria is very extensive, covering two large areas in the central and eastern parts of the state and encompassing thousands of individual localities. A relatively small number of sites of greater scientific importance could be considered for individual listing. These sites could include the following as well as others:

1. Lancefield Quarry. A small quarry north of the disused Mt William railway station, 5 kilometres north-east of Lancefield. Very fossiliferous and possibly the type locality for the Lancefieldian Zone La2 (Early Ordovician), which can be correlated throughout the Pacific faunal province (see Vandenberg and Cooper 1992: p. 51). Only a relatively thin interval (about 0.5 metres) has been sampled in detail and there is potential for finding more horizons. The site is on private land.
2. Cuttings in the Chewton–Castlemaine railway near Chewton (Harris 1916: p. 59). Good exposure of the Bendigonian Stage (Early Ordovician) that is widely distributed in central Victoria and has been recognised in sequences around the world. However, collecting of specimens is difficult due to the strong cleavage in the rocks.
3. The so-called 'good bed' locality of Thomas Smith, a farmer of the Sandon district who made extensive collections of graptolites from approximately 1898 to his death in 1944. His collection is now housed in Museum Victoria. The site is located on private property. Many characteristic and excellently preserved graptolites of the Bendigonian Zone Be1 have been described from the site by Harris (1933), Harris and Thomas (1938b, 1939, 1940) and others. The site is already listed by Heritage Victoria (VHR no. H1413).
4. The butts, Wellsford Rifle Range, approximately 4 kilometres east of Bendigo. Small exposure of fragile, soft, very weathered shale containing well preserved graptolites of the late Middle Ordovician Darriwilian Zone Da3 (see Vandenburg *et al.* 2002: p. 51).
5. Wellington River, at and near its confluence with the Dolodrook River, approximately 12 kilometres north-east of Licola, eastern Victoria (Harris and Thomas 1954, 1955). Complete record of well preserved Late Ordovician graptolites representing all zones up to and including mid-Bolindian Zone Bo3. The currently used Australasian zonation for the Late Ordovician is based mostly on the collections from this river section made in the 1950s by W.J. Harris and D.E. Thomas and subsequently by Thomas and a party from the Geological Survey of Victoria. The site is remote, requiring 4WD transport.

Comparison with similar sites in Australia

The Ordovician graptolite faunas of Victoria are more abundant, more widely distributed and more complete than those occurring in other states (NSW, Qld, Tas). Graptolites are the most important fossils for biocorrelation in the Ordovician. The Victorian faunas are the basis for the correlation of all Australasian sequences of this age, and have also played an important role globally in the development of graptolite taxonomy and the subdivision of the Ordovician.

Stories

The dioramic potential of the site is probably limited because of the deep-water, offshore depositional environment. However, there is a story to tell on the process of stratigraphic correlation and the relative dating of rock sequences, and the critical role played by graptolites in this process. Sir Frederick McCoy first recognised that some Victorian graptolites were identical with forms from other places, and provided some of the earliest descriptions and illustrations of many important species. Other historically noteworthy Victorian scientists achieved their reputations through their work on graptolite faunas (T.S. Hall, W.J. Harris, R.A. Keble, D.E. Thomas).

Graptolites also played a key role in the geological mapping of Bendigo and other goldfields at the turn of the nineteenth century because the lithologically monotonous and tightly folded sediments could be subdivided and mapped only on the basis of their graptolite faunas. There are historical associations with the Victorian gold rush, as some of the earliest discoveries of graptolites were by gold miners in Bendigo in the 1850s.

Silurian Baragwanathia site, Yea

Site overview

The Yea Silurian *Baragwanathia* macroflora (the Yea Flora Fossil Site) is located at Barclay's Cutting, Limestone Road, Yea district, Victoria. The beds exposed at the locality are brown to grey siltstones, belonging to the Yea Formation. This site contains the earliest land plants preserved in Australia, of apparent Late Silurian (Ludlovian) age, approximately 419–423 Ma (Garratt *et al.* 1984; Rickards 2000), and notably the lycopod *Baragwanathia longifolia* (Lang and Cookson 1935).

Scientific importance and research potential

Although there is some controversy relating to the precise age of the sediments, which is based on the co-occurrence of ancient marine invertebrates known as graptolites (see Douglas and Holmes 2006), *Baragwanathia* is indisputably of international botanical significance as it shows morphological complexity related to early adaptations for life on land. These extinct vascular plants probably attained heights of up to 1 metre, with stout stem axes of up to 5 centimetres wide covered with spirally arranged, simple leaves. They most likely occupied moist habitats and occurred in low herbfield communities in areas such as coastal estuaries and river delta swamps. Associated with *Baragwanathia* are algae, rhyniophytes and other simple plants, and several other invertebrates in addition to the graptolites. Douglas and Holmes (2006) concluded that there is considerable scope for further descriptive work on *Baragwanathia* fossils.

Fossil plants (including *Baragwanathia*) and graptolites from the site were first listed by Couper (1965: locality 4 and 62), who recognised that the beds there belong to the stratigraphically lower of two plant–graptolite horizons in the Yea–Molesworth district. Garratt (1978) identified the graptolites as Late Silurian species and illustrated two specimens of *Baragwanathia* from the site. Revised identifications of the graptolites were provided by Garratt and Rickards (1984) and Garratt *et al.* (1984), and the Late Silurian age was confirmed. Specimens of *Baragwanathia* and other plants from the site were illustrated by Douglas (1983), who also gave a provisional list of species present.

There is still controversy among scientists about whether the lower plant–graptolite horizon is of Late Silurian or Early Devonian age, as the *Baragwanathia* flora of Yea appears much more modern than floras that existed in the northern hemisphere in Silurian times. The Barclay's Cutting site is critical in resolving this controversy.

None of the plants from the site have been described scientifically, and some of them probably belong to new species. Work has been limited to provisional listing of the flora and identification of the graptolites. All of the specimens described by Lang and Cookson (1935) when they proposed *Baragwanathia* came from stratigraphically higher beds belonging to the upper plant–graptolite horizon and its equivalents of Early Devonian age (though at that time these beds were thought to be Silurian). There is thus a need for research on the Barclay's Cutting material to document the flora scientifically, establish whether it differs from that of the higher horizon, and possibly to confirm the Late Silurian age on the basis of more complete graptolite evidence.

Fossil significance

The *Baragwanathia* flora of Yea is one of the most important ancient terrestrial plant floras in the world. It occurs at a number of localities in central and east-central Victoria, but the critical feature of the Barclay's Cutting site is that it lies at a lower stratigraphical level than many of the other localities, and that the flora occurs here with Late Silurian graptolites. The site is one of only two known where the Baragwanathia flora occurs together with Late Silurian graptolites. The other site (Ghin Ghin, north of Yea) is not as favourable, as the rocks there are slightly metamorphosed, the graptolites are somewhat deformed, and the plants collected are more fragmentary. There is also some evidence that the beds there are younger than at Barclay's Cutting.

An important part of research on the site should be reconstruction of the environment in which these very early land plants lived.

Taphonomy and condition of the site

The site was originally a cutting on Limestone Road, approximately 2.5 kilometres south-east of Yea. In 1979 the site was bulldozed by Yea Shire Council and the cutting removed. *In situ* fossils can now only be collected by excavation of subsurface material, which presents some difficulties because of the proximity of the road. Fortuitously, extensive collections were made by the Geological Survey of Victoria in 1977 and 1978 before the bulldozing, and some of these are housed in Museum Victoria.

Comparison with similar sites in Australia

The site is critical internationally in establishing whether the *Baragwanathia* flora arose during the Silurian. No other known site is as promising in providing an answer to this question.

Stories

There is good potential for the development of a diorama depicting the ancient environment in which the *Baragwanathia* flora grew. A reconstruction of the environment was given by Jack Douglas (1983: scene 1, p. 25).

The site is of global significance in reconstructing the early evolution of land plants. A story could be told about how scientific views on the age of the *Baragwanathia* flora have changed, from Silurian as initially believed, then to Devonian, and finally to the recognition that the flora occurs at two stratigraphic levels of differing age. Changing opinions on the age of the *Baragwanathia* flora could be used to illustrate the formulation of scientific views on the early evolution of land plants.

Isabel Cookson (1893–1973), the co-author of the paper that described *Baragwanathia* (Lang and Cookson 1935), was one of Australia's great scientists, and an important pioneering role model for women in science. She recognised the importance of fossil plant research in Australia and greatly raised its profile through the publication of numerous papers on the subject and the training of many students. Her name is perpetuated in several ways, including through the genus *Cooksonia* (another key early land plant) and the Botanical Society of America's Isabel Cookson Award for the best student paper on palaeobotany presented at their annual meeting.

Torquay Oligocene whale and invertebrate site

Site overview

The Torquay Oligocene sites are divided into the vertebrate and invertebrate fossil localities for the purposes of this report.

Vertebrate fossil locality

The Torquay marine vertebrate fossil site is in reality not one geographically limited locality, but consists of several disjunct localities along a stretch of coastline in central coastal Victoria (see Fitzgerald 2004; Piper *et al.* 2006 for details). The stretch of coastline occurs immediately to the south-south-west of the township of Torquay, situated on Victoria's Surf Coast. The coastline in question is bounded to the north-east and the south-west by two well known geographic landmarks: in the north-east, Bird Rock (a rock stack) on Jan Juc Beach; in the south-west, Bells Headland, immediately south-west of the famous Bells Beach. In all, this coastline is about 6–7 kilometres long.

This stretch of coast has major national and international importance in the fields of geology, invertebrate palaeontology, and vertebrate palaeontology. The age of the fossil site spans the Late Oligocene through Early Miocene (about 27–21 Ma). However, almost all of the vertebrate fossils have been recovered from the Jan Juc Formation, which is entirely Late Oligocene in age (27–23 Ma). Taken as a whole, this fossil site has a rich and long history of research, and is canonical within the realms of Australian Cenozoic stratigraphy and marine invertebrate palaeontology. Indeed, the north-east part of this coastline forms the stratotype of the Janjukian Southeast Australian Regional Stage. This report focuses on the vertebrate palaeontology of the Torquay fossil site, which deserves separate treatment due to its singularity. More specifically, it is the fossil cetaceans (whales and dolphins) recovered from this site that make it especially significant from the perspective of vertebrate palaeontology.

Invertebrate fossil locality

The Jan Juc invertebrate site is located near Torquay, in southern Victoria. Beds exposed low in the cliffs at Bird Rock and Fisherman's Steps, south-west of Torquay, belong to the upper part of the Jan Juc

Formation, comprising silty glauconitic marls and sandy glauconitic limestones of Late Oligocene (Janjukian) age. The conformably overlying Puebla Formation consists of calcareous claystones and siltstones of Early Miocene (Longfordian) age. The Jan Juc Formation is coarser grained, more calcareous and more richly fossiliferous than the Puebla. The faunas of both formations are dominated by bivalves, gastropods and bryozoans. Shells from near the top of the Jan Juc Formation yielded a strontium isotope stratigraphic age of 25.6 Ma and those from about 10 metres above the base of the Puebla Formation gave an age of 21.6 Ma (Holdgate and Gallagher 2003).

Scientific importance and research potential

Vertebrate fossil locality

The vertebrate palaeontology of Torquay has a rather negligible history of study. It is important to note that this is due to a lack of research effort, not a lack of fossils to study. The principal limiting factor in the advancement of knowledge in vertebrate palaeontology is the rate and amount of collecting of new fossils. The poor record of scholarship on the Torquay fossil vertebrates is directly related to a lack of directed, systematic field prospecting and collecting. Thus, the vertebrate palaeontology of Torquay, as currently understood, is inherently biased by the somewhat casual, hit-and-miss, approach to its study.

We may conveniently divide the history of study, and prospectus of research potential, into categories based on vertebrate taxa recorded from the site.

Chondrichthyes (sharks and rays): The sharks and rays represent the bulk of the fossil assemblage, both in terms of numbers of specimens collected and in diversity of taxa. Kemp (1991) reviewed the fossil chondrichthyans recovered from Torquay in an authoritative analysis of the fossil sharks and rays of Australia, and pending significant additions of fossils via renewed field collecting, it is unlikely that much can be added to his conclusions. It is noteworthy that since Kemp's (1991) study, there have been no additions to knowledge on the fossil sharks and rays of Torquay. Older references to the sharks and rays collected from

Figure 6.7 Cliffs at Torquay. Photograph Nicholas Caire © National Library of Australia

Torquay are listed in the extensive bibliography in Kemp (1991).

Osteichthyes (bony fish): A moderately diverse assemblage of bony fish has been recorded from Torquay, largely on the basis of otoliths (see Stinton 1958, 1963; Long 1991b; Long in Warne *et al.* 2003; Fitzgerald 2004). Other remains of osteichthyan fish include isolated dentitions, vertebrae, ribs, fragmentary skulls, and jaw elements. The determination of the taxonomic identity of the non-otolith elements remains as a challenge for future researchers. Because virtually no basic research has been done on the osteichthyan fossils from Torquay, apart from otoliths, it is an understatement to say that this area represents a wide-open field for discovery. References to fossil osteichthyans from Torquay are listed in Fitzgerald (2004), Long (1991b), and Long and Turner (1984).

Aves (birds): One avian bone has recently been recorded from the Jan Juc Formation at Torquay. This specimen, noted by Fitzgerald (2004), represents an incomplete tibiotarsus. Due to its incompleteness, it is not possible to identify this fossil beyond Aves indeterminate. However, it is clearly not a penguin. This unique fossil demonstrates that avian fossils occur at Torquay, and that future prospecting could result in the discovery of more complete marine avian material.

Cetacea (whales and dolphins): Fossil cetaceans from Torquay were first recorded in the scholarly literature in Dennant and Kitson (1903), and have figured sporadically in discussions of the fossil record and evolution of cetaceans since that time. The published specimens, for the most part, represent fragmentary, non-diagnostic material. The unpublished fossil material includes some exceptionally complete and well preserved specimens. The cetaceans represent by far the most scientifically significant fossils, vertebrate or otherwise, collected from Torquay. Because of this, and the enormous potential for future work, this report is focused on the cetaceans from Torquay.

Three cetaceans have been described from Torquay: *Parasqualodon wilkinsoni* McCoy 1866 (recorded in Hall 1911); *Mammalodon colliveri* Pritchard 1939 (see Fitzgerald 2010); and *Janjucetus hunderi* Fitzgerald 2006. Of these *Parasqualodon wilkinsoni* is now referred to *Prosqualodon* sp. (see Fitzgerald 2004). Other unpublished specimens are currently under study (Fitzgerald 2003, 2004, 2005a, 2005b, 2005c, 2006, 2010). Key references to fossil cetaceans lie in the published secondary literature (Fitzgerald 2004; Fordyce 1982, 1984, 1991).

With more intensive study and publication, Torquay is likely to represent the canonical Paleogene marine vertebrate site in Australia.

Invertebrate fossil locality

The stratigraphy of the invertebrate site has been well studied from the 1890s (e.g. Tate and Dennant 1893), but most work was carried out in the 1950s and later (e.g. Raggatt and Crespin 1952, 1955; Abele 1979). There has been a long history of research on the fossil invertebrates, beginning in the second half of the nineteenth century with Sir Frederick McCoy (mainly in the Prodromus of the Palaeontology of Victoria), R. Tate, J. Dennant, J. Tenison Woods, P.H. MacGillivray and G.B. Pritchard. More recent work has been by F. Chapman, J. Dennant, G.B. Pritchard and F.A. Singleton in the first half of the twentieth century, and A.G. Beu, T.A. Darragh, C.A. Fleming, K.J. McNamara, G.M. Philip, W.F. Ponder and others in the second half. However, many elements of the fauna remain poorly known (e.g. sponges, some corals and bryozoans).

This is regarded as one of the most significant Cenozoic sites in Australia because of the richness and diversity of the invertebrate fossil faunas, the exposure of the Oligocene–Miocene boundary in a continuous section, and the lateral facies change with the Jan Juc Formation passing into the Point Addis Limestone towards the south-west. It is the type section for the Australian Janjukian Stage.

There is a need for updating the taxonomy for many invertebrate groups, including molluscs, while other groups remain essentially unstudied (sponges, corals, bryozoans, etc.).

Fossil significance

Vertebrate fossil locality

Some context is needed to address the issues of the rarity, unusualness, and representativeness of the fossil cetaceans of Torquay, and of Torquay itself. Fossil whales are extremely poorly known from Australia, reflecting the long history of a lack of interest and effort by palaeontologists. Thus, although there are hundreds of important fossil cetacean specimens residing in museum collections, perhaps less than a dozen have been recorded in the published literature. The lack of effort in searching for and collecting fossil cetaceans in the field means there are only a few reasonably complete fossil skulls and skeletons in museum collections. In all, only 12 associated skulls and partial skeletons of cetaceans are known to exist in museums, and one of these, the holotype skull of *Prosqualodon davidis* from Tasmania, has subsequently been lost (see Mahoney and Ride 1975 for details).

Of those 12 exceptional specimens, six were collected from Torquay. In addition, among those six specimens from Torquay are two of the three reasonably complete fossil whale skulls discovered in Australia. Therefore, from the perspective of numbers of relatively complete skeletons discovered, Torquay is by far the most important and exceptional fossil cetacean, indeed marine mammal, site in Australia. Of the six more complete specimens from Torquay, two are named (*Mammalodon colliveri* and *Janjucetus hunderi*) and one other is complete and distinctive enough to be the holotype specimen of a new species, in a new genus. None of these taxa have been recorded from anywhere else in the world. Indeed, based on the fossil record as currently understood, the Mammalodontidae are endemic to Torquay, having not been found elsewhere (Fitzgerald 2010).

From the perspective of past cetacean biodiversity, Torquay is extremely unusual. At least two toothed baleen whale families are represented at Torquay. At all other fossil sites in the Pacific and Atlantic realms where toothed baleen whales have been discovered, only one family is ever present. The distinctiveness of Torquay as a fossil cetacean locality is further enhanced by the fact that as proportions of the known fossil whale assemblage, toothed baleen whales dominate. Elsewhere, this is not known to be the case. As such, Torquay is an unusual fossil cetacean site nationally and internationally.

The scientific importance of the fossil whales from Torquay is unquestionable. The toothed baleen whales from this site, due to their archaic nature, are critical to understanding the origins and early evolution of one of the major marine mammal groups alive today. Indeed, *Mammalodon colliveri* and *Janjucetus hunderi* from Torquay are arguably two of the top ten most internationally significant (scientifically) fossil vertebrate specimens discovered in Australia.

Being components of a diverse fossil marine ecosystem, there is enormous potential for reconstructing the palaeoenvironmental context of the Torquay fossil cetacean site. Due to the large amount of research carried out on the geology and invertebrate palaeontology of Torquay, most of the data required for palaeoenvironmental reconstruction is already available.

Invertebrate fossil locality

The invertebrate fauna at the site ranges from some taxa that are very common to others known only from a few specimens. The common faunal elements (e.g. many of the molluscs) belong to the same species found at other localities in south-eastern Australia where strata of the same age are exposed (Late Oligocene and Early Miocene, Janjukian and Longfordian). However, many of these other localities are no longer accessible due to weathering or human activity.

The site is notable because of the abundance and diversity of invertebrate fossils and the excellent preservation of the specimens. It is also the only site in Australia where the position of the Oligocene–Miocene boundary has been established in a continuous section.

The palaeoenvironment of the site has been studied as part of a PhD project at the University of Melbourne but the results remain unpublished.

Taphonomy and condition of the site

Vertebrate fossil locality

The vertebrate fossils are found scattered along the Torquay coastal section of this site. Fossils are either found *in situ* within the cliffs, or eroded out onto the beach in boulders. Fossils are usually in excellent condition when found due to the highly indurated nature of the host sediment, and some material includes partially articulated or associated skeletal elements.

Invertebrate fossil locality

The site encompasses coastal cliffs extending from Torquay in the north-east to Bell's Headland in the south-west (see Holmes 1997: fig. 1).

Comparison with similar sites in Australia

Vertebrate fossil locality

Put simply, there are no other sites like Torquay yet known from Australia. Nowhere else has yielded fossil cetaceans of similar exceptional preservation, completeness, unusualness, or scientific value. It is arguable that Torquay represents the geologically earliest complete picture we have of the Cenozoic marine ecosystem off south-east Australia.

Invertebrate fossil locality

There is no other site in Australia where the Oligocene–Miocene boundary is known to be exposed in continuous section.

Stories

There is good potential for dioramic depiction of the different depositional environments in which the laterally equivalent Jan Juc Formation and Point Addis Limestone were deposited, and also the facies change through time represented by the vertical change in the sequence from the shallower water marls of the Jan Juc Formation to the deeper water clays of the Puebla. Evolutionary changes in the faunas (e.g. molluscs) can be documented vertically through the section (see Darragh 1985).

There is excellent potential for demonstrating and depicting the diversity and evolution of early baleen whales based on the fossil material found at Torquay. This is a prominent feature in Museum Victoria's current Evolution Gallery and is also featured in the Surf Coast Shire's Surf World Museum, Torquay.

Many important discoveries of vertebrate fossil material at Torquay have been made by members of the local community, including amateur fossil collectors and surfers. Their contributions are extremely important and deserve to be recognised as such.

Historically noteworthy scientists who have published on the invertebrate faunas from the site include Sir Frederick McCoy, Professor Ralph Tate, Julian Tenison Woods, G.B. Pritchard, F.A. Singleton and Frederick Chapman on molluscs; P.H. MacGillivray on bryozoans; and John Dennant on corals. More recent work of note has been by T.A. Darragh, A.G. Beu and W.F. Ponder on molluscs, and K.J. McNamara and G.M. Philip on echinoids.

The site was also the location of some of the first oil wells (unsuccessful) drilled in Victoria.

Hamilton Miocene–Pliocene invertebrate sites

Site overview

The Muddy Creek and Grange Burn invertebrate sites are located near Hamilton, western Victoria. The lowermost rock unit, exposed in Grange Burn and the lower part of Muddy Creek, is the Bochara Limestone, of late Early Miocene (Batesfordian) age. It contains abundant bryozoans as well as pectinid bivalves (scallops), echinoids, sponges and the large foraminifer *Lepidocyclina*. The conformably overlying Middle Miocene (Balcombian to Bairnsdalian) Muddy Creek Marl contains an abundant fossil fauna of which molluscs (mainly gastropods and bivalves) are the dominant element. Strontium isotope stratigraphic dating of shells from the unit gives an age of 10.8 Ma (Dickinson *et al.* 2002). In Grange Burn and the upper reaches of Muddy Creek the Muddy Creek Marl is disconformably overlain by the Grange Burn Formation, consisting of richly fossiliferous marls and sandy to pebbly limestones with a phosphatic nodule horizon at its base. Molluscs and foraminifers from the formation indicate a Kalimnan (very late Miocene to Early Pliocene) age, and strontium isotope dating of shells at the base gives an age of 5–4 Ma.

Scientific importance and research potential

The fossil invertebrate faunas of the site have been extensively studied from the 1860s to the present, but many faunal elements remain poorly known or completely undescribed, including some bryozoans, corals, echinoderms, and smaller types of molluscs. The research has been carried out by well-known specialists on Cenozoic faunas and is generally of a high standard. The site is regarded as a classic in the study of Cenozoic invertebrates in eastern Australia, containing one of the richest and most diverse invertebrate faunas of late Early to Middle Miocene and early Pliocene ages. Although the site has been extensively studied, much of the work was carried out in the nineteenth and early twentieth centuries. There is a need for updating the taxonomy for many groups, including molluscs, while other groups remain essentially unstudied (e.g. sponges and some corals).

Fossil significance

The fauna at the site ranges from some taxa that are very common, to others known from only a few specimens. The common faunal elements (e.g. many of the molluscs) belong to the same species found at other localities in south-eastern Australia where strata of the same age (late Early Miocene to Early Pliocene) are exposed. However, many of these other localities are no longer accessible due to weathering or human activity. The site is notable because of the abundance of fossils, the diversity of species (including very rare taxa), and the excellent preservation of specimens. The presence of an ancient rocky shoreline is unique in Cenozoic sites in south-eastern Australia.

The only published palaeoenvironmental investigation of the site, carried out 50 years ago (Gill 1957), concluded that the marine sequence was deposited in progressively shallowing waters as the climate gradually cooled. The uppermost Cenozoic beds exposed are terrestrial and include lake deposits and fossil soils of Late Pliocene age. A more recent palaeoenvironmental study as part of a BSc (Hons) thesis has not been published. There is a need for modern re-interpretation, perhaps including oxygen isotope analysis to document climate change.

Taphonomy and condition of the site

The Miocene–Pliocene sedimentary sequence is covered by younger basalts and is only exposed where Muddy Creek and Grange Burn have cut through the volcanics. Outcrop in the creeks is generally good but parts of the sequence may be covered by alluvium or by plant growth.

Comparison with similar sites in Australia

There is no other single invertebrate fossil site in south-eastern Australia where a Cenozoic sequence is exposed of such age range and lithological and faunal diversity.

Stories

There is potential for dioramic depiction of depositional environments, ranging from the Miocene shallow seas to the late Pliocene terrestrial environments.

Historically noteworthy scientists who have published on faunas from the site include Sir Frederick McCoy, Professor R. Tate and J. Tenison Woods on molluscs; C.M. Maplestone and P.H. MacGillivray on bryozoans; I. Crespin, W. Howchin and W.J. Parr on foraminifers; J. Dennant and P.M. Duncan on corals; and T.H. Withers on barnacles. More recent work of note has been by T.A. Darragh (molluscs); A.G. Beu (molluscs), J.V. Neil (ostracods), and J.S. Buckeridge (barnacles). The site is also very well known to amateur fossil collectors and has long been subject to exploitation by them.

Early Cretaceous (Aptian) Koonwarra site

Site overview

This Early Cretaceous site is located at Koonwarra in the Gippsland Basin. It contains a variety of fossilised remains, including fishes and insects, but the flora from the site is the subject of this report.

Scientific importance and research potential

In general, Early Cretaceous floras resemble those of the Late Jurassic, with strong representation of conifers including Podocarpaceae and Araucariaceae, pteridosperms, cycads, bennettitaleans, ferns, cryptogams and sometimes *Ginkgo* (Douglas 1994). In Australia, flowering plant (angiosperm) pollen is first recognised in sediments of the Eromanga and Gippsland Basins during the Early Cretaceous, making the floras from these basins of great significance. Australian Early Cretaceous sites are relatively well represented and studied, and include several floras containing cuticle-bearing fossils, notably the Lower Strzelecki Group of the Gippsland Basin (McLoughlin *et al*. 2002). Although the site at Koonwarra in the Gippsland Basin unfortunately does not exhibit such high quality preservation, in 1990 it received global botanical attention for containing the world's oldest known fossil flowers, which were interpreted to have been derived from a herbaceous wetland plant (Taylor and Hickey 1990). Older flowers have subsequently

Figure 6.8 Fossilised *Tarwinia australis* (flea) found at Koonwara. Photograph John Broomfield © Museum Victoria

Figure 6.9 Fossil Fish *Leptolepis koonwarri*.
Photograph © Museum Victoria

been reported from China (Sun *et al.* 2002), but the Koonwarra flowers are still Australia's most ancient.

Fossil significance

Overall, more than 60 plant taxa are recognised from Koonwarra (Drinnan and Chambers 1986), including other possible aquatic angiosperms represented by seeds (Douglas 1994). It also contains excellent fossils of many animals, including several fish, crustacean and arachnid species, and an amazing total of more than 70 insects (Jell and Duncan 1986). This detailed site reconstruction thus offers a fascinating window into Early Cretaceous life and climate. At this time, much of the Australian continent was covered in ocean, but southern Australia remained connected to Antarctica at very high latitudes (75–80°S). It is a continuing topic of global interest as to how the biota functioned under the extremely seasonal conditions that would have prevailed (e.g. Royer *et al.* 2005).

Taphonomy and condition of the site

The Koonwarra site is one of the few *Konservat-Lagerstätten* localities in Australia. Many fossils are preserved complete, and there is a high level of detail retained, such as the delicate structures of flowers, feathers and soft tissue.

Comparison with similar sites in Australia

The Bauhinia Downs (NT) and Dampier Peninsula (WA) macrofloras are better preserved Cretaceous plant fossil sites but the Koonwarra site contains Australia's oldest known fossil flowers and feathers. It is comparable in mode of preservation to the Belmont insect beds (NSW) and Emu Bay, Kangaroo Island (SA), but the age of the Koonwarra site elevates its significance. While Devonian fish fossil sites are very well known in Australia, Mesozoic sites are rarer. Koonwarra is comparable to the Jurassic-aged Talbragar beds in New South Wales.

Stories

As a *Lagerstätte*, the site has very high dioramic potential. The presence of fish, insects and plant fossils *in situ*, and their excellent preservation, allows detailed environmental reconstruction.

Middle Eocene Anglesea site, Eastern View Formation

Site overview

This Middle Eocene site is a macroflora fossil site located in Anglesea, Victoria.

Scientific importance and research potential

Middle to Late Eocene floras are characterised by a rise in prominence of *Nothofagus* (southern beech) pollen. Leaf, fruit and flower fossils representing over 100 species have been recovered in mummified form from clay lenses in the Anglesea brown coal mine (Christophel *et al.* 1987). A large collection of these fossils has been made with the majority collected during the 1980s. Typical macrofossil taxa include many species of Lauraceae, Proteaceae, Myrtaceae, Cunoniaceae, Elaeocarpaceae, *Brachychiton* (Sterculiaceae) and *Gymnostoma* (Casuarinaceae), and of conifers such as *Podocarpus* and *Agathis* (Christophel 1994).

Fossil significance

The Middle to Late Eocene was an important time for Australia as it marks the beginning of a long period of effective isolation for the continent after it separated from Antarctica and drifted northward. The plant fossils from the Anglesea site are significant because of the abundance and diversity of taxa that appear to represent a rainforest such as that currently found in New Caledonia and the Atherton Tablelands of the Wet Tropics (Qld).

Significant increases in our understanding of the development of modern eastern Australian rainforest communities, including that of the Wet Tropics, will ensue with further study of the fossils collected from this site.

Taphonomy and condition of the site

Many leaf, fruit and flower specimens have been recovered from the clay in mummified form, and are thus able to be studied in great detail. The environment in which the fossils were deposited is likely to have been complex, with each lens perhaps representing a different transient oxbow lake in a large river system. The quality of the preservation sets the site apart from others of similar age. Described fossils include delicate, three-dimensionally preserved floral structures and the earliest record of mites in leaf domatia (O'Dowd *et al.* 1991).

Figure 6.10 Steraula leaf fossil from Angelsea, Victoria. Photograph © Museum Victoria

Comparison with similar sites in Australia

Numerous macrofloras of similar age are known from across the continent, including the Cape Van Diemen Flora, Melville Island (NT); the Stuart, Poole and Nelly Creek sites (SA); and the Eastern Shore of Macquarie Harbour (Tas). However, among several contemporaneous Middle to Late Eocene macrofloras across Australia the Anglesea site is the standout, due to its impressive diversity, with over 100 species represented, and the quality of preservation, which includes many mummified samples and delicate three-dimensional fossils.

The Latrobe Valley coal measures span a much longer time period, from the Eocene to the Miocene.

Stories

The site has high dioramic potential due to the quality of preservation, the preservation of depositional settings and the diversity of fossils, and also to the number of other macrofloras of similar age across Australia, which allow correlation. It can also be compared to other late Gondwanan sites, and hints at parallels with modern Australian and Gondwanan rainforests.

Latrobe Valley coal measures

Site overview

The Latrobe Valley coal measures contain a number of late Middle Eocene to late Middle Miocene plant fossil sites, including Traralgon South/Loy Yang, Morwell and Yallourn seams. These sites are all found in the Gippsland Basin, Victoria.

Scientific importance and research potential

Brown coals and associated sediments in the Latrobe Valley preserve an unparalleled 30 million year continuous record of climatic and floristic change in south-eastern Australia. This region is thus considered to be a plant fossil site of international palaeobotanical significance. The palaeobotany, stratigraphy and palynology of the Latrobe Valley sediments have received attention from researchers for many years, and microfossil studies of the Gippsland Basin (especially Stover and Partridge 1973) underpin our understanding of the terrestrial biostratigraphy of south-eastern Australia and beyond (Macphail *et al.* 1994).

However, considering the exceptional quantity and quality of the material, much of the vast potential of the region is so far untapped. For instance, there are no monograph-style taxonomic reports on the macrofossil floras of any of the coal seams. Perhaps the most comprehensive work on the site was a detailed vegetation and environmental reconstruction by Blackburn and Sluiter (1994). This is of profound significance for understanding the development of the modern Australian vegetation, in part because environmental change such as swamp to dry land transitions and the effects of fires could be tracked over time, with consequent changes in vegetation associations. Holdgate *et al.* (2007) recently outlined these events for the Middle Miocene. This was a warm period of relatively high sea levels and high water tables, and a huge (565 square kilometre) area of peat swamp developed at the western end of the Latrobe Valley to form the Yallourn coal seam. Stratigraphic units of the seam contain a high sclerophyll plant component and abundant charcoal, suggesting alternating seasonally dry and wet warm periods characterised by abundant rushes and dense kauri (*Agathis*) and podocarp forests.

Fossil significance

The coals are also significant as being the subjects of several papers by the eminent Australian palaeobotanist Isabel Cookson and her co-workers. Notable among these was a landmark study describing several species of *Banksia*-like leaves from the Yallourn sequence (Cookson and Duigan 1950).

Comparison with similar sites in Australia

The Latrobe Valley coal measures are unparalleled among Australian Cenozoic palaeobotanical sites for the virtually continuous record of climate and floristic change from the Early Eocene to the late Middle Miocene.

Stories

Besides being a noted plant fossil site, the huge open-cut coal mines at Yallourn, Morwell and Loy Yang are obviously also of great economic and social importance and are used to generate electricity for Victoria through associated power stations formerly run by the State Electricity Commission (SEC) and now by several private operators.

The sites have high dioramic potential because of their geographic and temporal extent, and because of the detailed picture offered by the stratigraphy, palynology and palaeobotany about changes in vegetation distribution and climate patterns over time during the Cenozoic Era.

Western Australia

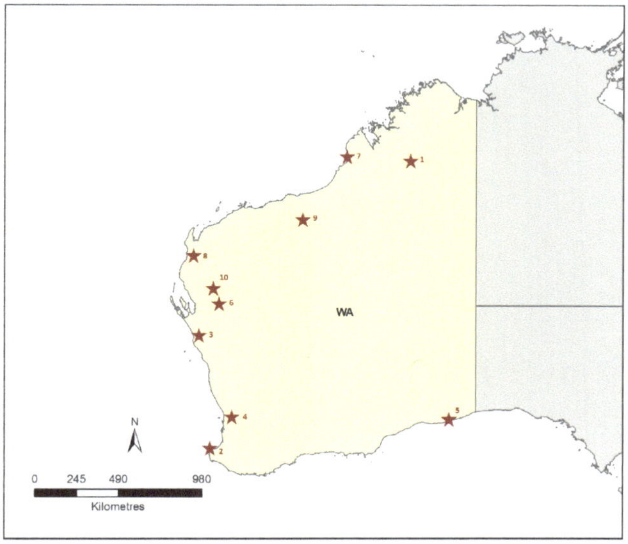

Fossil Sites ★

1 Devonian 'Great Barrier Reef' and Gogo Fish Fossil Site, Fitzroy Crossing
2 Devil's Lair, Cape Leeuwin – Cape Naturaliste Region
3 Tumblagooda Sandstone, Kalbarri and Murchison River Gorge Trace Fossils
4 West Dale, Darling Plateau
5 Roe Calcarenite Invertebrate Fauna, Roe Plains
6 Callytharra Formation, Callytharra Springs (Permian Invertebrates and Foraminifera)
7 Broome Sandstone Dinosaur Trackways and Macro Flora, Broome
8 Miria Marl, Giralia Anticline
9 Warrawoona Group (North Pole and Strelley Pool Sites), Proterozoic Early Life Fauna
10 Cundlego Formation Permian Invertebrate Fauna, Gascoyne Junction

Western Australia locality map © SEWPaC

Devonian 'Great Barrier Reef' and Gogo fish fossil site, Fitzroy Crossing

Site overview

The Lennard Shelf, a structural subdivision of the Canning Basin, contains the remains of a 'great barrier reef' that fringed an ancient Kimberley island landmass during the Devonian. The reef developed off the shore of the Kimberley block, which lies to the north-east of this shelf and is bounded on the south-west by the Pinnacle Fault. The main reef may have been as long as 1400 kilometres. Outcrops now extend for 350 kilometres along the northern margin of the Canning Basin and are up to 50 kilometres wide (Playford *et al.* 2009). The reef is exposed in the limestone rocks of the Napier, Oscar, Emmanuel and Pillara ranges that run from near Derby to Fitzroy Crossing, extending almost to Hall's Gap (Long 2006). These ranges are parallel to the King Leopold Ranges and reach heights of up to 300 metres above sea level. The Ningbing range north of Kununurra is a smaller remnant of the reef (Tyler 2000).

One of the limestone formations within the reef complex, the Gogo Formation, has produced spectacular fossils of Devonian fishes. This unit is mostly subsurface, intertonguing with the Sadler Limestone. The fossils are found within concretions, 'Gogo nodules', that weather out of the formation (Playford *et al.* 2009). The preservation of these fishes is exceptional; many specimens are near-complete, three-dimensional skeletons that have recently been found also to preserve soft tissue features. These fossils represent the fauna of the pelagic, deep-water basin associated with the reef.

Scientific importance and research potential

The Devonian reef was built from 390 million years ago (the Givetian stage of the middle Devonian period) to 370 million years ago (the Famennian stage of the late Devonian) (Playford 1980; Tyler 2000). The extent and duration of this deposit, as well as the time period it represents, is unique on this continent. It spans the Frasnian–Famennian mass extinction and is among only four sites in the world where reef environments survived this event, although the major reef-building organisms themselves changed (Wood 2002). The Devonian reef outcrops are relatively undisturbed by burial, taphonomic or tectonic processes and offer an *in situ* sequence through this time (Playford *et al.* 2009). Famennian reefs are rare throughout the world and none is present elsewhere in Australia. In addition, valleys cut through the reef at Windjana and Geike Gorge by the Lennard and Fitzroy rivers provide sections through the deposit that give palaeontologists and geologists a unique window on this sequence.

A Late Devonian continental shelf can be reconstructed from the outcrops found on the Lennard Shelf. These outcrops represent island archipelagos with their fringing reefs, and outcrop as forereef accumulations, bioherms (mud mounds) that grew on pinnacle reefs rising from the shallow sea floor of the backreef lagoon, and limestone nodules that preserve entire fish and crustaceans. The Gogo fossil localities provide an integrated picture of a proto-Australian continental shelf environment from 390–370 million years ago (Johnson and Webb 2007; Playford and Lowry 1966; Playford *et al.* 2009; Webb 2001). A suite of diverse environments is represented, allowing a large-scale picture to be reconstructed. The minimal post-deposition deformation that the area has experienced is integral to this reconstruction (Johnson and Webb 2007; Playford *et al.* 2009). As noted by Playford *et al.*

Figure 7.1 Bugle Gap Geological Monument, Fitzroy Crossing. Image © SEWPaC

(2009) 'the present-day topography mimics that of the Devonian sea floor, so that from the air it might appear that the sea has only recently withdrawn from the area'.

The Gogo fossil sites found within the reef sequence preserve up to 50 different fish species and have produced the most complete Devonian tetrapodomorph fish yet discovered: *Gogonasus andrewsae* (Holland and Long 2009; Long *et al.* 2006). This specimen provides important information about the transition from fish to tetrapods, the first terrestrial vertebrates (and as such, early ancestors of humans) (Long *et al.* 2006). The complete nature of the fish fossilised at Gogo has meant that many specimens preserve features not seen on fish preserved elsewhere; for example, *Gogonasus* is the only known Devonian fish that shows a complete pectoral limb (Long *et al.* 2006). Preservation of delicate, tiny bones has even shown the presence of embryos in some of the fish fossils, while others have revealed anatomical details that provide the earliest evidence of internal fertilisation and sexual dimorphism within vertebrates, as well as other details of soft tissue anatomy (Long *et al.* 2008; Long *et al.* 2009; Ahlberg *et al.* 2009a). For these reasons, Gogo is acknowledged as one of the world's most important early vertebrate fossil localities. One of the Gogo placoderms, *Mcnamarapsis kaprios*, is now the official fossil emblem of Western Australia.

Fossil significance

Reef-building species in the Devonian are dominantly corals (over 40 species of rugose corals), stromatoporoids (some 25 species) and algae, in association with many other major invertebrate groups. Sea level changes occurred rapidly at the end of the Frasnian and at other reefs around the world; this boundary marks their extinction and burial by deeper-water sediment (Wood 2000). At the Devonian 'Great Barrier Reef' at Fitzroy Crossing, there are records of the sea level changes, as well as the extinction of a major component of the reef-building organisms (large, calcified animals) at this time. However, the reef itself withstood the extinction and was wither-colonised by different taxa, mostly microbial, taking up the role of reef building in the Famennian (Wood 2000). Studying this sequence can provide information about how reef communities react to climate change and to changes in sea level, both of which are key issues facing the Great

Barrier Reef today (Wood 2000; Wood 2002; Vernon 2008).

A diverse array of up to 50 species of Devonian fishes has been found at Gogo, including placoderms (armoured fish), sarcopterygians (lobe-finned fish) and actinopterygians (ray-finned fish). The placoderms, which are now extinct, are represented by numerous arthrodire species and one antiarch. The sarcopterygians are tetrapod ancestors and the actinopterygians are now represented by the teleosts, the most common type of fish living today.

Recently, Gogo fossils have been discovered to preserve soft tissue structures along with bone (Trinajstic *et al.* 2007; Trinajstic and Long 2009; Ahlberg *et al.* 2009a, 2009b). This has allowed researchers to examine fine details of anatomy that are usually lost during the fossilisation process. Most remarkable of these finds has been evidence for viviparity (live birth) within the placoderms. Embryos, an umbilical cord and a possible yolk sac have been found preserved. This represents the earliest evidence for internal fertilisation and live birth within the vertebrates (Long *et al.* 2008). Extensive remains of soft tissue have allowed reconstruction of the body musculature in an early vertebrate that is ancestral to tetrapods (Ahlberg *et al.* 2009b). Despite the large amount of work done on the Gogo sites since the 1960s and the large number of published studies, significant new finds are still common.

Taphonomy and condition of the site/fossils

The fossil reef forms a limestone range that extends for 350 kilometres in an arc that flanks the present day Kimberley ranges; it is also found in the Ningbing range north of Kununurra. The reef sequences have been subject to little significant tectonic deformation or dolomitisation (Playford *et al.* 2009). The fish fossils found at the Gogo sites are spectacularly preserved, with both skeletal and soft tissue material present. Lime-rich mud entombed animals that drifted into the area from the reefs as decomposing, bloated carcasses. The extraordinary preservation of the Gogo fish fossils (which is without equal in fossil fish of this age) is a result of a high rate of sedimentation in the inter-reef basins; early diagenesis (with formation of limestone nodules around the fossils); and a tectonically stable region. Low global oxygen levels (about 17% compared to the current 22%) resulting in slower rates of decay contributed to the preservation of the Gogo fish.

The Gogo fish and other fossils (mainly crustaceans) are extracted from the limestone concretions using acetic acid. Collecting involves walking the outcrop equipped with a hammer and splitting open the nodules, which litter the plain. A few of the nodules contain fish or other fossils. However, the relative abundance of fish fossils within the nodules varies considerably from one local site to another. The main threat to the site is unauthorised removal of the limestone nodules.

Comparison with similar sites in Australia

Other areas within Australia that preserve Devonian reefs are Buchan (Vic) and Taemas–Wee Jasper (NSW). These outcrops date from the Early Devonian, preserving reefs that are tens of millions of years older than that of the Middle–Late Devonian Lennard Shelf (Basden *et al.* 2000). The outcrops at Taemas are equivalent in age to the Buchan Group, while those at Wee Jasper continue for a few million years longer (Basden *et al.* 2000). Both the Taemas and Buchan sites have been deformed through tectonic movement, folded by the Middle Devonian Tabberabberan Orogeny and often subject to extensive dolomitisation (Yeates 2001). Neither of these formed at the right time period to provide any evidence relating to the Frasnian–Famennian extinction or any other significant extinction events.

The Gogo fossil site, in preserving early tetrapodomorph fishes, has provided new evidence on the evolution of vertebrates; it is one of the world's most important early vertebrate fossil localities. The story of early vertebrates in Australia is also told by the Devonian fish fossil site of Wee Jasper (NSW), which is similar to Gogo but has not yet produced fossils as complete in fine details. There are four other known Devonian fish sites in Australia that preserve articulated skeletons (Long 1991b). Two of these, Canowindra and Wee Jasper, are located in New South Wales, and two, Mt Howitt and Buchan, are in Victoria. Preservation at these sites varies, with Canowindra and Mt Howitt preserving fish in two dimensions, while Buchan, Wee Jasper and Gogo preserve fish in three dimensions. Two-dimensional preservation at Mt Howitt consists mainly of impressions of the bones left behind in the mudstone, as any bone preserved is usually badly weathered. Canowindra's preservation is mostly natural moulds filled with hard sandstone casts, some of which preserve internal structure of braincases and gills. There are eight species of fish found in Canowindra, but the fauna is overwhelmingly dominated by just two species. Note that, for both of these types of preservation, while fine detail can be retained, soft tissue cannot.

Both Buchan and Wee Jasper represent Devonian coral reef facies. At these sites, the fish fauna is preserved in limestone rather than shale, which has made three-dimensional preservation of the actual bone possible. Buchan's fossil fish are rare, and as the most

Figure 7.2 Reconstuction of Devonian marine tropical reef from the Kimberley region. Peter Schouten © SEWPaC

fossiliferous part of the facies has been mined for many years, the deposit may be all but removed (Inan *et al.* 1992). Wee Jasper, by contrast, has nearly as diverse a fish fauna as Gogo and apparently similar preservation (Long 1991b). However, as yet, there is no reported fossil from Wee Jasper that preserves soft tissue structure.

Stories

The story of the Lennard Shelf's ancient barrier reef, responding to climate change on the west coast of a proto-Australia, resonates with the story of the modern Great Barrier Reef on the east coast of Australia. In contrast to the Lennard Shelf reef, which grew for 20 million years, the Great Barrier Reef has been in place in its current structure for approximately 6000 years, although some researchers argue that it has been present on-and-off for as much as 30 million years (Vernon 2008). While the taxa involved in reef building during the Devonian were markedly different from today's reef organisms, the structures are analogous and the Lennard Shelf allows us to study a reef that existed for a period of time comparable to the Great Barrier Reef. The Devonian was the last major reef-building period in Earth's history prior to the Holocene, so comparing the two great barrier reefs of Australia provides fertile ground for research.

The Gogo fish fossil site has been important in the development and history of global and Australian geosciences. A major part of the story of Gogo studies is the development of new technology for the preparation and study of specimens, which has allowed new layers of information to be revealed. When the site was first found by Curt Teichert in the 1950s, he disregarded the fossil fish because, encased in hard limestone; the delicate bones of the fish would not have survived mechanical preparation. In the 1960s, when the site was visited by Harry Toombs, the newly developed method of dissolving the limestone in an acid, chemically 'etching' the bones, allowed entire, delicate skeletons to be revealed, often from fossils that had been discovered some time earlier (Long 2006). The most recent advances use high-resolution scanning electron microscopy, high-resolution computer tomography, X-ray and synchrotron CT scanning to reveal details of the soft tissue morphology that might otherwise be obscured by bone and buried within the supporting matrix (Ahlberg *et al.* 2009a, 2009b; Trinajstic and Long 2009). Gogo's history is that of a site producing remarkable specimens with a potential for study that increases with each new technological development. It is reasonable to expect that it has a high potential to yield further information on the evolution of early Australian fish.

Devil's Lair, Cape Leeuwin–Cape Naturaliste region

Site overview

Devil's Lair is a single chamber cave formed in Quaternary limestone of the Leeuwin–Naturaliste Ridge. Well stratified cave sediments reach a depth of 6.6 metres, containing abundant archaeological material in the upper half and an extensive, well dated and stratified vertebrate faunal sequence throughout. Initial radiocarbon dates of the lowest layers containing extinct megafauna species were between 35 000–31 000 years BP (Balme *et al.* 1978). Subsequent, more refined dating methods indicate that the sequence exceeds a radiocarbon age of 55 000 years BP (Turney *et al.* 2001) and that human occupation dates back to approximately 48 000 years ago.

Devil's Lair is a remnant of a once much larger cave that formed within a lithified aeolian calcarenite of the Tamala Limestone. The latter comprises a series of Pleistocene sand dunes running parallel to the coastline in the south-western part of the state. The name of the cave relates to the findings of skeletal remains of *Sarcophilus harrisii*, the Tasmanian devil. It is thought that devils were responsible for the fragmentation of some of the bones found in the cave. Devil's Lair is located in the south-west corner of the state, half-way between the towns of Augusta and Margaret River.

Scientific importance and research potential

The scientific work on the site is multidisciplinary and is held in high esteem. Excavation of the deposit in the Devil's Lair cave commenced in 1955 when an American Fulbright Scholarship student, E.L. Lundelius, explored the Pleistocene mammal faunas of Western Australia. The extraordinary importance of the cave was realised when re-examination of presumed kangaroo teeth revealed a human tooth. This discovery resulted in a series of systematic excavations in the 1970s by palaeontologists and archaeologists from the Western Australian Museum. The fragmented bones found in the cave were originally thought to be predominantly the leftovers from Tasmanian devils, hence the name of the cave.

The locality is one of the best researched caves in Australia and certainly the most thoroughly studied in Western Australia. It records about 40 000 years (49 000 to around 8000 years ago) of late Pleistocene and early Holocene vertebrate faunal dynamics of the region. Artefacts, a human deciduous tooth, radiocarbon-dated charcoal bands, and prehistoric pits dug into the cave, indicate that humans used the cave at times from about 30 000 to 8000 years ago. This is among the oldest evidence of human presence in Australia.

The site continues to have considerable research potential. Its excellent stratigraphic sequence and organic content are attractive to new dating techniques.

Fossil significance

This site provides extremely rare evidence of dated Pleistocene human association with megafauna. The excavations yielded fragmentary remains from 35 species of late Pleistocene to early Holocene mammals in addition to numerous species of birds, snakes, lizards and fishes. The fauna is typical of the timeframe represented and contains many well preserved specimens of small to medium-sized species (bandicoots, macropodoids, thylacinids and dasyurids) as well as some specimens of late Pleistocene megafauna (*Sthenurus*, *Zygomaturus*, and *Protemnodon*).

The very detailed bed-by-bed sampling, conducted by field parties of the Western Australian Museum, revealed a high diversity and surprisingly stable late Pleistocene community of vertebrates, fluctuating only moderately in composition over time in response to the drastic changes in regional climate that characterises the Pleistocene. The cave is located about 5 kilometres from the sea, but during the last peak glaciation the distance to the shore grew to 40 kilometres. The increased distance to the shore may in itself have had an effect on the local vegetation and fauna. The transition to the Holocene marks a drop in diversity of mammals brought into the cave by humans and other predators, to a level comparable to the modern fauna of the Devil's Lair region.

Taphonomy and condition of the site/fossils

Initially it was thought that Tasmanian devils were the primary agency for the introduction of bones to the cave (hence its name), but with the discovery of archaeological material, the pendulum swung towards humans. The consensus is that both humans and other predators, including Tasmanian devils, thylacines, owls and bats, contributed to the accumulation of bones in the cave.

The strata do not yield complete, articulated skeletons but rather the fragmented leftovers from what is thought to be human meals, and bones from rodents, lizards and other small vertebrates originating from droppings from owls and kestrels. Devils may have worked over some of the leavings of humans.

Comparison with similar sites in Australia

Few localities containing extinct late Pleistocene fauna and evidence of human occupation are in a finely documented and dated stratigraphic sequence. Devil's Lair is certainly one of the most complete and least ambiguous of late Pleistocene human and faunal associations on the continent.

Recent discoveries of 'pit fall trap' caves on the Nullarbor Plain by speleologists have enabled palaeontologists at the Western Australian Museum to collect Pleistocene mammals of a state of preservation and significance superior to those of the Devil's Lair mammals. These new vertebrate localities are not included in this report as the research is still at a very early stage. In comparison to the Nullarbor caves, the high scientific importance of the Devil's Lair locality is largely related to the recorded prehistoric activity of humans.

Stories

Devil's Lair has excellent potential for museum exhibition, with abundant well preserved fauna and association with Australia's earliest humans. There are readily reconstructed scenarios showing how humans and other animals brought bones into the cave and how each agency has been discriminated. The site is also linked to an important evolutionary story. Devil's Lair Cave has an extensive and detailed successional record of species, showing times of extinction of local fauna. The site continues to have an important bearing on the riddle of Pleistocene megafaunal extinctions. Devil's Lair is among the first of the human-associated faunal sites to have been studied in detail with a multidisciplinary approach.

Tumblagooda Sandstone, Kalbarri and Murchison River Gorge trace fossils

Site overview

The Tumblagooda Sandstone crops out extensively in the Murchison River gorge and in coastal cliffs south of Kalbarri. It stretches some 70 kilometres between Kalbarri and the Pre-Cambrian Northampton Block and is a strongly jointed, red-bed sandstone with a composite thickness of about 1 kilometre in the type area. It rests on Proterozoic basement rocks and is overlain by the Early Cretaceous siliciclastic units of the Winning Group in the lower Murchison River area. In the absence of age-diagnostic body fossils, the precise age of the deposit is unknown. To the north the Tumblagooda is overlain by the Dirk Hartog Formation, which has yielded Late Silurian conodonts in cored samples from a borehole approximately 100 kilometres north of the Kalbarri outcrops. A late Silurian age was advocated by Hocking (1991), McNamara and Trewin (1993), and Trewin and McNamara (1995). More recent data points to an older, Ordovician age (Braddy 2003).

The deposit comprises a variety of terrestrial to nearshore marine facies, including fluvial channel deposits, aeolian sandsheet and ponds, aeolian dunes and nearshore sandstones. Recognisable body fossils in the type area are restricted to a single specimen of the euthycarcinoid *Kalbarria brimmellae*, one of the oldest body fossils of this enigmatic arthropod group.

Scientific importance and research potential

The locality is renowned for its spectacular trackways of early arthropods in the lower, non-marine part of the sandstone. The antiquity, diversity, abundance and state of preservation of the Tumblagooda arthropod trackways are almost unparalleled. The significance of the locality cannot be overstated, especially as recently published data indicate a Late Ordovician age of the sandstone, approximately 30 million years older than previously assumed.

Fossil significance

Myriapods, eurypterids, euthycarcinoids, xiphosurids and scorpionids are thought to have been the source of the trackway assemblage. Eurypterids ('sea scorpions') are known from body fossils elsewhere reaching up to 2 metres in length and some of the trackways, almost 20 centimetres wide, do indeed indicate arthropods of a very large size. Imprints of the ichnogenus *Diplichnites* provide evidence of arthropods walking out of the water and represent some of the earliest examples of animal activity on land. The upper part of the red-bed sequence indicates a more marine influence with *Skolitos-Diplocraterion* burrows.

Taphonomy and condition of the site/fossils

The site contains excellent exposures of the Tumblagooda Sandstone covering hundreds of square kilometres. Much of the strata are located within Kalbarri National Park and there are no major threats to the two main localities (i.e. the sea cliffs and the gorge), other than possible vandalism of the fossil trackways.

Stories

The trackways at Tumblagooda have significant dioramic potential. They tell a fascinating story about the evolution of terrestrial organisms, providing some of the earliest evidence of organisms walking on land. The trackways consist of a variety of marks that seem to have formed mostly at the edge of a wind-blown pond. The marks have been interpreted as showing feeding behaviour; it is suggested that the organism dined on organic matter that was either blown out of the pool, or that was deposited on land as the pool's water levels dropped. Further tracks show travel across dunes: the organism walks up the slope of the dune then slides down the other side and into a pool. Another set of trackways show the paths of two organisms converging and then separating again.

West Dale, Darling Plateau

Site overview

West Dale is a small Middle Eocene–Oligocene macroflora deposit on the Darling Plateau, 76 kilometres east-south-east of Perth and 32·5 kilometres west-south-west of Beverley, Western Australia. This diminutive exposure of siltstone has yielded an extraordinary and diverse assemblage of leaves, stems, wood and infructescences. The diversity and particularly the state of preservation of the fossil leaves puts West Dale in a class of its own among the Mesozoic and early Cenozoic plant localities in Western Australia. Surface cell detail is often comparable to that of living plants (as observed during scanning electron microscopy), allowing examination of individual cells.

Scientific importance and research potential

The West Dale flora is a window into the past of a region that has preserved extraordinary and diverse vegetation with several thousand endemic species. The unique preservation of minute details on the fossil leaves provides us with more palaeoenvironmental data than is typically available from fossil vegetation (Hill and Merrifield 1993; McLoughlin and McNamara 2001).

Fossil significance

The flora comprises four species of conifers, one fern, and a multitude of flowering plants (angiosperms). The last group includes both broad-leaved, rainforest types and sclerophyllous species at West Dale. Scleromorphy is manifested by thick but often small and narrow lignin-rich leaves, an adaptation to a deeply weathered soil, deficient in phosphorus and other trace elements. The fossil leaves from West Dale show no well developed xeromorphic characters other than those also associated with scleromorphy (e.g. small and thick leaves). Desert plants have evolved a whole range of water-conserving strategies. One of these involves the configuration of the stomata on the underside of the leaves. The stomata would typically have protective hairs covering them or be situated in a deep pit within the leaf. In both cases the stomata are protected from open exposure to the hot and dry air. The remarkable leaf surface preservation of the West Dale flora enable us to interpret the local climate with some accuracy. The stomata show no protective measures and therefore indicate a rather humid climate without the dry summers of today's Mediterranean climate of the region. This observation agrees well with the presence of drip tips (a rainforest indicator) on several of the angiosperm leaf types.

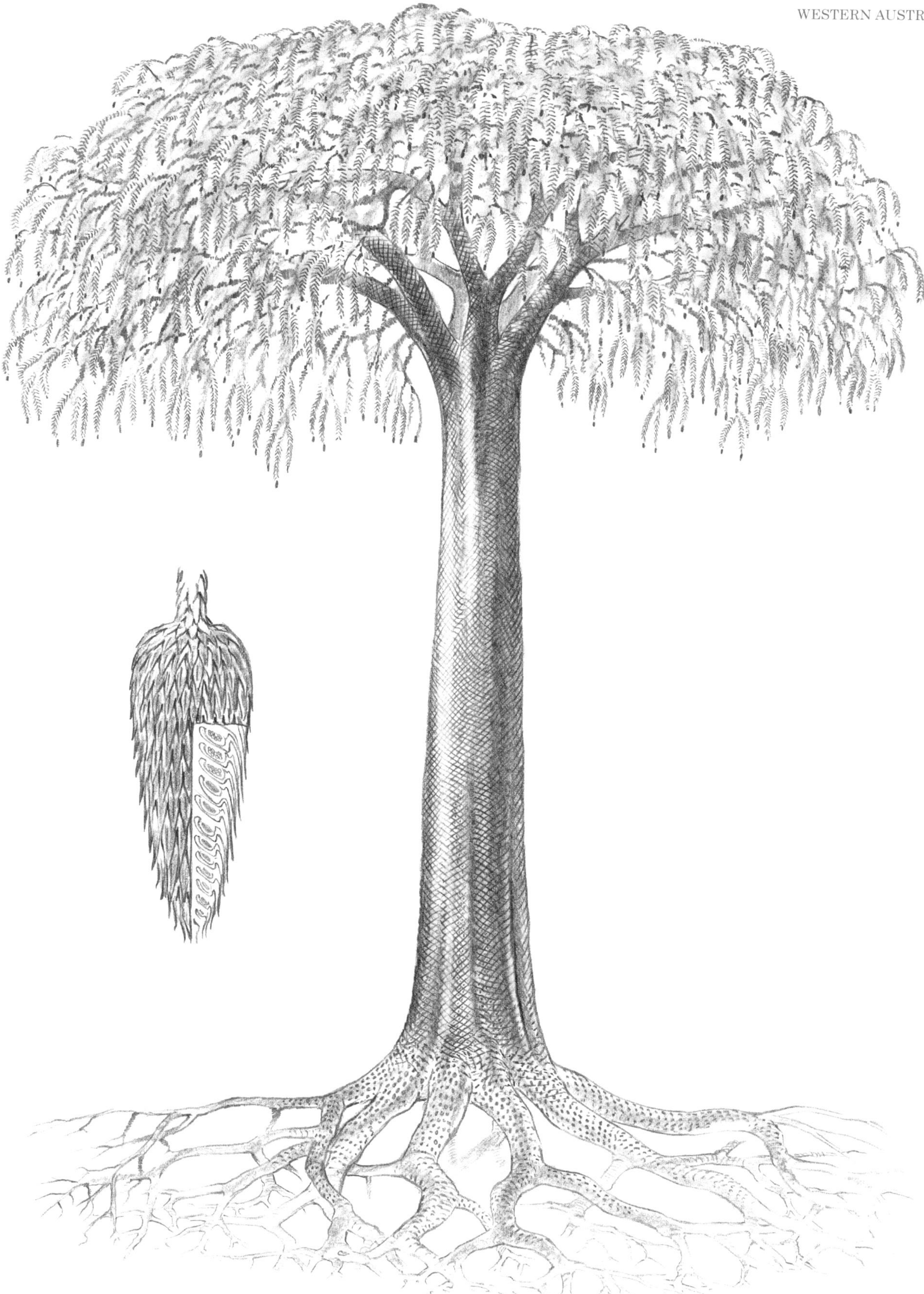

Figure 7.3 *Lepidodendron veltheimianum*. Drawing Peter Schouten © SEWPaC

It is thought that scleromorphy might have provided local species with a pre-adaptive advantage over rainforest species as more arid conditions with extensive laterite formation developed in the later part of the Cenozoic. However, few (probably *Agonis* and *Banksia/Dryandra*) of the West Dale scleromorphic lineages survived in south-western Australia by evolving stomatal protection. Most became locally extinct and their surviving descendants are today found in wet climates, typically on nutrient-deficient soil. The modern scleromorphic vegetation in south-western Western Australia appears to have originated largely from more or less xeromorphic scleromorphs growing elsewhere, presumably in drier inland areas from which we have no fossil record.

Taphonomy and condition of the site/fossils

The West Dale siltstone was deposited in a low energy, meandering river system that drained the peneplain of the Darling Plateau. The fossil plant material is preserved in iron oxides, which indicates an anaerobic depositional environment such as a stagnant billabong. Pliocene uplift of the Darling Plateau rejuvenated the river system, and new streams, cutting through the older flood plain deposits, exposed the plant-bearing strata at West Dale.

Roe Calcarenite invertebrate fauna, Roe Plains

Site overview

The Roe Calcarenite is a 2–3-metre thick carbonate that accumulated in shallow water at the centre of the Great Australian Bight (now the Roe Plains, Western Australia) during the early Pleistocene. There are only two natural exposures of the Roe Calcarenite on the Roe Plains: the Madura and Nurina Caves. The Roe Calcarenite contains a diverse and superbly preserved Pliocene invertebrate fauna.

Scientific importance and research potential

The deposits of the Roe Calcarenite provide a profusely rich record of communities living in a shallow, nearshore or intertidal subtropical setting. These lower Pleistocene deposits have many parallels with Holocene grass-bank facies in western and southern Australia and probably represent accumulation in a slightly warmer ocean than today, with the depositional setting heated by solar radiation. Units within the Roe Calcarenite provide an important conceptual bridge into the older Cenozoic rock record (James and Bone 2007).

Fossil significance

The current interpretation holds that the Roe Calcarenite represents one of the high-stand sheets of calcareous sands but is preserved as a result of a regional uplift (Megahiatus 3 tectonic event). The deposit is unique within Western Australia as it contains a beautifully preserved and extremely diverse and rich invertebrate fauna from an interglacial interval in the early Pleistocene. The molluscs alone are represented by 265 species and subspecies, of which two-thirds are extinct. Other groups include foraminifera, bryozoans, corals, brachiopods, annelids, arthropods, echinoderms, calcareous algae and even fish (vertebrates).

Taphonomy and site condition

Unlike the indurated older Cenozoic units of the area, the beautifully preserved fossils are easily extracted from the relatively unconsolidated Roe Calcarenite. Several abandoned quarries and temporary, shallow pits preserve more complete sections of the unit, in particular the Main Roads quarry, 16 kilometres south of Madura.

Stories

The Roe Calcarenite contributes to a story of the formation of the southern edge of Australia and cooling climates of the early Pleistocene. The Nullarbor Plain forms the eroded top surface of a marine sequence of Cenozoic limestones and minor sandstones. Coastal sedimentation in this part of Australia commenced with the opening of a rift graben as Australia began to separate from Antarctica in the Eocene. Uplift and seaward tilting, probably in the Late Miocene, exposed the strata to subaerial erosion with widespread karstification. Renewed transgression in the Pliocene by the high energy Southern Ocean hammered away at the vertical cliffs formed in the indurated limestones of the Eucla Group. The cliffs receded by up to 85 kilometres over a period of roughly three million years, as the sea worked its way deeper into the tilted layers of the Upper Oligocene to Early Miocene Abrakurrie Limestone. Orbitally driven cycles of transgressions and regressions, which culminated in the Pleistocene (and are still in effect), intensified in magnitude towards the end of the Pliocene. Thin veneers of calcareous sands were laid down on the more-or-less indurated flat surface of the Abrakurrie Limestone basement during peak transgressions as the Antarctic icesheet and Arctic sea ice retreated. As the sea withdrew again from the Roe Plain when solar radiation levels began to drop off at the onset on the next Milankovich cycle, the shallow marine sand was subjected to subaerial erosion. The remnants of the Pliocene cover would later be completely destroyed as the surf zone once again worked its way up the gentle slope of the plain to complete the cycle.

Callytharra Formation, Callytharra Springs (Permian invertebrates and foraminifera)

Site overview

The Callytharra Formation is located 500 metres west of Callytharra Spring, on the south bank of Wooramel River in the Carnarvon Basin. It is a richly fossiliferous Sakmarian (Early Permian) unit deposited in an interior rift basin.

Scientific importance and research potential

The Callytharra Formation, together with the underlying Carrandibby Formation, record the initial stages of Early Permian marine transgression in eastern Gondwana. The sequence is among the most fossiliferous sedimentary rocks of any kind in Western Australia. Skwarko (1993) reported 64 species of foraminifera, eight corals, 29 different bryozoans, 42 species of brachiopods, 27 bivalves, three taxa of ammonoids, five nautiloids, three gastropods, one conulariid, five blastoids, 15 crinoids, two different plants, and one annelid; in total 205 fossil species. The formation is also known to yield rare shark teeth and more common ostracods. The crinoid component of the fauna is the most diverse of its kind in the Australian Palaeozoic.

The Callytharra Formation at Callytharra Spring was listed by Carter (1987) as one of just over 100 major geological localities in Western Australia in need of protection. At the time of publication of the 1987 report, about 100 fossil species of organisms were known from the formation. Just six years later the figure had doubled to over 200 species. It indicates the untapped potential of most major fossil sites in Western Australia, considering that this is one of the best known and most researched Palaeozoic sites in the state.

Fossil significance

The underlying Carrandibby Formation represents the initial stage of the Early Permian marine transgression into the rift system, as the Gondwanan ice sheet began to recede near the Carboniferous/Permian boundary. The formation yields a low diversity, cool-water invertebrate fauna. The overlying Callytharra Formation marks a faunal change that has long been thought to reflect a continuation of local and global warming following the retreat of the icesheet. However, recent work on foraminifera indicates that the faunal turnover was controlled by a complex interaction of water temperature and salinity. With increasing distance from the receding ice sheet, the input of meltwater into the rift basin decreased. Overlying the brackish-water Carrandibby Formation, the Callytharra sequence thus records the first fully marine biota in the central part of the western rift basins.

Taphonomy and condition of the site/fossils

The type section of the Callytharra Formation comprises 146 metres (composite section; Dixon and Haig, 2004) of thin limestone beds separated by mudstones. The middle and upper part of the formation is rather poorly exposed in the type area.

Broome Sandstone dinosaur trackways and macroflora, Broome

Site overview

The Broome Sandstone of the northern Canning Basin is a nearly 300-metre-thick sequence representing a variety of mostly non-marine, coastal environments during the Cretaceous Period. Dinosaur trackways are ubiquitous within the Broome Sandstone outcrops, which are exposed discontinuously along the coast near Broome for up to 200 kilometres. The trackways are most easily seen in the intertidal zone but occur throughout the outcrop.

Scientific importance and research potential

Footprints of dinosaurs found near Broome were first mentioned in the scientific literature in 1952. Fifteen years later Colbert and Merrilees (1967) published a description of the theropod tracks, naming the ichnospecies *Megalasauropus broomensis*. More recently, Thulborn *et al.* (1994) and Thulborn (pers. comm. 2009) described enormous sauropod footprints measuring 1.75 metres in diameter. Their size suggests they were made by sauropods comparable in size to the largest known from anywhere in the world. Thulborn has also reported the presence of the world's smallest sauropod tracks, making the size range here the largest in the world. Up to 15 types of tracks are now recognised as occurring within the Broome Sandstone, including thyreaphoran (possibly stegosaurid) tracks, which would be the first and only record of a stegosaurid in Australia and the first occurrence of stegosaurid tracks in the world.

In addition to the dinosaur footprints and trackways, the Broome Sandstone contains one of the best Cretaceous floras in Western Australia. The type section has yielded a great variety of plants, including lycophytes, ferns, seed-ferns and conifers. Some plant fossils appear in their growth positions, with roots descending into the substrate and dinosaur trackways weaving around them (Thulborn pers. comm. 2009).

Fossil significance

The ichnofossils indicate a diverse dinosaur fauna with very small to gigantic sauropods, small to medium-sized theropods, small to large ornithopods and possible stegosaurs. These dinosaur trackways are of world-class standard, and together with previously described tracks at Broome (Gantheaume Point), provide the only substantial insight into the world of these long-gone giants. There are virtually no dinosaur body fossils recovered from Mesozoic rocks in Western Australia other than a few isolated bone fragments that put together would fit in a human hand. The number of dinosaur tracks drawn from a single stratigraphic layer is an otherwise unobtainable census of the dinosaur community that occurred here. Some tracks are so well preserved they even retain evidence of skin impressions and manus claws (Thulborn pers. comm. 2009).

Taphonomy and condition of the site/fossils

Isolated sections of Broome Sandstone surfaces were exposed along an 80–100 kilometre stretch of the Broome coast in 1997 after a cyclone washed away the beach sand covering the strata. Dr Tony Thulborn and members of his field party discovered remnants of forests, swamps, lagoons and fluviatile/deltaic facies along with associated dinosaur footprints and trackways in an abundance and quality never seen before in Western Australia. Since a much-publicised theft of these footprints, researchers will no longer reveal the precise location of the dinosaur tracks to the public. Movement of the sand in response to storms and tides as well as the location of most of the known tracks in the intertidal zone makes it difficult to access the prints for more than brief periods.

Stories

The dramatic visual evidence provided by such dinosaur trackways has great public appeal. The presence of both a flora and trackways of several herbivorous dinosaurs at a single site is intriguing as it gives an indication about the diet of the local dinosaurs. One could imagine the sauropods raking off the leaves in the conifer canopy while the smaller, bipedal ornithopods browse the lycophytes below.

Miria Marl, Giralia Anticline

Site overview

The Miria Formation is a thin (0.6–2.1 metres), mid-shelf, latest Maastrichtian (Late Cretaceous) foraminiferal calcarenite and calcisiltite with its main surface exposure in the Giralia Range. The Miria represents the only late Maastrichtian surface unit in Australia, and contains a rich and diverse macrofossil assemblage.

Scientific importance and research potential

The end of the Maastrichtian marks one of the most severe global extinction events known to science (the end-Cretaceous mass extinction event). Virtually all organism groups were affected and many became extinct, including dinosaurs, mosasaurs, plesiosaurs, pterosaurs, ammonites and larger planktic foraminifera. Some or all of the upper part of the Miria can be referred to the very last planktic foraminiferal zone (*Abathomphalus mayaorensis* Zone) of the Maastrichtian. The significance of the site is thus closely tied to the age of the formation at the very end of the Mesozoic. Monographs on several of the major invertebrate groups have been published, including ammonites, bivalves, scaphopods and gastropods.

The contact between stratigraphic layers representing the actual K/T boundary is vividly exposed in the Miria Marl of the Giralia Range, but nowhere else in Australia: a thin bed of the Boongerooda Greensand overlies 2 metres of soft fossiliferous limestone. The greensand was almost certainly formed in a cold ocean, but the immediately underlying limestone must have formed in much warmer waters. This contact demonstrates that in a virtual geological instant, the marine fauna changed on a scale not seen since the Permo-Triassic extinction 180 million years earlier, at the close of the Palaeozoic era 250 million years ago (Craig 2002; McNamara 1997).

Fossil significance

Ammonites are a formerly abundant group of molluscs akin to the living pearly nautilus. These important and widely collected global biostratigraphic markers suddenly disappear from the fossil record all over the world at the close of the upper Maastrichtian. Their presence in the Miria Formation helps to date it to the Maastrichtian, 71–65 million years ago. Palaeontologists have described more than 60 other species of marine mollusc in addition to nearly 30 species of ammonite. The site also includes Australia's most diverse late Cretaceous shark tooth assemblage and the most diverse assemblage of corals from the Mesozoic era in Australia. These coral species only occur elsewhere in Cretaceous rocks in Antarctica and southern India, which during the Mesozoic era were connected to Australia as part of Gondwana.

Taphonomy and condition of the site/fossils

The Miria Formation crops out over a strike distance of some 80–100 kilometres in the Giralia Anticline (see Henderson and McNamara 1985: fig. 1). It rests with a disconformity on the late Campanian to early Maastrichtian Korojon Calcarenite and is succeeded disconformably by the mid-Palaeocene (upper Selandian – lower Thanetian) Boongerooda Greensand. In many places, weathering has reduced the formation to a residual surface, littered with phosphatic steinkerns of molluscan macrofossils, mainly ammonites and a nautiloid. Partial and complete sections are present in the northern part of the anticline. The Giralia Range is very scenic with some stunning cliffs off the Korojon Calcarenite.

Comparison with other similar sites in Australia

The Miria Marl is one of very few Maastrichtian surface rocks in the country (the other few are all also in Western Australia) and the only one with a rich and diverse macrofossil assemblage. Ammonites, a group of fundamental importance in global biostratigraphy, are the most common macrofossils in the Miria, and are far more abundant here than anywhere else in the world in rocks of Maastrichtian age.

The end-Palaeozoic (Permo-Triassic or P/T) extinction horizon (250 million years ago) is nowhere represented in Australia as dramatically as the Giralia Range exposure presents the end-Mesozoic boundary (65 million years ago). P/T boundary deposits are rare in Gondwana in general, because the boundary is difficult to correlate with the marine standard sections of northern hemisphere deposits, hampered by the rarity of marine index fossils (Foster *et al.* 1999). P/T boundary sites like Coalcliff and Wybung Head near Sydney are identified predominantly by isotopic analysis rather than their specific fossil content, and do not provide the dramatic contrast of pre- and post-extinction sediments and fossil ecosystems displayed so strikingly at the contact of the Miria formation with the Boongerooda Greensand in Giralia Range. Furthermore, the K/T boundary is irrevocably associated with the extinction of charismatic megafauna like the dinosaurs and marine reptiles, and among invertebrate fossils with historically important, iconic and beautiful ammonites and belemnites.

Stories

The formation provides us with a snapshot of life on a shelf in the southern hemisphere at the very end of the Cretaceous period. The invertebrate diversity is among the highest known for latest Maastrichtian strata and provides us with invaluable data on the health of a marine ecosystem that was just about to be obliterated by the violent cosmic (and/or volcanic) event that terminated the Mesozoic.

Warrawoona Group (North Pole and Strelley Pool sites), Proterozoic early life fauna

Site overview

The Warrawoona Group localities contain unique sites, including the North Pole site, located near Marble Bar, which provides the oldest established evidence of life on Earth. Nearby stromatolites in the Warrawoona Group represent the first biogenic structures known, and may be interpreted as the first evidence of fossil reefs. Western Australia is perhaps the most important region in the world for the study of stromatolites, the dominant fossil group in the Archaean and Proterozoic. These stromatolites relate well at gross scale to modern living relatives in Shark Bay (WA).

Scientific importance and research potential

Ever since the discovery of fossilised early life forms in the Pilbara in 1959 by L.E. de la Hunty, the region has been the focus of intense research by palaeobiologists. Arguably the oldest known fossils on Earth were found in the Dresser Formation and Strelley Pool Chert of the Warrawoona group in the late 1970s. They include stromatolitic structures at North Pole and Strelley Pool and have been dated at 3.49 Ga (North Pole; Dresser Formation) and 3.42 Ga (Strelley Pool; Strelley Pool Chert). Alternative hypotheses explain the structures as abiotic and derived from precipitation from hydrothermal solutions.

The series of reports describing 11 species of 3.465-billion-year-old microfossils in chert dykes have met more resistance in the scientific literature. Schopf and Packer (1987) described what they believe are cellularly preserved filamentous and colonial microfossils from carbonaceous chert dykes in the Apex Basalt and Towers Formation at Chinaman Creek near Marble Bar, 30–50 kilometres east of the North Pole and Strelley Pool sites. A bacterial and cyanobacterial affinity was suggested, which would indicate oxygen-producing photosynthesis at a very early stage in the evolution of our planet. Brasier and colleagues (2002) challenged Schopf and Packer's interpretation of their findings by suggesting that the 'microfossils' were instead formed from amorphous graphite in metalliferous hydrothermal veins. The controversy is far from over, and both sides have gained support to a variable degree from other research groups.

The Warrawoona group localities are without a doubt the best known Western Australian fossil sites internationally, as indicated by numerous publications by national and overseas research groups in the most prestigious scientific journals (e.g. *Nature* and

Science). The area has played and will continue to play a vital role in our understanding of early life on Earth. Ongoing research has been facilitated by the renewed interest in nanofossil and microfossil forms purportedly found in a Martian meteorite, which has highlighted the need for greater examination of this site.

Fossil significance

The site contains a unique assemblage of Archaean fossils for Australia and the oldest representation of life on Earth. It has extremely high potential for palaeoenvironmental reconstruction, and may prove to be critically important in understanding Archaean atmospherics, sea water, geological conditions, and, through these variables, the origin of life.

Taphonomy and condition of the site/fossils

The site contains metamorphosed cherts and altered carbonates that preserved fine detail. The fossils themselves are microscopic but stromatolites are visible in outcrop. Indiscriminate collecting at North Pole has removed most stromatolitic structures. The most important localities are the North Pole, Strelley Pool and the Chinaman Creek area (chert dykes in the Apex basalt).

Comparison with similar sites in Australia

This is among the most important geological sites on the globe. It cannot be easily compared with other sites given its unique nature. It is as unusual as the living stromatolites in Shark Bay and can only be compared with stromatolite localities in the much younger Fortescue Group (WA), which are better preserved but less important scientifically.

Stories

The classic Smithsonian Institution poster (Archaean Earth) is based on these sites, and demonstrates their dioramic potential. Warrawoona Group helps tell the critical story of the beginnings of life in the shallow Archaean sea. It may become historically important as the site of discovery of the first record of life.

Cundlego Formation Permian invertebrate fauna, Gascoyne Junction

Site overview

The Cundlego Formation is exposed for about 10 kilometres along the Gascoyne River at Gascoyne Junction. The formation is an early Permian marine unit of quartz greywacke and carbonaceous siltstone. At Gascoyne Junction it reaches its maximum thickness of 412 metres.

Scientific importance and research potential

The uniqueness of the locality lies in the abundance and exceptional preservation of crinoids and starfish specimens preserved. It is without doubt one of the best localities in the world for these echinoderms. Crinoids like the common *Jimbacrinus bostocki* may occur in large and dense congregations of specimens on bedding surfaces.

Fossil significance

The formation has yielded 16 species of foraminifera, two corals, one bryozoan, 13 brachiopods, one bivalve, one nautiloid, two gastropods, one conulariid and three crinoids, in total 40 fossil species or just one-fifth of the number of taxa that are present in the slightly older Callytharra Formation. There is, however, a diverse trace fossil assemblage identified from the Cundlego Formation.

Taphonomy and condition of the site/fossils

The site is in need of protection from commercial collectors who pillage the locality on a regular basis. Mining leases are held over the area by commercial collectors, and slabs of rocks with beautifully preserved crinoids and starfish are removed by the truck load. Crinoids from the Gascoyne Junction locality have found their way to many museums around the world as they make a very attractive display.

Stories

The rich marine fauna and depostional environment of the Cundlego formation preserves the actions of storms which periodically mantled benthic marine animals like sea stars, brittle stars and crinoids in silt and sand. The site records a storm-swept coastal shelf in the Early Permian, around 280 million years ago, covered by warming seas, as the supercontinent Gondwana emerged from a major ice age (Western Australian Museum 2011).

The 280-million-year-old Cundlego Formation was deposited on a storm-swept coastal shelf. The southern supercontinent Gondwana, of which the Australian continent was part, had just emerged from a major ice age. A relatively rich marine fauna flourished in the cool but warming waters (Western Australian Museum 2011).

List of sites by type

Early life sites

Ediacara sites, Nilpena, Flinders Ranges, South Australia

Mount Skinner fauna, Northern Territory

Warrawoona Group (North Pole and Strelley Pool sites), Proterozoic early life fauna, Western Australia

Landscapes sites

Bitter Springs Formation, Ross River Homestead, Northern Territory

Devonian 'Great Barrier Reef' and Gogo fish fossil site, Fitzroy Crossing, Western Australia

Ellery Creek Section, Northern Territory

Fletcherview – Big Bend 'Leichhardt Cliffs' Devonian reef complex, Queensland

Jack Hills Gorge and Fish Hill, Broken River Province, Queensland

Taemas–Cavan–Wee Jasper, New South Wales

Tumblagooda Sandstone, Kalbarri and Murchison River Gorge trace fossils, Western Australia

Trace fossil sites

Broome Sandstone dinosaur trackways and macroflora, Western Australia

Ellery Creek Cambrian–Ordovician trace fossil locality, Northern Territory

Lark Quarry Dinosaur Stampede National Monument (Winton Formation), Queensland

Mount Morgan Jurassic dinosaur footprint site, Queensland

Myponga Beach, Fleurieu Peninsula, South Australia

Tumblagooda Sandstone, Kalbarri and Murchison River Gorge trace fossils, Western Australia

Plant fossil sites

Bauhinia Downs flora, Balbarini, Northern Territory

Broome Sandstone dinosaur footprints and macroflora, Western Australia

Cape Van Diemen flora, Northern Territory

Dinmore Quarry, Queensland

Early Oligocene macroflora sites of north-western Tasmania

Eastern shore of Macquarie Harbour, Tasmania

Fennell Bay (Kurrur-Kurran) fossil forest, New South Wales

Homevale Station, Queensland

Lake George, New South Wales

Late Cretaceous (Cenomanian) Winton Formation, Queensland

Latrobe Valley coal measures, Victoria

Lune River Jurassic plant site, Tasmania

Silurian Baragwanathia site, Yea, Victoria

Stuart, Poole and Nelly Creek fossil plant sites, South Australia

West Dale, Darling Plateau, Western Australia

Invertebrate fossil sites

Ajax Mine achaeocyathid locality, South Australia

Belmont insect beds, New South Wales

Bitter Springs Formation, Ross River Homestead, Northern Territory

Black Mountain Cambrian–Ordovician Boundary Section, Queensland

Callytharra Formation, Callytharra Springs (Permian invertebrates and foraminifera), Western Australia

Chatsworth Limestone echinoderm and trilobite localities, Queensland

Christmas Hills, Tasmania

Cliefden Caves–Belubula River Valley, New South Wales

Cundlego Formation Permian invertebrate fauna, Gascoyne Junction, Western Australia

Ediacara sites, Nilpena, Flinders Ranges, South Australia

Ellery Creek Section, Northern Territory

Emu Bay (Cape d'Estaing to Big Gully), Kangaroo Island, South Australia

Fossil Cliffs, Maria Island, Tasmania

Hamilton Miocene–Pliocene invertebrate sites, Victoria

Miria Marl, Giralia Anticline, Western Australia

Mount Crosby Triassic insect locality, Queensland

Ordovician faunal succession of the Florentine Valley, Tasmania

Roe Calcarenite invertebrate fauna, Roe Plains, Western Australia

The shoreline near Point Hibbs, Western Tasmania

Victorian Ordovician graptolite province, Victoria

Vertebrate sites

Alcoota vertebrate fossil beds, Northern Territory

Bullock Creek vertebrate fossil site, Northern Territory

Canowindra fish fossil site, New South Wales

Chinchilla Rifle Range, Queensland

Cuddie Springs, New South Wales

Devil's Lair, Cape Leeuwin – Cape Naturaliste region, Western Australia

Devonian 'Great Barrier Reef' and Gogo fish fossil site, Fitzroy Crossing, Western Australia

Early Cretaceous (Aptian) Koonwarra site, Victoria

Genoa River Devonian tetrapod site, Victoria

Grange Burn Pliocene mammal site, Victoria

Green Waterhole Cave, South Australia

Hines Quarry Diprotodon site, Bacchus Marsh, Victoria

Jack Hills Gorge and Fish Hill, Broken River Province, Queensland

Lake Callabonna, South Australia

Lake Frome Basin, Namba Formation, South Australia

Lake Palankarinna, Lake Eyre and Tirari Desert, South Australia

Middle Eocene Anglesea site, Eastern View Formation, Victoria

Mt Howitt fish fossil site, Victoria

Mt Watt and Mt Charlotte, Northern Territory

Otway and Strzelecki Ranges, Victoria

Rewan, The Crater, Queensland

Riversleigh, Queensland

Torquay Oligocene whale and invertebrate site, Victoria

Triassic vertebrate faunas, Knocklofty Formation, Tasmania

Wellington Caves, New South Wales

Microfossil sites

Bitter Springs Formation, Ross River Homestead, Northern Territory

Darwin Meteorite Crater, Tasmania

Lake George, New South Wales

Mixed sites

Lake Ngapakaldi Leaf Locality, South Australia

Lightning Ridge, New South Wales

Moon Plain Cretaceous fossil site, Coober Pedy, South Australia

Redbank Plains Formation, Queensland

Table Cape – Fossil Bluff, Wynyard, Tasmania

Talbragar fossil beds, New South Wales

Torquay Oligocene whale and invertebrate site, Victoria

Walsh River and Elizabeth Creek, Queensland

Geological timeline

Eon/era/period/epoch				Million years
Phanerozoic Eon	Cenozoic Era	Quaternary Period	Holocene epoch	0.0117–0
			Pleistocene	2.588 ± 0.01–0.0117
		Neogene	Pliocene	5.33 ± 0.005–2.588 ± 0.01
			Miocene	23.3 ± 0.05–5.33 ± 0.005
		Palaeogene	Oligocene	33.9 ± 0.1–23.3 ± 0.05
			Eocene	55.8 ± 0.2–33.9 ± 0.1
			Palaeocene	65.5 ± 0.3–55.8 ± 0.2
	Mesozoic	Cretaceous	Late Cretaceous	99.6 ± 0.9–65.5 ± 0.3
			Early Cretaceous	145.5 ± 4–99.6 ± 0.9
		Jurassic	Late Jurassic	161.2 ± 4–145.5 ± 4
			Middle Jurassic	175.6 ± 2–161.2 ± 4
			Early Jurassic	199.6 ± 0.6–175.6 ± 2
		Triassic	Late Triassic	228 ± 2–199.6 ± 0.6
			Middle Triassic	245 ± 1.5–228 ± 2
			Early Triassic	251 ± 0.4–245 ± 1.5
	Palaeozoic	Permian	Late Permian	260.4 ± 0.7–251 ± 0.4
			Middle Permian	270.6 ± 0.7–260.4 ± 0.7
			Early Permian	299 ± 0.8–270.6 ± 0.7
		Carboniferous	Late Carboniferous (Pennsylvanian)	318.1 ± 1.3–299 ± 0.8
			Early Carboniferous (Mississippian)	359.2 ± 2.5–318.1 ± 1.3
		Devonian	Late Devonian	385.3 ± 2.6–359.2 ± 2.5
			Middle Devonian	397.5 ± 2.7–385.3 ± 2.6
			Early Devonian	416 ± 2.8–397.5 ± 2.7
		Silurian	Late Silurian	443.7 ± 1.5–422.9 ± 2.5
			Early Silurian	422.9 ± 2.5–416 ± 2.8
		Ordovician	Late Ordovician	460.9 ± 1.6–443.7 ± 1.5
			Middle Ordovician	471.8 ± 1.6–460.9 ± 1.6
			Early Ordovician	488.3 ± 1.7–471.8 ± 1.6
		Cambrian	Late Cambrian	501 ± 2–488.3 ± 1.7
			Middle Cambrian	513 ± 2–501 ± 2
			Early Cambrian	542 ± 0.3–513 ± 2
Proterozoic	Neoproterozoic 1000–542	Ediacaran		635–542 ± 0.3
		Cryogenian		850–635
		Tonian		1000–850
	Mesoproterozoic			1600–1000
	Palaeoproterozoic			2500–1600
Archaean				3800–2500
Hadean				4567.17 ± 0.1–3800

Key references used for this table: Gradstein *et al.* (2004), International Commission on Stratigraphy (2009), Ogg *et al.* (2008), Rohde (2005).

Glossary

Abiotic: Not living, non-biological, usually describing factors in an ecosystem such as atmospheric gases, inorganic salts, mineral soil particles and water. This expression is also used to describe the chemical and physical factors such as salinity and humidity that influence organisms. An abiotic environment is one without any life.

Acanthodian: A group of extinct, jawed fishes, distinguished by bony spines projecting in front of their fins and by having small, diamond-shaped scales. Most had relatively large eyes set near the front of short blunt heads, and, like most early fishes, they possessed tail fins with the top longer than the bottom (heterocercal). They are among the earliest jawed vertebrates known, extending some 160 million years from as early as the Late Ordovician to about the Early Permian, with greatest diversity from the Late Silurian and the Devonian.

Acritarch: In general, acritarchs are any small, non-carbonate or non-siliceous organic structure that cannot otherwise be taxonomically classified. Acritarchs are found in sedimentary rocks that range in age from the Proterozoic through to the present-day. They include the remains of a wide range of different kinds of organisms, ranging from the egg cases of small metazoans to resting stages or cysts of many different kinds of chlorophyta (green algae). It is likely that some acritarch species represent the resting cysts of algae that were ancestral to the dinoflagellates (one-celled, chiefly marine organisms that have two dissimilar flagella and characteristics of both algae and protozoans). The nature of the organisms associated with older acritarchs is generally not clear, though many are probably related to unicellular marine algae. In theory, when the biological source (taxon) of an acritarch does become known, that particular microfossil is removed from the acritarchs and classified with its proper group.

Actinia: Sea anemones and related animals. Marine polyps that resemble flowers but have oral rings of tentacles, differing from corals in that they form no hard skeleton.

Actinopterygians: Ray-finned fishes, possessing fins consisting of webs of skin supported by bony or horny spines ('rays'). These fin rays attach directly to the proximal or basal skeletal elements (the radials), which link the fins to the internal skeleton (e.g. pelvic and pectoral girdles). They range in age from the Late Silurian (approximately 420 Ma) to the present, and are the dominant class of modern vertebrates with nearly 30 000 species. They are ubiquitous throughout freshwater and marine environments.

Adaxial surface: Of or relating to the side or surface that faces, or is nearest to, the axis of an organ such as the upper surface of a leaf.

Aeolian: Relating to, or caused by, the wind (Aeolus = Greek god of the winds).

Agnatha: Jawless fishes (sometimes called cyclostomes), which have existed since the Cambrian Period. Agnatha contains two extant groups – lampreys and hagfish – with about 100 species in total. In addition to the absence of jaws, agnathans are characterised by absence of paired fins; presence of a notochord (a primitive, flexible, rod-like 'backbone' that in higher vertebrates is present in embryos and develops into the spinal column) in both larvae and adults; and seven or more paired gill pouches. There is a light-sensitive pineal eye (homologous to the pineal gland in mammals). All living and most extinct agnathans do not have an identifiable stomach or any appendages. They are ectothermic (cold-blooded: body temperature is controlled externally), with a cartilaginous skeleton, and the heart contains two chambers.

Agnostida: An order of small trilobites. They had tails (pygidia) that were similar in size and shape to their heads (cephalons), and most agnostid species were eyeless. It is not certain whether the agnostids lived a pelagic (middle–upper zone of the open ocean) or a benthic lifestyle. Agnostids first appeared in the late Early Cambrian, thrived in the Middle Cambrian, and went extinct in the Late Ordovician.

Albian: The last stage of the Early epoch of the Cretaceous Period. Extends from 112 ± 1–99.6 ± 0.9 Ma.

Allochthonous: Components of a rock or deposit that originate from a place other than where the rock or deposit formed.

Allometry: The study of the relative rates of change of two variables of a system. In biology it is an important method for describing morphological evolution. It is the relation between the size of an organism and the size of any of its parts, and can be studied during the growth of a single organism, between different organisms within a species, and between organisms in different species.

Alluvium: Sediments deposited by a river or other running water. Alluvium is typically made up of a variety of materials, including fine particles of silt and clay and larger particles of sand and gravel.

Ammonites: An extinct group of marine cephalopod molluscs that are classified in part on the ornamentation and structure of the septa that divides the internal chambers of the shell. Ammonites are split into three orders and eight known suborders. This makes them excellent index fossils, and it is often possible to link the rock layer in which they are found to specific geological time periods. Their fossil shells usually take the form of planispirals, although there were some helically spiralled and non-spiralled forms. Their name came from their spiral shape, as their fossilised shells resemble the tightly coiled horns of rams (the Egyptian god Ammon was typically depicted wearing rams' horns). Ammonites first appeared in the Late Silurian to Early Devonian (about 400 million years ago) and became extinct at the close of the Cretaceous (65 Ma) along with the dinosaurs.

Amorphous: A term applied to material having no regular arrangement of atoms. In mineralogy it refers to a mineral without definite crystalline structure.

Angiosperms: Flowering plants. They have a fossil record extending back at least 130 million years and they are the most widespread group of land plants today.

Annelids: A large phylum of segmented worms. They occur in marine environments, in freshwater, and in moist terrestrial environments. They are generally poorly preserved as fossils. The annelid fossil record ranges from the Early Cambrian, or possibly the Ediacaran, to the present.

Antiarchs: The second most diverse group of placoderm fishes. They possessed a jointed head and thoracic shields that formed a heavily armoured box with a small anterior opening for two protruding eyes. Each pectoral fin was completely encased in bone and in the more advanced genera it was jointed in a fashion more reminiscent of arthropods than of vertebrates.

Aptian: A late stage of the early epoch of the Cretaceous Period. Extends from 125 ± 1–112 ± 1 Ma.

Archaean: The earliest formally recognised eon in the geological record. Archaean rocks and fossils are the oldest known on Earth and the eon ranges from 3800–2500 Ma. It is preceded by the informal Hadean division.

Archaeocyatha: Sessile, reef-building marine organisms that lived in warm tropical and subtropical waters during the Early Cambrian period. First known from about 530 million years ago, they quickly diversified into over 100 families and formed the first reef-building animals.

Archosaurs ('ruling lizards'): A large animal group that included modern birds and crocodilians and extinct non-avian dinosaurs, pterosaurs, and relatives of the crocodilians. They are characterised by having teeth set in sockets (thecodont), although birds are secondarily toothless; openings (fenestrae) in the skull and jaws to reduce weight; and a fourth ridge (trochanter) on the femur for attaching muscles. They probably evolved during the Late Permian but quickly became the dominant land vertebrates in the Early Triassic following the Permian–Triassic extinction event.

Archostemata: Among the earliest known beetles, with a fossil record extending from the Permian (when they were relatively common and dispersed over a wide portion of the globe) through to the present. Modern archostematans have a number of primitive characteristics and are the smallest suborder of beetles, consisting of fewer than 50 known species organised into five families; the species that remain are probably scattered vestiges of their former population.

Arkose: A sandstone rock with a significant proportion (>25%) of feldspar grains. Thought to form under conditions of minimal chemical weathering, due to aridity and/or rapid erosion and burial.

Arthrodires: The most diverse group of placoderm fish, distinguished by having two pairs of tooth-like plates (gnathal) extending from the upper jaw. The more advanced arthrodires also have an unusual neck joint that considerably increases gape size. Most placoderms were benthic but several of the more advanced arthrodires have reduced armour and were probably adept swimmers. Arthrodires are typically less than one metre in length, but some became the giants of their day.

Arthropods: A large phylum of invertebrate animals that have an external skeleton (exoskeleton), a segmented body, and jointed appendages. The phylum includes the insects, arachnids and crustaceans. They first appeared in the Cambrian and have over one million described modern species (>80% of all described living species).

Assemblage zone: A biostratigraphic unit or level of strata that is characterised by a particular assemblage of animals and/or plants.

Aves: The class name for birds: winged, bipedal, warm-blooded (endothermic), vertebrate animals that lay eggs and inhabit ecosystems across the globe. They evolved from theropod dinosaurs during the Jurassic Period around 150–200 million years ago. Birds are regarded as the only clade of dinosaurs that survived the Cretaceous–Cenozoic extinction event that occurred approximately 65.5 million years ago. There are around 10 000 living bird species, making them the most numerous of tetrapod vertebrates.

Bairnsdalian: A middle Miocene Australian Stage, extending from 15–10.5 Ma, roughly contemporaneous with the Serravalliam ISC Stage.

Balcombian: A middle Miocene Australian Stage, extending from 15.5–15 Ma, roughly contemporaneous with the late Burdigalian and Langhian ISC Stages.

Basalt: A fine-grained, black or grey extrusive igneous (volcanic) rock, produced when molten rock cools rapidly.

Batesfordian: An early Miocene Australian stage, extending from 16.5–15.5 Ma, roughly contemporaneous with the mid Burdigalian ISC Stage.

Bedding: In sedimentary rocks, a bed is the smallest lithostratigraphic unit or division of a geologic formation or stratigraphic rock series, usually ranging from 1 centimetre to several metres in thickness. It is separated from layers above and below by well defined bedding planes and differentiated by characteristics such as rock or mineral type, colour, or particle size. Bedding planes are not necessarily originally horizontal.

Belemnites: An extinct group of marine cephalopods, very similar in many ways to the modern squid and cuttlefish but possessing ten arms of roughly equal length and no tentacles. Belemnites originated in the Devonian and were abundant in Mesozoic marine rocks of the Jurassic and Cretaceous periods. They became extinct, along with the ammonites, at the end of the Cretaceous.

Bennettitaleans: A plant group, originally mistaken for cycads because of their cycad-like growth form of woody stems heavily armoured with persistent leaf bases, and entire simple leaves or pinnate foliage. However, the reproductive structures of these plants and their stomatal morphology (syndetocheilic) unequivocally differentiates Mesozoic cycad and bennettitalean foliage. Bennettitaleans first appeared in the Triassic and became extinct towards the end of the Cretaceous.

Benthic: The benthos includes all aquatic organisms that live on, or within, the sediment at the bottom of a water body. Benthic organisms can be free-living (mobile) or non-mobile (sessile).

Biochron: The length of time represented by a biostratigraphic zone, which is an interval of geological strata defined on the basis of its characteristic fossil taxa. Biochrons are named after particular fossil organisms or taxa that characterise that interval in time.

Biochronology: The relative age-dating of rock units based on their fossil content, particularly index taxa.

Bioclastic packstone: A sedimentary carbonate rock. The granular material consists of broken fragments of organic skeletal material arranged in a self-supporting framework that also contains some matrix of calcareous mud.

Biocorrelation: The use of fossils or fossil assemblages to demonstrate correspondence in character and/or stratigraphic position.

Biogenic: Material produced by life processes. It may be either constituents or secretions of plants or animals.

Biogeography: The study of the geographical distributions of organisms and their habitats and the environmental or historical factors that produce them.

Bioherm: A mound-like or circumscribed mass of rock built by sessile organisms such as corals, stromatoporoids or algae, and enclosed in rock of different lithological character. Reefs are typical examples of bioherms.

Biostratigraphy: The organisation of rock bodies into distinctive, mappable units based on their fossils or fossil assemblages. Biostratigraphy is useful for establishing the distribution and relationship in space of rock bodies and their succession in time, and for interpreting their geologic history.

Biostrome: A horizontally bedded stratum of fossilised remains of sessile organisms.

Bioturbation: The disturbance of soils and sediments by animals or plants, such as burrowing, tracks and other traces.

Biozone: An interval of a geological stratum defined on the basis of its characteristic fossil taxa. Equivalent to a biostratigraphic zone.

Bivalve: Descriptively, refers to an animal with a shell in two parts, hinged together. Taxonomically, the mollusc class Bivalvia consists of marine or freshwater organisms having a soft body with plate-like gills enclosed within two shells hinged together. Typically the valves are symmetrical along the hinge line.

Blastoids: An extinct class of echinoderms, they were high-level, stemmed, suspension feeders that inhabited clear-to-silty, moderately agitated ocean waters from shelf to basin. Like most echinoderms, the blastoid main body (theca) was protected by a set of interlocking carbonate plates and attached to a stalk or column of stacked disc-shaped plates, which was attached to the ocean floor by a holdfast. The mouth was located at the top of the theca and food brought to it by many long, thin, fine structures called brachioles. Blastoids originated, along with many other echinoderm classes, in the Ordovician. They reached their greatest diversity in the Carboniferous and became extinct at the end of the Permian about 250 million years ago.

Body fossils: The anatomical components of an organism preserved in rock. Body fossils most commonly consist of hard parts such as exoskeletons, bones and teeth, but can occasionally include soft tissue parts.

BP: An abbreviation for 'before present' used when reporting radiocarbon dates, which by convention must be reported in the form 11 500 ± 250 years BP. By convention, BP refers to years before the year 1950 AD. BP should only be used in reporting uncalibrated radiocarbon dates.

Brachiopoda: A phylum of benthic (bottom dwelling), marine, bivalved organisms characterised by the two shells being dissimilar but both bilaterally symmetrical. Brachiopods are filter feeders and gather microscopic organisms and bits of organic matter by a specialised tube-like structure (lophophore) with hair-like projections (cilia), which move food particles to the mouth. Brachiopods evolved in the Cambrian, were highly abundant in genera and numbers through the Palaeozoic, and were severely reduced by the end of the Permian extinction. Several orders survived the extinction to the present day.

Braided channel: A fluvial (or river) system that consists of a network of small channels separated by small and often temporary islands called braid bars. Braided channel morphology is promoted by an abundant supply of sediment, high stream gradient, rapid and frequent variations in water discharge, and easily eroded banks. The channels and braid bars are usually highly mobile, with the river layout often changing significantly during flood events.

Branchiopod: A group of primitive and primarily freshwater aquatic crustaceans, mostly resembling shrimps. They have flattened leaf-like abdominal appendages, to which the gills are attached. These appendages also have hairs for filtering food from the water, and in some groups are used to aid in swimming. Their fossil record extends from the Cambrian to the present day.

Bryophytes: A group of land plants that have tissues and enclosed reproductive systems but lack vascular tissue that circulates liquids. They do not have flowers or produce seeds; instead they reproduce via spores. The bryophytes consist of three groups: the liverworts (Marchantiophyta); the hornworts (Anthocerotophyta); and the mosses (Bryophyta). Bryophytes evolved in the Devonian and continue to the present day.

Bryozoans: Small, colonial, mostly shallow-water marine animals, superficially similar to corals. They generally build stony carbonate skeletons, although some species lack any calcification in the colony and instead have a mucilaginous structure. Colonies range from millimetres to metres in size, but the individuals (zooids) are usually less than 1 millimetre long and individuals assume different functions such as food gathering, structural strengthening, and cleaning the colony. They feed with a specialised, ciliated structure (lophophore), which is a crown of tentacles surrounding the mouth. The beating of the cilia drives water, together with entrained food particles (mainly phytoplankton), towards the mouth. Bryozoans are known from the Early Ordovician through to the present, and consist of about 8000 living species that generally inhabit warm, tropical waters worldwide.

Calcarenite: An indurated rock composed mainly of sand-sized detrital reworked calcium carbonate fragments. The term is commonly used for indurated coast-marginal foredunes formed by a mixture of calcareous shell fragments and quartz grains.

Calcareous: Describes sediment, sedimentary rock, soil type or biotic secretion, that is formed from, or contains a high proportion of, calcium carbonate in the form of calcite or aragonite (two polymorphs of the calcium carbonate mineral).

Calcification: The process of sedimentation and solidification of calcium carbonate into a mineralised, hardened form.

Calcimicrobes: Calcareous colonial microfossils, which include many morphologically dissimilar organisms. These organisms created the earliest recognisable reef systems by secreting calcium carbonate into a range of tubules, threads, chambered structures, and other solid forms. They are known from the Neoproterozoic and the Cambrian.

Calcimudstone: An indurated rock composed mainly of mud-sized (clay and silt) calcium carbonate fragments.

Calcirudites: An indurated rock composed mainly of calcium carbonate fragments larger than sand-sized (>2 mm).

Calcisiltite: An indurated rock composed mainly of silt-sized calcium carbonate fragments.

Calcium carbonate: A chemical compound with the equation $CaCO_3$. Typically forms the mineral calcite, but can also form aragonite and vaterite.

Calyx (pl. calices): The body part of a crinoid that contains the digestive and reproductive organs.

Cambrian explosion (Cambrian radiation): A major evolutionary diversification of organisms that appear in the fossil record around 530 million years ago. This includes the seemingly rapid appearance of most major groups of complex animals.

Cambrian: The first period of the Palaeozoic Era. Extends from 542.0 ± 1.0–488.3 ± 1.7 Ma.

Campanian: A late ISC Stage of the Late Cretaceous. Extends from 83.5 ± 0.7–70.6 ± 0.6 Ma.

Carbonaceous: Describes a material or substance characterised by being carbon-rich.

Carbonate: The carbonate ion is a polyatomic anion consisting of one central carbon atom surrounded by three identical oxygen atoms and has the empirical formula CO_3^{2-}. Mineral salts formed by combinations of carbonate ions and a wide variety of cations are commonly referred to as carbonate minerals. The most common is calcite or calcium carbonate, $CaCO_3$, the chief constituent of limestone.

Carboniferous: The penultimate period of the Palaeozoic Era. Extends from 359.2 ± 2.5–299.0 ± 0.8 Ma.

Cenozoic: The most recent era of the geological time scale consisting of the Palaeogene, Neogene and Quaternary Periods. Extends from 65.5 ± 0.3 Ma to the present day. May also be spelled 'Cainozoic'.

Cephalopods: A group of entirely marine molluscs with the foot modified to a ring of tentacles around the mouth. It commonly possesses a univalved aragonitic (a form of calcium carbonate) shell, usually containing chambers, which are gas-filled for buoyancy. The group includes modern squids, octopuses and argonauts, and fossil nautiloids (Cambrian to Holocene), ammonoids (Silurian to Cretaceous) and belemnites (Devonian to Cretaceous).

Cetacea: The order of marine mammals that includes whales, dolphins and porpoises.

Chaetetids: A small group of organisms originally known from Ordovician to Miocene fossils and presumed to be related to the tabulate anthozoan corals. However, recently discovered living representatives indicate a closer relationship to the sponges (phylum Porifera). Their rigid skeleton is calcareous and consists of a cluster of closely packed tubes with floors or tabulae. They are mostly warm, shallow-water organisms, often associated with photosynthetic algae, and apparently required hard substrates for settlement and the onset of growth. They contributed to the development of reefs during the Palaeozoic and Mesozoic, both overgrowing and serving as substrate for other reef-associated organisms.

Chalcedony: A cryptocrystalline (made up of minute crystals) form of silica, composed of very fine intergrowths of the silica minerals quartz (trigonal) and moganite (monoclinic). Chalcedony has a waxy lustre, may be semitransparent or translucent, and can assume a wide range of colours.

Chancellorides: An extinct family of animals common in Early Cambrian to Late Cambrian sediments. Fossils commonly consist only of spines and other fragments, presumed to belong to the same type of organism. Other specimens appear to be more complete and to represent sessile, bag-like organisms with a soft skin armoured with star-shaped, calcareous, hard body parts (sclerites) from which sharp spines radiate.

Chert: A fine-grained, microcrystalline, cryptocrystalline or microfibrous silica-rich opaque sedimentary rock, which can vary greatly in colour.

Chitinozoa: A group of flask-shaped millimetre-scale marine microfossils produced by animals of unknown affinities. They were common from the Ordovician through to the Devonian. Because of their wide distribution and rapid evolution, they are valuable biostratigraphic markers.

Chondrichthyans: Jawed fish with paired fins, paired nares, scales, two-chambered hearts, and skeletons made of cartilage rather than bone. They include the cartilaginous sharks, rays and skates and ghost sharks (chimaeras, which are sometimes separated into their own class). They range in age from at least the Early Silurian (perhaps Late Ordovician) to the present.

Clade: A taxonomic term used for a group of taxa consisting of a single common ancestor and all its descendants.

Cladogram: A branching diagram representing the most parsimonious distribution of derived characters within a set of taxa (or clade). The branching pattern of a cladogram is intended to show the relative relationships among taxa; it is not a true 'evolutionary tree' of how those relationships came to be, but a hypothesis of relationships.

Class: See Taxonomy.

Clast: A grain of sediment, silt, sand, gravel, etc. formed from the breakdown of pre-existing rocks, as distinguished from a chemical or biogenic component.

Clastic: A sedimentary rock composed of fragments, or clasts, of pre-existing rock, which are produced by weathering and erosion and transported to a site of deposition.

Clay pan: Ephemeral lacustrine features in low-relief arid lands that form where floodwater carrying fine sediment slows and deposits clay particles and then dries to a cracked clay surface. Clay pans usually occur as remnants of old drainage systems or as areas that receive localised runoff from surrounding sandy or rocky rises. They are similar to playas but are usually smaller and are not closed to groundwater, so do not accumulate salts as a surface crust.

Cleavage: In minerals, it refers to flat planes of breakage parallel to planes of weakness in the atomic lattice of the crystal structure. In rocks it refers to the capacity to split along certain parallel surfaces more easily than along others. It is a secondary structure produced by rock deformation and is generally better developed in fine-grained rocks.

Clubmoss: Small, creeping, terrestrial or epiphytic vascular plants, which lack flowers and reproduce sexually by spores, either clustered into small cones or borne in the axils of the small scale-like leaves. They often inhabit moist places, especially in tropical and subtropical forests, and were a dominant plant group in the Carboniferous period, when they attained the size of trees and contributed to the coal deposits then being formed.

Coleoptera: The order of insects that includes the beetles and weevils. They are generally characterised by a particularly hard exoskeleton and hard forewings (elytra) that are not used for flight but tend to cover the hind part of the body and protect the hind pair of wings (alae) used for flight. They extend from the Early Permian to the present day and are the insect group with the largest number of modern species. They occur in all non-polar terrestrial ecosystems.

Colonial organisms: A group of single-celled organisms living closely together, usually for mutual benefit. Individual organisms (cells) removed from a *colonial organism* can survive on their own, while cells separated from a multicellular life form cannot. Colonial organisms were probably the first evolutionary step towards multicellular organisms.

Comatulida: An order of the Echinoderm class Crinoidea. Comatulids (feather stars) include most living crinoids. They shed their stalk at an early growth stage and retain only the topmost stalk segment, which usually bears numerous hooks for anchoring the animal. Comatulids first appeared during the Jurassic and have radiated substantially since. They occur throughout most marine environments from the intertidal zone to abyssal depths.

Conchostracans: Branchiopod crustaceans with a bivalved carapace. The two valves are joined by a dorsal hinge and controlled by a strong adductor muscle. They are known from the Lower Devonian to the present and occur on in ephemeral waters on all continents except Antarctica.

Conglomerate: A rudaceous (having clasts larger than sand-sized) sedimentary rock with rounded or sub-rounded fragments (clasts), indicating a significant degree of transport prior to deposition.

Coniferales: The plant order that includes all extant conifers, although some fossil conifers belonged to other orders. It has been renamed Pinales.

Conifers: Profusely branching and chiefly evergreen trees with some shrubs. They have narrow or needle-like leaves, and are characterised by a reproductive structure known as a cone. They arose in the Carboniferous, diversified in the Triassic, and are well represented in the present.

Conodonts: Microscopic (0.5–2 mm), phosphatic, tooth-like structures, which ranged from the Cambrian to the Jurassic. Originally of unknown affinities, they were used for biostratigraphic division of the Devonian and the Carboniferous periods. Since the 1980s, conodonts are known to have formed a bilaterally symmetrical array of 15 or (more rarely) 19 elements, which constitute the feeding apparatus in the head of an extinct type of chordate resembling an eel.

Conulariids: Poorly understood fossil organisms preserved as shell-like structures. They are made up of rows of calcium phosphate rods, resembling an ice-cream cone, with four-fold symmetry. Soft tentacles protruded from the wider end of the cone, and a holdfast from the pointed end attached the organisms to a hard substrate. Their fossil record begins abruptly in the Middle Cambrian and extends without significant break through numerous major mass extinctions until they finally disappeared during the Early Triassic.

Coprolites: Fossilised excrement, faeces or droppings of ancient animals. They are classified as trace fossils rather than body fossils and provide evidence of animal behaviour (in this case, diet) rather than morphology.

Coquina: A limestone composed chiefly of reworked, sorted and cemented fossil debris, usually shells and shell fragments.

Cordaites: A genus of extinct gymnosperms, including large trees and shrub-like plants, which grew generally in wet conditions. They had characteristic, mostly long, strap-shaped leaves. Fossils of the group are known from the Late Carboniferous through to the Late Permian.

Corystosperms: A Mesozoic seed-fern order (Pteridospermatophyta) that forms a dominant component of most southern hemisphere Triassic palaeofloras. The widespread and highly variable corystosperm leaf genus *Dicroidium* has been used extensively in biostratigraphic, phylogenetic, and palaeoecological research. However, the group is poorly understood due to the paucity of fossils containing attached ovulate organs, pollen and leaves.

Cretaceous: The final period of the Mesozoic Era. Extends from 145.5 ± 4.0–65.5 ± 0.3 Ma.

Crinoid: Marine echinoderm animals. They are characterised by a mouth on the top surface that is surrounded by feeding arms, and a U-shaped gut with the anus located close to the mouth. Basic echinoderm five-fold symmetry can be recognised, but most crinoids have many more than five feeding arms. Crinoids live in shallow water to depths as great as 6000 metres. Fossils are known from the Ordovician through to the present, and crinoids are still alive today. However, their diversity was greatly reduced by the extinction event at the end of the Permian.

Cross-bedding: A series of inclined bedding planes in arenaceous sedimentary rocks that have some relationship to the direction of current flow, the angle of the rest of the sediment, and the rate of sediment supply. Also often called current bedding or cross stratification.

Cryptogams: Plants and plant-like organisms that lack flowers or are not reproduced by seeds, including ferns, mosses, fungi and algae. It is not a taxonomic classification.

Cuticle: Any tough but flexible non-mineral protective outer covering of an organism or part of an organism. Cuticles differ in their origin, structure, and chemical composition. Plant cuticles are a protective waxy covering produced only by the epidermal cells.

Cyanobacteria: A bacterial phylum, also known as blue-green algae, blue-green bacteria or Cyanophyta, that obtain their energy through photosynthesis. They are important primary producers in marine, freshwater and hypersaline inland lake environments, and are a major component of arid-zone biological soil crusts. The production of oxygen by cyanobacterial photosynthesis converted the Earth's early reducing atmosphere to an oxidising atmosphere, dramatically changing the composition of the planet and its life forms. Fossil cyanobacterial stromatolites are known from the Archaean to the present.

Cycadophyte: A group of evergreen, dioecious (having separate sexes), gymnosperm seed plants characterised by a large crown of compound pinnate leaves and a stout trunk. They resemble, but are not related to, palms and ferns. They arose in the Late Permian or perhaps the Late Carboniferous and continue to the present.

Cyclic sediments: Sequences of sedimentary rocks that are characterised by repetitive patterns of different rock types. They are also called rhythmic sediments or rhythmites.

Dendroidea: An order of graptolites with a many-branched form of growth and characterised by three types of cups (thecae) arranged in regular triads and connected by internal tubes (stolons). They existed from the Middle Cambrian to the Early Carboniferous.

Devonian: A middle period of the Palaeozoic Era. Extends from 416.0 ± 2.8–359.3 ± 2.5 Ma.

Diagenesis: Any chemical, physical, or biological change to sediment after its initial deposition and during and after its lithification. These changes, which alter the original mineralogy and texture of the rock, happen at relatively low temperatures and pressures. There is a gradational boundary between diagenesis and metamorphism, which occurs under conditions of higher temperature and pressure.

Dicotyledons: One of the two major groups of flowering plants (angiosperms), characterised by a seed with two seed leaves (cotyledons). The other group, monocotyledons, have a seed with one seed leaf.

Dinosaurs: An extinct group of archosaurian reptiles consisting of mostly large or extremely large carnivorous and herbivorous taxa. Two main groups of dinosaurs, the Saurischia and Ornithischia, are recognised on the basis of their pelvic structure. Dinosaurs lived from the Late Triassic to the end of the Cretaceous Period.

Diorama: A three-dimensional scenery display, common in museums, that often consists of models arranged in front of a painted background.

Diploid: Of an organism or cell, having two sets of chromosomes. A haploid organism or cell has one set of chromosomes.

Dipnoans: Freshwater fish belonging to the subclass Dipnoi, which includes lungfish, and characterised by the ability to breathe air, lobed fins, and a well developed internal skeleton. Their fossil record extends from the Devonian to the present. Modern lungfish live only in Africa, South America and Australia.

Diprotodontia ('two front teeth'): A large group of herbivorous graviportal marsupial mammals that includes kangaroos, wallabies, possums, koalas, wombats, and many others. Extinct diprotodontians include the rhinoceros-sized *Diprotodon*. All species in this group possess a pair of large, procumbent incisors on the lower jaw, and the second and third digits of the hind foot are fused by a sheath of skin (syndactyly). The earliest known fossils are from the Late Oligocene, but the group is likely to have evolved earlier, during the Palaeogene.

Disarticulation: A taphonomic condition of fossils in which the bones are found separated or not contacting each other as they would in life.

Dolerite: A medium-grained igneous rock that forms in the earth crust as dykes, sills and other intrusions. It has the same composition and mineralogy as basalt.

Dromornithidae: An extinct group of large, flightless birds from Australia. They are classified as a family of Anseriformes, with waterfowl such as ducks and geese being their closest living relatives. One Miocene species, *Dromornis stirtoni*, was 3 metres tall and may have been the largest bird that ever lived. Their known fossil record extends from the Oligocene to the Late Pleistocene. The last surviving species, *Genyornis newtoni*, was an element of the Australian Pleistocene megafauna, which became extinct soon after the arrival of humans.

Dropstones: Isolated fragments of rock found within finer grained water-deposited sedimentary rocks and ranging in size from small pebbles to boulders. They were transported by ice rather than normal water currents, and dropped vertically through the water column after melting out of the ice.

Dyke: A tabular body of intrusive igneous rock, crosscutting the host strata at a high angle.

Echinodermata: A large group (phylum) of radially symmetric, exclusively marine animals, commonly with spiny skins. Echinoderms are found at every ocean depth, from the intertidal zone to the abyssal zone. They can be extremely abundant and their lithified skeletons are major contributors to many limestone formations. They range in age from the Cambrian (or earlier) to the present.

Ediacaran: The last period of the Neoproterozoic Era of the Proterozoic Eon. Extends from 635–542 Ma. Ediacaran biota include the oldest definite multicellular organisms with tissues (cells differentiated according to function). It is the first geological period defined in the last 120 years and the first from an Australian type section (Ediacaran Hills, South Australia).

Edrioasteroids: An extinct class of echinoderms that had a main body (theca), which was composed of many small plates, circled by a peripheral rim for attachment, and (in some species) a pedunculate zone for extension and retraction. Five arms (ambulacra) radiated in a spiral pattern from the mouth in the centre of the body. Edrioasteroids were sessile and often attached to the substrate via a stalk. They lived from the Ediacaran to the Permian (if *Arkarua* was indeed an edrioaster).

Edrioblastoids: A group of small (approximately 2.0 cm), primitive, attached echinoderms of uncertain affinities, which have been grouped with blastoids, cystoids and edrioasteroids. They are only known from the Middle Ordovician of Canada and Australia.

Elytra: Hardened forewings of beetles (Coleoptera) that serve to protect the functional posterior wings.

Endemic: Unique to a particular geographic location (e.g. specific island, habitat type, nation, or other defined zone) and found nowhere else.

Entomofauna: An insect or terrestrial invertebrate fauna.

Eocene: The middle epoch of the Palaeogene Period of the Cenozoic Era. Extends from 55.8 ± 0.2–33.9 ± 0.1 Ma.

Eocrinoid: Sessile benthic suspension feeders that were the earliest known group of stalked, arm-bearing echinoderms. Early eocrinoids with a short holdfast and irregularly structured plates evolved to later forms with a fully developed stalk and regular rows of plates. Their fossil record ranges from the Early Cambrian to the Late Silurian and they were the most common echinoderms during the Cambrian.

Ephemeral: Lasting for a short period of time. Ephemeral streams and lakes hold water after rain but dry up during subsequent arid periods.

Erratic boulders: Boulders and other large rock fragments transported by glacial ice from their place of origin to an area where the bedrock is different. Similar to dropstones.

Eukaryote: An organism characterised by cells that are organised into complex structures (such as a nucleus), enclosed within membranes. Most living organisms, including all animals, plants, fungi, and protists, are eukaryotes. Bacteria and Archaea are prokaryotes and lack these features.

Eurypterids: An extinct group of arthropods related to arachnids, and which included the largest known arthropods. They were predators that thrived in warm shallow water, with near-global distribution from the Ordovician to the Permian. They went extinct during the Permo-Triassic extinction event.

Euthycarcinoid: An extinct group of amphibious freshwater arthropods that extended from the Cambrian to the Triassic mass extinction. They may have produced the earliest arthropod trackways on land.

Exuvia: The remains of a cast-off or moulted exoskeleton from an arthropod (insect, crustacean or arachnid).

Facies: The features of a sedimentary rock, such as lithology, mineral content, grain size, sedimentary structures, bedding characteristics, fossil content etc. that characterise it as having formed in a given environment.

Family: See Taxonomy.

Faulting: Earth movement that produces relative displacement of adjacent rock masses along a fracture.

Feldspar: A common silicate mineral that forms in igneous (volcanic) rocks.

Fenestellid: A group of extinct, fenestrated, cryptostomatous bryozoans with a large range of genera that were abundant during the Silurian.

Fenestrate bryozoans: Lace-like bryozoan colonies with individual bryozoan animals living in microscopic tubes or pores on the lace branches.

Ferns: A large group of flowerless and seedless vascular plants that reproduce from spores. They have true roots from a rhizome, and fronds that uncurl upward. Ferns range from the Middle Devonian to the present.

Filamentous organisms: Organisms that grow in a thread-like or filamentous form, and including Thiothrix, Actinomycetes, and Cyanobacteria (blue-green algae). Cyanobacteria can form a symbiotic relationship with some varieties of algae.

Fluoroapatite: A hexagonal mineral with the formula $Ca_5(PO_4)_3F$ (calcium halophosphate). Vertebrate bones and teeth are commonly rich in hydroxylapatite ($Ca_5(PO_4)_3OH$), which in fossils converts to fluoroapatite by diagenetic uptake of environmental fluorine. The relative amount of fluorine in bones in an assemblage is the basis of the fluorine relative dating technique.

Fluviatile: The processes associated with rivers and streams and the deposits and landforms created by them.

Foliar cuticles: Plant cuticles of the leaves or foliage.

Footprint positives: A natural cast (positive relief) of a footprint formed when a natural footprint imprint (impression, negative relief, or mould) is later filled in by accumulating sediments that are subsequently indurated.

Foraminifera: A large group of protozoan, benthic and

planktonic, mostly marine organisms. They have reticulating pseudopods, which are fine strands of cytoplasm that branch and merge to form a dynamic net. They typically produce a test, or shell, which can have either one or multiple chambers, often with elaborate structure. These shells are made of calcium carbonate or agglutinated sediment particles. They are usually less than 1 millimetre in size, but some are much larger. They range from at least the Cambrian to the present.

Fossil: The preserved remains or traces of animals, plants, and other organisms from the remote past.

Fossiliferous: Bearing or containing fossils.

Frasnian: An early ISC stage of the latest epoch of the Devonian Period. Extends from 385.3 ± 2.6–374.5 ± 2.5 Ma.

Ga: Billion years.

Gastropoda: A class of molluscs that typically have a one-piece (usually coiled) carbonate shell, a head bearing stalked eyes, and an unsegmented soft body that forms a flattened muscular foot. They live in terrestrial, marine and freshwater environments, and range in age from the Early Cambrian to the present.

Genus (pl. genera): See Taxonomy.

Geochronology: The determination of the numeric age of rocks, fossils, and sediments, within a certain degree of uncertainty inherent to the method used.

Geopetal structure: A marker that indicates an initial horizontal plane that can be used to determine the original orientation (up or down) of a rock unit and can therefore act as a precise measure of post-depositional tilting. They most commonly consist of a void or cavity in a rock that has been part-filled with sediments or mineral deposits.

Geothermometry: A group of techniques designed to determine the temperatures at which various geological processes took place.

Ginkophytes: A gymnosperm plant group that appeared in the Early Jurassic and continues to the present. *Ginkgo biloba* is the only extant representative.

Givetian: A late ISC Stage of the middle epoch of the Devonian Period. Extends from 391.8 ± 2.7–385.3 ± 2.6 Ma.

Glacial: The processes associated with glaciers and the deposits and landforms created by them.

Glendonites: Calcite pseudomorphs after ikaite (hexahydrate of calcium carbonate, $CaCO_3 \cdot 6H_2O$), associated with coldwater depositional systems, including glaciomarine and deepwater settings. Ikaite precipitation is favoured by elevated alkalinity and dissolved phosphate, conditions encountered commonly in association with organic-rich marine sediments where methane oxidation is occurring. Rapid recrystallisation of ikaite to calcite during slight warming or pressure release results in considerable solid volume loss, producing a highly porous crystal mesh.

Glosselytrodea: An extinct order of small insects.

Glossopterids: An extinct group of Gondwanan Pteridospermatophyta (seed-fern) plants that arose in the early Permian and were extinct by the end of the Permian. The best known genus is *Glossopteris*.

Gondwana: A southern supercontinent that separated from Laurasia 180–200 million years ago during the break up of the Pangaea supercontinent that had existed from about 500 Ma. Gondwana included most of the current southern hemisphere landmasses, including Antarctica, South America, Africa, Madagascar, Australia–New Guinea, and New Zealand, as well as Arabia and the Indian subcontinent, which have now drifted into the northern hemisphere.

Granule: A sedimentary particle or clast with a diameter ranging 2–4 millimetres in size.

Graptolites: Fossil colonial animals believed to be related to Hemichordates. Each colony (rhabdosome) has a variable number of branches (stipes), and each individual (zooid) was housed within a tubular or cup-like structure (theca), which varied in size within individuals. The number of branches and the arrangement of the thecae are important features in the identification of graptolite fossils. They are important index fossils for dating Palaeozoic rocks as they evolved rapidly and had a worldwide distribution. Their fossil record extends from the Late Cambrian to the Early Carboniferous.

Greensand: A marine sandstone rock with an olive-green colour derived from the presence of the mineral glauconite (an iron potassium silicate).

Greywacke: An impure grey or dark-coloured sandstone generally characterised by its hardness and by poorly sorted angular grains of quartz, feldspar, and small rock fragments set in a compact, clay-rich fine matrix. The formation of greywacke is attributed to submarine avalanches or strong turbidity currents.

Ground ferns: Pteridophyte (fern) habit reflects the environment and substrate, and three classes are commonly recognised: those that grow on the ground; on trees; or on water. Terrestrial (ground) ferns have erect or creeping stems and the leaves are held more or less upright, either vertical or spreading and arching. Erect stems are generally unbranched, radial with a more or less terminal rosette of fronds, and may be stout and woody, stout and fleshy, or arborescent.

Gymnosperm: A group of spermatophyte or seed-bearing plants with unenclosed ovules on scales, which are usually arranged in cone-like structures. They originated in the Late Carboniferous and continued to the present.

Haploid: Of an organism or cell, having one set of chromosomes. A diploid organism or cell has two sets of chromosomes.

Hardground: A surface or synsedimentarily cemented carbonate layer that has been exposed on the seafloor. A hardground is essentially a lithified seafloor.

Heterochrony: A developmental change in the timing of events, leading to changes in the size and shape of organisms. There are two main components: first, the onset and cessation of a particular process, and second, the rate at which the process operates.

Heteromorph ammonites: Ammonites with irregularly coiled shells, in contrast to homomorph ammonites, which have regularly coiled shells.

Heterostracan: An extinct class of jawless vertebrates that lived primarily in marine and estuary environments and differed from other Palaeozoic agnathan classes in both the arrangement of their scales and the structure of their scales. They had two plates that formed a large dorsal shield and a large ventral shield, and a series of scales arranged in various patterns on the sides of their bodies. Scale structure consisted of three layers, composed of dentine and aspidine and a middle layer honeycombed with small voids (cancella). They arose during the Ordovician, and finally became extinct during the extinction event at the end of the Devonian.

Holocene: The latest epoch of the Quaternary Period of the geological timescale. It extends from 11.7 ka (thousand years ago) to the present day.

Homoptera: An insect group having membranous forewings and hind wings; includes aphids, leafhoppers, cicadas, scale insects, psyllids, and whiteflies.

Horsetails: The Equisetaceae, a family of vascular plants that reproduce by spores rather than seeds and are characterised by jointed stems with a ring of long, pointed leaves and branches at each joint. Horsetails first appeared in the Devonian and dominated the understory of late Palaeozoic forests. *Equisetum* is the only type of horsetail alive today.

Hox genes: A group of related genes that specify the anterior–posterior axis and segment identity of metazoan organisms during early embryonic development. These genes are critical for the proper number and placement of embryonic segment structures (such as legs, antennae, and eyes).

Hydrothermal: Processes associated with heated or superheated water associated with igneous activity. Water at very high temperatures is highly reactive.

Hyoliths: Small (1–4 cm), extinct, enigmatic, benthic animals of indeterminate affinity with calcareous conical shells. Their shells have a cover (operculum) and two curved supports (helens) and are triangular or elliptical in cross section. Some species have rings or stripes. They are known from the early Cambrian to the end of the Permian.

Hypsilophodontid dinosaurs: Small, fast, herbivorous, biped, ornithopod dinosaurs, ranging in size from 1–2 metres. The group traditionally has included almost all bipedal bird-hipped dinosaurs but their phylogenetic relationships are uncertain. Their fossils are known from Middle Jurassic to Late Cretaceous rocks from Asia, Australia, Europe, North America and South America.

Ice-house conditions: The state of the Earth's climate that prevailed on a timescale of millions of years. During icehouse periods, glaciers are present in fluctuating amounts due variations in solar insulation caused by variations of the Earth's orbit. This may result in many ice ages, consisting of alternating glacials and interglacials. During a greenhouse Earth period, the planet's atmosphere contains sufficient greenhouse gases such as carbon dioxide and methane for ice to be entirely absent from the planet's surface.

Ichnofauna: Living organisms that change the depositional fabric of sedimentary rocks by burrowing in them, walking on them, or constructing edifices within the sediment. They are the animals responsible for producing trace fossils.

Ichnogenus: An ichnotaxon at the generic level.

Ichnotaxa: Names used to identify and distinguish morphologically distinctive trace fossils. The taxa are based on the fossilised traces of an organism and are assigned genus and species ranks much like organisms in Linnaean taxonomy. Ichnotaxa include trace fossils such as burrows, borings and etchings, tracks and trackways, coprolites, gastroliths, regurgitaliths, nests, leaf mines, and bite and gnaw structures, as well as secretions modified by organismal activity, such as cocoons, pupal cases, spiderwebs, embedment structures and plant galls.

Ichthyosaurs: Large marine reptiles that resembled fish and dolphins, with a porpoise-like head, a long-toothed snout, a dorsal fin and a tail. They arose from as-yet unidentified land reptiles in the Middle Triassic, were particularly abundant in the Jurassic Period, and became extinct in the Late Cretaceous, about 25 Ma before the dinosaurs became extinct.

Ilariidae: A poorly known extinct marsupial family from the Late Oligocene in South Australia and Late Oligocene–Early Miocene in Queensland. Their tooth structure is unique among Diprotodontians in having a complicated folding pattern.

Immature sandstone: Maturity of a sandstone is a measure of how nearly it approaches an ideal stable state, in terms of its mineral content, sorting and roundness of grains. An immature sandstone contains a high abundance of unstable mafic and/or feldspar minerals, is poorly sorted, and has angular grains. It is indicative of minimal chemical weathering and abrasive transport.

In situ: Describes materials found in their original position of formation, deposition or growth.

Index fossils: Taxa that are distinct, widespread and temporally restricted to a relatively narrow time range and are used to define geological periods and biochronological units.

Induration: A process by which soft sediment becomes hard through pressure and cementation; it may include baking by proximity to an igneous rock emplacement.

Infauna: Aquatic animals that live within the substrate of a body of water, especially in soft sandy or muddy sea bottom sediment.

Infructescence: The fruiting stage of a plant inflorescence (a group or cluster of flowers arranged on a stem).

Insects: A large group of arthropods characterised by having a hard exoskeleton, a three-part body (head, thorax and abdomen), three pairs of jointed legs, compound eyes, and two antennae. The insect fossil record extends from the Middle Devonian to the present, and they are the most diverse group of modern animals on the planet.

Intercalary lineage: A lineage inserted by interpolation between lineages in a phylogenetic scheme.

Intercalations: Interpolations or insertions into a sequence.

Interglacial: A geological interval of warmer global average temperature that separates cold glacial periods within an ice age. The current Holocene interglacial has persisted since about 11.7 ka.

Intraclast: A sedimentary particle (clast) formed by reworking of eroded materials formed within the same depositional environment.

Intrusion: A body of igneous rock that has forced itself into pre-existing rocks, either along an existing structure or by deforming, displacing or cross-cutting the existing rock.

ISC: International Commission on Stratigraphy.

Janjukian: An Australian stage of the Oligocene Epoch of the Palaeogene Period. It overlapped the Rupelian and Chattian ISC stages and extends from 30–27.5 Ma.

Jurassic: The middle period of the Mesozoic Era. Extends from 199.6 ± 0.6–145.5 ± 4.0 Ma.

ka: Geochronological abbreviation for thousands of years.

Kalimnan: A stage in the Late Neogene (Pliocene) of south-eastern Australia, underlain by the Cheltenhamian, overlain by the Yatalan, and roughly contemporaneous with the late Zanclian and early Piacenzian ISC stages. Extends from 4.3 to 3.4 Ma.

Karst: Landscape topography shaped by the dissolution of soluble, usually carbonate, bedrock, such as limestone or dolomite.

Kingdom: See Taxonomy.

Konservat-Lagerstätten: A *Lagerstätte* (German: literally 'place of storage'; plural *Lagerstätten*) is a sedimentary deposit that exhibits extraordinary fossil richness or completeness. *Konservat-Lagerstätten* ('conservation' *Lagerstätten*) are deposits known for the exceptional preservation of fossilised organisms, where the soft parts are preserved in the form of impressions or casts. This is caused by incompleteness of biological recycling, for example where anoxic conditions, as in oxygen-free mud, have suppressed common bacterial decomposition long enough for the initial casts of soft body parts to register.

Labile: Describes rocks and minerals that have unstable constituents which are easy to mechanically or chemically decompose.

Lacustrine: Of deposits or environments that pertain to lakes.

Lamprophyre: An uncommon, small-volume, potassium-rich igneous rock primarily occurring as small intrusions such as dykes, lopoliths, laccoliths, and stocks. They are alkaline, silica-undersaturated, ultramafic rocks with a primary mineralogy consisting of amphibole or biotite in a feldspar groundmass.

Leaf domatia: Small structures on the lower leaf surface in some woody dicotyledons, located in the axils of the primary veins and usually consisting of depressions partly enclosed by leaf tissue or hairs.

Lignin: A complex biopolymer, most commonly occurring in wood where it is an integral part of the secondary cell walls and provides strength and support. It is one of the most abundant organic polymers on Earth, exceeded only by cellulose, and constitutes 25% to 35% of the dry mass of wood.

Lingulids: A group of bivalved brachiopods with separate valves that do not articulate about a hinge, but possess a complex arrangement of muscles. They are semi-sessile, suspension-feeding, benthic infaunal marine animals. They range in age from the Early Cambrian to the present. Two living genera of this group originated in the Ordovician Period, when the group achieved maximum diversity.

Lithification: The process of formation of a hard rock from a soft sediment.

Lithostratigraphy: The organisation of rock bodies into distinctive, mappable units based on their lithology or rock type, used to establish their distribution and relationship in space, their succession in time, and to interpret their geologic history.

Longfordian: An Australian Stage of the Late Oligocene and the Early Miocene, which followed the Janjukian and preceded the Batesfordian stages and extends from 27.5–16.5 Ma. It overlaps the Chattian, Aquitanian and Burdigalian ISC stages.

Lutite: A rock consisting of clay- and silt-sized particles.

Lycophytes: A group of primitive vascular plants that reproduce by shedding spores. They differ from all other vascular plants in having microphylls, leaves that have only a single vascular trace (vein) rather than the much more complex megaphylls found in ferns and seed plants. Their fossil record extends from the Silurian to the present.

Ma: Geochronological abbreviation for millions of years.

Maastrichtian: The last stage of the Cretaceous Period. Extends from 70.6 ± 0.6–65.5 ± 0.3 Ma.

Macropodoidea: Marsupial group (superfamily) that includes kangaroos, wallabies, tree-kangaroos, pademelons, bettongs, and the musky rat-kangaroo. The first three categories are usually grouped as the family Macropodidae (macropods). They vary in size and most have very large hind legs and a long, powerfully muscled tail, with locomotion achieved by energy-efficient hopping. Macropodoids are browsing and grazing herbivores and are equipped with appropriately specialised teeth for cropping and grinding up fibrous plants. Their fossil record extends from the Late Oligocene to the present.

Magnetostratigraphy: The organisation of rock bodies into distinctive mappable units based on their magnetic polarity signature (normal or reversed), used to establish their distribution and relationship in space and their succession in time, and to interpret geologic history.

Malacostracans: The largest group of crustaceans, including crabs, lobsters and shrimp (decapods); mantis shrimp (stomatopods); krill (euphausiids); amphipods; and the terrestrial wood-lice (isopods). Their fossil record extends from the Cambrian to the present.

Marl: A calcium carbonate or lime-rich mud or mudstone, which contains variable amounts of clay. Often used to describe indurated marine deposits and lacustrine sediments.

Mass kill: Any sudden, catastrophic mass mortality event in animal populations. Mass-kill events are most commonly seen in fish populations, where they are caused by abrupt changes in water body properties such as salinity, temperature and dissolved oxygen.

Megafauna: Animals roughly larger than 5 kilograms, including mammals, reptiles and birds. The term is specifically associated with large animals that existed on all continents during the Pleistocene and mostly became extinct in the late Pleistocene in Australia, North America, South America, Madagascar and New Zealand after the arrival of humans. Megafaunal extinctions were less general in Eurasia and minimal in Africa.

Megalodontid bivalves: An extinct group of medium to large heterodont bivalves with a fossil record that extends from the Ordovician to the Jurassic.

Mesic: Characterised by intermediate moisture conditions, neither very wet (hydric) nor very dry (xeric).

Mesozoic: An era of the Phanerozoic Eon. Extends from 251.0 ± 0.4–65.5 ± 0.3 Ma.

Metamorphism: The solid-state recrystallisation of pre-existing rocks, due to heat, pressure, and the introduction of chemically active fluids. Mineralogical, chemical and crystallographic changes can occur during this process.

Metazoans: Multicellular animals having cells differentiated into tissues and organs and usually a digestive cavity and nervous system. The group includes all animals except protozoans and sponges.

Microfossils: Small fossils (generally <1–4 mm), which require the use of microscopy for study.

Miocene: The earliest epoch of the Neogene Period of the geological time scale. Extends from 23.03–5.332 Ma.

Molluscs: A very large group of soft-bodied invertebrates that typically have a hard shell of one or more pieces. They occur in a very wide range of marine, freshwater, and terrestrial environments. Their fossil record extends from at least the Early Cambrian to the present.

Monocotyledon: One of the two major groups of flowering plants (angiosperms), characterised by a seed with one seed leaf (cotyledon). The other group, dicotyledons, have a seed with two seed leaves.

Monoplacophorans: A poorly known mollusc group thought to be extinct (Cambrian to Devonian) until a living specimen was collected in 1952 from the deep ocean.

Monotremes: A primitive mammal group that lays eggs instead of giving birth to live young like all other marsupial and placental mammals. Modern representatives (platypus and echidna) are restricted to Australia and New Guinea, although their fossil record extends from the Early Cretaceous, suggesting a formerly wide distribution.

Monotypic: A taxonomic group with only one type or species.

Mosasaurs: Snake-like predatory marine reptiles considered to be the closest relatives of snakes. They appeared in the Early Cretaceous and became the dominant marine predators during the last 20 Ma of the Cretaceous, following extinction of the ichthyosaurs and pliosaurs.

Moulds (internal, external): A mould fossil is essentially a hardened impression of an organism made in the sediments. An external mould is an imprint of the outside of an organism and an internal mould is an imprint of the inside of an organism.

Mucilaginous: Thick, viscose, and sticky.

Myriapod: A terrestrial arthropod group characterised by an elongated body composed of many similar segments. Includes centipedes and millipedes.

Nautiloids: A group of cephalopod molluscs with an external, chambered shell that is either straight or coiled. They are ancestral to all cephalopods and their fossils are known from the Cambrian to the present, with their former diversity (Ordovician to Devonian) reduced to only a single living genus today.

Nektic: Non-benthic aquatic organisms that are free-swimming rather than floating (planktic).

Neogene: The middle period of the Cenozoic Era. Extends from 23.03–2.59 Ma.

Neoproterozoic: The last era of the Proterozoic Eon. Extends from 1000–542 Ma.

Neoptera: The animal group that includes most winged insects, characterised by the ability to flex their wings over their abdomens.

Niche: The role or functional position of a species within the community of an ecosystem. How an organism or population responds to the distribution of resources and competitors.

Oligocene: The latest epoch of the Palaeogene Period. Extends from 33.9 ± 0.1–23.03 Ma.

Opalised: Replacement by or conversion to opal, a hydrated amorphous form of silica.

Order: see Taxonomy.

Ordovician: An early period of the Palaeozoic Era. Extends from 488.3 ± 1.7–443.7 ± 1.5 Ma.

Organosedimentary: Sedimentary rocks that form from the accumulation and lithification of organic debris, such as leaves, roots, and other plant or animal material including microbialites.

Ornithopods: A group of bird-hipped dinosaurs that

started out as small, bipedal running grazers, and grew in size and numbers and became highly successful during the Cretaceous Period. Their fossil record extends from the early Jurassic until the end of the Cretaceous.

Orogeny: An episode of mountain building by large-scale tectonic activity associated with intense deformation of the rocks in a region. Generally accompanied by metamorphism and intrusive igneous activity.

Orthoceratite nautiloids: A group of extinct nautiloid cephalopods with slender, elongate shells. Their fossil record extends from the Ordovician to the Triassic.

Orthoptera: An insect group characterised by leathery forewings, membranous hind wings, and chewing mouthparts. Includes locusts, cockroaches, crickets, and grasshoppers. Their fossil record extends from the Carboniferous to the present.

Osteichthyan: Bony fish, a group that includes the ray-finned fish (Actinopterygii) and lobe-finned fish (Sarcopterygii). Their fossil record extends from the Early Devonian to the present.

Osteolepidae: An extinct family of lobe-finned fish from the Devonian. *Osteolepis* was about 20 centimetres long, and covered with large, square scales.

Osteolepiforms: A group of extinct sarcopterygians, including the Devonian genus *Osteolepis*, which was characterised by a long streamlined body, large median fins, and an epicercal tail (having a longer upper lobe). Osteolepiforms survived the major extinction event at the end of the Devonian, but became extinct at the beginning of the Permian.

Ostracods: A very diverse group of small crustaceans, mostly only a few millimetres long, occurring in salt and fresh water. Enclosed by a usually calcareous bivalve carapace, their fossil record extends from the Early Cambrian to the present.

Overlying: Of sedimentary strata, stratigraphically higher than a particular stratum.

Ovules: In seed plants, the structure that gives rise to and contains the female reproductive cells.

Palaeocene: The lower epoch of the Palaeogene Period. Extends from 65.5 ± 0.3–55.8 ± 0.2 Ma.

Palaeoclimate: Refers to past states of the Earth's climate at the temporal scale of the entire history of the planet (thousands to billions of years). It uses proxy records from ice sheets, tree rings, pollen records of vegetation associations, fossil records of faunal associations, sediments and rocks, and trace element and isotope geochemistry etc.

Palaeoecology: The reconstruction of past ecosystems from the study of fossil organisms, their traces, life cycle, living interactions, natural environment, and manner of death and burial.

Palaeoenvironment: The physical and ecological characteristics of an area in the geologic past (including the climate, vegetation, air, water, geology, geomorphology, organisms, etc.).

Palaeoflora: The plants of a specific region in the geologic past.

Palaeogene: The earliest period of the Cenozoic Era. Extends from 65.5 ± 0.3–23.03 Ma.

Palaeomagnetism: The study of the record of the Earth's magnetic field through geologic time. The record preserved in various magnetic minerals demonstrates that substantial variations in the polarity and strength of the magnetic field have occurred through past geological times.

Palaeoniscoid: A diverse group of generally small, ray-finned fish (actinopterygians), with scales covered by a shiny material (ganoine). Palaeoniscoids superficially resemble the bony fish of today and ranged from elongated eel-like forms to compressed forms resembling living angelfish. They were the dominant ray-finned fishes from the Carboniferous to the Triassic, and were extinct by the end of the Mesozoic.

Palaeozoic: The earliest era of the Phanerozoic Eon. Extends from 542.0 ± 1.0–251.0 ± 0.4 Ma.

Pandemic species: A pandemic species lives in all places in a specific geographic region, as opposed to an endemic species which lives in only one place. Cosmopolitan species inhabit the whole world, though not necessarily in all places.

Paraneoptera: An abundant and diverse group of insects that includes true bugs, hoppers, cicadas, aphids, scale insects, thrips, lice, and others. Many of these insects have beak-like mouthparts specialised for piercing and/or sucking.

Patch reef: A small, isolated, often circular or oval reef. Usually develops in calm, protected waters behind barrier reefs or within a lagoon or embayment.

Pectenoids: A group of bivalve molluscs including the modern scallop that occurs worldwide on banks of clean, firm, near-coastal sands in waters up to 100 metres deep. They can be attached by means of a secreted filament, or are simply recumbent and can be propelled through the water by their valves. Their fossil record extends from the Triassic to the present.

Permian: The last period of the Palaeozoic Era. Extends from 299.0 ± 0.8–251.0 ± 0.4 Ma.

Permineralisation: A type of fossilisation involving deposits of minerals within the pores or spaces of the organic tissues of organisms, forming a mineral cast. Permineralisation is different from petrification in that the organic material is only filled with minerals and not completely replaced.

Petrification: A type of fossilisation involving complete replacement of the organic remains by new minerals. Despite the complete replacement, there can be excellent preservation of micro-structures when the process is gradual and fine-scale. The term is mostly used with reference to silicification of wood.

Phyllolepids: A group of flattened, freshwater, Devonian placoderms that were bottom-dwelling ambush predators found throughout the world. They were sightless, possessed armour of whole plates, and had extraordinarily wide mouths.

Phylogeny: The evolutionary relationships among a group of organisms. It effectively details evolution through time and expresses the ancestor–descendant relationships between non-contemporaneous populations and the common ancestry of distinct contemporaneous populations.

Phylum: see Taxonomy.

Phytogeography: The study of all aspects of the geographic distribution of plants, from the controls on the distribution of individual species ranges to factors that govern the composition of entire communities and floras.

Placoderm: Early jawed fishes that first appeared in the Early Silurian and diversified dramatically during the Devonian, but became extinct at the end of that period. About 200 genera of placoderms evolved during this interval, with the vast majority restricted to the Devonian. Placoderms (= 'plate skin') are armoured by an extensive dermal skeleton of bones (or plates) that form head and thoracic shields that are either articulated by distinctive joints or fused into a single unit. Pectoral fins are typically well developed. True teeth are absent and substituted by bony shearing or crushing structures on the jaws. The jaw joint is simple. Most placoderms are known only from plates of the head or thoracic shields, but some exceptionally preserved specimens reveal heterocercal (shark-like) tails, a pair of pelvic fins, and single dorsal fins.

Platform carbonates: Carbonate precipitation in the form of autochthonous biogenic skeletal precipitates or microbially mediated precipitates that produce a sedimentary body with topographic relief referred to as a carbonate platform.

Pleistocene: The first epoch of the Quaternary Period. Extends from 2.59 Ma–11.7 ka.

Plesiosaurs: Large, carnivorous, aquatic (mostly marine) reptiles that were characterised by having a long neck, in contrast to the short-necked pliosaurs, a related group. The plesiosaur fossil record extends from the Late Jurassic to the end of the Cretaceous.

Pliocene: The last epoch of the Neogene Period. Extends from 5.33–2.59 Ma.

Pliosaur: Large (4–15 metres), carnivorous marine reptiles that were characterised by having a short neck and an elongated head, in contrast to the long-necked plesiosaurs, a related group. The pliosaur fossil record extends from the Late Triassic to the Late Cretaceous.

Podocarp: A large group of mainly southern hemisphere conifers, considered to be of Gondwanan origin.

Pollen: Small, light, dry protein particles that contain the male gametes and are produced in the stamens of flowering seed plants. They can be preserved for geological periods of time under certain conditions and have taxon-specific morphology.

Polymerida: The most common (95%) and familiar group of trilobites which appeared in the Early Cambrian and went extinct at the end of the Permian. Unlike the agnostids, polymerid trilobites typically have more than two or three thoracic segments, the tail (pygidium) is usually smaller than the head (cephalon), and most had remarkably well developed compound eyes.

Polymict: A term for sedimentary rocks that have particles or clasts of many different materials.

Polyneoptera: An insect group including the cockroaches.

Polyplacophorans: A small group of entirely marine molluscs, including the chitons, that are highly adapted to life in the rocky intertidal zone. Chiton shells are divided into eight transverse overlapping shell plates or valves. The polyplacophoran fossil record extends from the Ordovician to the present.

Pre-Cambrian: The portion of the geological time scale which precedes the Cambrian (i.e. older than 542.0 ± 1.0 Ma). It includes the Proterozoic and Archaean Eons, which extend back to 4000 Ma, as well as the informal Hadean time period which precedes the Archaean.

Procolophonids: A group of small reptiles that were shaped like stocky lizards with broad-cheeked skulls and a stout backward-pointing spike(s) on their cheeks. Their fossil record extends from the Middle Permian to the Late Triassic.

Protactinoceratids: An early, primitive group of Cambrian cephalopods characterised by small, straight or gently curved, laterally compressed shells, with close-spaced septa, and transverse partitions present in the siphuncle.

Proterosuchids: An early group of basal archosauriform reptiles that were slender, medium-sized (approximately 1.5 metres), long-snouted and superficially crocodile-like animals. However, they lacked the armoured scutes of true crocodiles, and their skeletal features are much more primitive. Their fossil record is known to extend from the Late Permian to the Early Triassic.

Proterozoic: The latest eon of the Pre-Cambrian part of the geological time scale. It extends from 2500–542.0 ± 1.0 Ma.

Protozoans: Micro-organisms classified as unicellular eukaryotes. The term is not strictly taxonomic and is mostly used to refer to a unicellular heterotrophic protist, such as an amoeba or a ciliate.

Pseudomorph: One mineral occurring in the crystal morphology of another due to replacement.

Pteridophytes: Vascular plants that reproduce and disperse via spores rather than by producing either flowers or seeds.

Pteridosperms: A very heterogeneous group of extinct plants, also called seed-ferns, with mostly fern-like foliage but with real seeds. They appear to have been predominantly small trees, but forms with a climbing growth habit have also been found. Pteridosperms evolved in the latest Devonian and survived until the Eocene.

Pterosaurs: A group of flying reptiles with wings formed by a membrane of skin, muscle, and other tissues stretching from the legs to a dramatically lengthened fourth finger. Pterosaurs are the earliest vertebrates known to have evolved powered flight. Their fossil record extends from the late Triassic to the end of the Cretaceous.

Quaternary: The last period of the Cenozoic Era. Extends from 2.59 Ma to the present.

Regression: A geological process whereby areas of submerged seafloor are exposed above sea level. The opposite event is transgression, whereby flooding from the sea covers previously exposed land.

Relictual: Refers to surviving remnants from former time periods. Relictual taxa are organisms that at an earlier time were abundant and widespread but now occur more sparsely.

Remanié: Organic skeletal fossils that are reworked and accumulate over a period of time, with a paucity of other sediment, before eventual burial and preservation.

Replacement: A type of fossilisation involving complete replacement of the organic remains by new minerals. Despite the complete replacement, there can be excellent preservation of micro-structures when the process is gradual and fine scale.

Rhizodontidae: An extinct family of often large (up to 7 metres), freshwater, predatory lobe-finned fishes. Their fossil record is known to extend from the late Middle Devonian to the Late Carboniferous.

Rhyniophytes: A group of Early Devonian vascular plants.

Rostroconchia: An extinct class of sedentary, semi-infaunal molluscs characterised by a single pseudo-bivalved shell. Their fossil record extends from the Early Cambrian to the Late Permian.

Rugose corals: An extinct coral group that was characterised by a unique horn-shaped chamber made of calcite with a wrinkled (rugose) wall. There were solitary and colonial forms, and they were benthic and reef forming. Their fossil record extends from the Early Ordovician to the Late Permian.

Sakmarian: An early middle stage of the Cisuralian (early) Epoch of the Permian Period. Extends from 294.6 ± 0.8–284.4 ± 0.7 Ma.

Sarcopterygians: Lobe-finned fishes; a group of bony fish with fleshy, lobed-paired fins, joined to the body by a single bone, which evolved into legs of the first tetrapod land vertebrates. They also possess two dorsal fins with separate bases, as opposed to the single dorsal fin of actinopterygians (ray-finned fishes). Their fossil record extends from the Late Silurian to the present.

Sauropods: A group of herbivorous, lizard-hipped (saurischian) dinosaurs notable for the enormous sizes attained by some species. The group includes many of the largest animals to have ever lived on land. Their fossil record extends from the Early Jurassic to the Late Cretaceous.

Scaphopods: A group of small to medium (up to 15 cm) marine molluscs that live on soft offshore substrates. They usually have whitish, aragonitic shells that are conical, curved (resembling miniature elephant's tusks), hollow, and open at both ends. Their fossil record extends from the Late Ordovician to the present.

Scleromorphy: The morphological adaptations of sclerophyll plants.

Sclerophyll: A type of vegetation with hard leaves and short internodes (the distance between leaves along the stem), that occurs worldwide. Sclerophyll plants are drought-tolerant and adapted to low nutrients and frequent burning. Many sclerophyllous woodland plants also produce leaves unpalatable to herbivores because they include toxic and indigestible compounds.

Scorpionids: Predatory arachnid arthropod animals with a two-part body, four pairs of legs, one pair of which (palps) are modified into pincer-like appendages used for grasping and defence, and an abdomen that tapers into a tail with a sting containing a poison gland. Their fossil record extends from the Early Silurian to the present.

Scutes: Chitinous or bony dermal plates or scales, such as the shell of a turtle, the skin of crocodiles, or the feet of some birds.

Sedimentary: Refers to basins, environments, processes, rocks etc. where eroded and weathered products of pre-existing rocks are transported and deposited.

Seed ferns: The pteridosperms are a very heterogeneous group of extinct plants with mostly fern-like foliage but real seeds. They usually occurred as small trees but forms with a climbing growth habit have also been found. Pteridosperms evolved in the latest Devonian and survived until the Eocene.

Semi-articulated: In fossil taphonomy, articulation refers to separate parts of an individual animal that are preserved in their correct anatomical position, such as two vertebrate bones or two bivalve valves. Full articulation refers to a whole fossil animal with all individual components in growth position. Semi-articulation usually refers to articulated portions of a whole animal, but some usage refers to parts of an animal, inferred to be from the same individual, that are preserved in close proximity to an articulated position. It can also be described as 'associated'.

Semi-deciduous: Deciduous plants lose their foliage seasonally, usually during cold winters. Semi-deciduous plants lose their foliage for a very short period, when old leaves fall off and new foliage growth is starting. This phenomenon occurs in tropical and subtropical woody species during a monsoonal dry season.

Sessile: Immobile, and regularly attached to a fixed substrate.

Shale: A fine-grained sedimentary rock consisting of clay- and silt-sized grains and abundant clay minerals. Shale is characterised by thin laminae and breakage often parallel to the bedding plane (fissility). Shale is the most common sedimentary rock.

Siliceous: Characterised by having silica (SiO_2) as the principal constituent.

Siliciclastic: Clastic sedimentary rocks that are almost exclusively silica-bearing, as either quartz or other silicate minerals. Siliciclastic rocks are deposited through some mechanical process and subsequently lithified.

Silicification: Secondary replacement or impregnation of original material with silica (SiO_2).

Silurian: A middle period of the Palaeozoic Era. Extends from 443.7 ± 1.5–46.0 ± 2.8 Ma.

Species: See Taxonomy.

Spicules: Small spines of silica (SiO_2) that form the skeletons of many sponges. In some species, it provides some defence against predators.

Spiriferidae: A family of extinct, articulate brachiopod fossils, known for their long hinge-line. The latter is often the widest part of the shell, and they often have a deep fold down the centre of the shell. A thin, tightly coiled ribbon of calcite (brachidium) within the shell is the internal support for the lophophore (the organ surrounding the mouth) and gives the spiriferids their name. They first appeared in the Early Ordovician, radiated dramatically during the Devonian, and finally became extinct during the Early Jurassic.

Sponges: Sessile, aquatic, mostly marine animals of the phylum Porifera. Their bodies consist of jelly-like material sandwiched between two thin layers of cells, and lack nervous, digestive or circulatory systems. They maintain a constant water flow through their bodies to obtain food and oxygen and to remove wastes, and the shapes of their bodies are adapted to maximise the efficiency of the water flow. Their fossil record extends from the Ediacaran to the present.

Spore: A reproductive structure that is adapted for dispersal and survival for extended periods of time in unfavourable conditions. Spores form part of the life cycles of many bacteria, plants, algae, fungi, and some protozoans. A chief difference between spores and seeds as dispersal units is that spores have very little stored food resources compared with seeds. Spores are usually haploid and unicellular, and are produced by meiosis in the sporangium by the sporophyte. Once conditions are favourable, the spore can develop into a new organism using mitotic division, producing a multicellular gametophyte, which eventually goes on to produce gametes.

Stegosaurs: A group of herbivorous, mostly northern hemisphere dinosaurs characterised by long, narrow heads with a horn-covered beak (rhamphotheca) covering the front of the upper and lower jaws, fore limbs much shorter than their hind limbs, and rows of special bones (osteoderms) that develop into plates and spines along the back and tail. Their fossil record extends from the Middle Jurassic to the Late Cretaceous periods.

Steinkern: The fossilised outline of a hollow organic structure, such as a skull or a mollusc shell, formed when mud or sediment consolidates within the structure and the structure itself later disintegrates or dissolves. Also called an endocast or an internal cast.

Stenoporidae: A family of bryozoan animals.

Sthenurines: An extinct group of robust, short-faced kangaroos that appeared in the late Miocene of central Australia and became most diverse (at least 14 species) in the Pleistocene faunas of southern and eastern Australia. All had broad faces, a single long toe with a hoof-like claw on their hind foot, a short, thick tail, and long arms. The largest was *Procoptodon goliah*, which stood 2.5 metres high and weighed up to 200 kilograms, the largest kangaroo known. They were an element of the Australian megafauna.

Stomata: Respiratory openings on plant leaves.

Stratified: Descriptive of rocks or sequences characterised by layering or bedding (strata).

Stratiform: Formed in layers. Materials that are distributed in a manner that resembles particles in a sedimentary rock.

Stratigraphy: The study of stratified rocks, their sequence in time, the character of the rocks, and the correlation of beds in different localities.

Stratotype section: An actual rock succession, chosen at a particular locality (type locality) to act as the standard for comparison with all other related chronostratigraphic or lithostratigraphic units. A Global Boundary Stratotype Section and Point (GSSP) is an internationally agreed upon stratigraphic section that serves as the reference section for a particular boundary on the geologic time scale.

Strike: Refers to the orientation of a bed, fault, or other planar geological feature and is a line representing the intersection of that feature with a horizontal plane.

Stromatolites: A type of microbialite, they are layered accretionary structures formed in shallow water by the trapping, binding and cementation of sedimentary grains by biofilms of micro-organisms, especially cyanobacteria (blue-green algae). They include some of the most ancient records of life on Earth, extending from the Archaean to the present, but are rare in modern environments and restricted to areas where particular conditions (such as high salinity) exclude grazers.

Stromatoporoids: An extinct group of reef-forming, colonial, aquatic invertebrates related to sponges, with spherical, branching or encrusting calcareous skeletons. Groups of fossil stromatoporoids lived the Palaeozoic (Ordovician to Early Carboniferous) and the Mesozoic, and were dominant reef builders in both time periods.

Successions: A group of rock units or strata that succeed one another in chronological order and that are readily seen in exposed geological sections.

Synecology: The branch of ecology that deals with the structure and development of entire ecological communities and the interrelationships of the plants and animals within them.

Tabulate corals: An extinct group of mostly colonial corals characterised by individual hexagonal corallites defined by a calcite skeleton, similar in appearance to a honeycomb. Their fossil record extends from the Early Ordovician to the Late Permian.

Taphonomy: The study of the conditions and processes by which organisms become fossilised.

Taxodiaceae: A group (family) of coniferous trees.

Taxonomy (taxon, taxa): The practice and science

of classification. Biological classification (Linnaean taxonomy) uses a hierarchy of taxonomic ranks, including among others: kingdom (broadest), phylum, class, order, family, genus, and species (narrowest).

Tectonics: A field of geological study concerned with rock structures and particularly with the forces and movements that have operated to create those structures.

Teleosts: A group of ray-finned fishes (Actinopterygii) characterised by a movable maxilla and premaxilla that allow the teleosts to protrude their jaws outwards from the mouth. Their fossil record extends from the Early Triassic to the present.

Temnospondyl amphibians: An extremely diverse group of small to giant primitive amphibians that adapted to a wide range of habitats worldwide, including freshwater, aquatic, semi-aquatic, amphibious, terrestrial and near-shore marine. They arose in the Early Carboniferous, flourished in the Carboniferous, Permian, and Triassic, and became extinct in Middle Cretaceous.

Terebratulids: A group of articulate brachiopods characterised by a biconvex, ovoid to circular, short hinge-lined shell that can be either smooth or have radial ribbing. The support for the lophophore (organ surrounding the mouth) is loop-shaped. There is a circular pedicle opening located in the beak. Their fossil record extends from the Early or Middle Silurian to the present.

Tertiary: A portion of geologic time extending from the end of the Cretaceous Period to the beginning of the Quaternary Period (65.5 ± 0.3–2.59 Ma). Tertiary is no longer included in the International Stratigraphic Chart (ISC) and is replaced by the Palaeogene and Neogene.

Tetrapods: Vertebrate animals with four feet, legs or leg-like appendages, and including amphibians, reptiles, dinosaurs, birds and mammals. The limbless snakes are tetrapods by descent. The earliest tetrapods radiated from the lobe-finned fish (Sarcopterygii). Their fossil record extends from the Late Devonian to the present.

Thanetian: The latest stage of the Palaeocene Epoch of the Palaeogene Period. Extends from 58.7 ± 0.2–55.8 ± 0.2 Ma.

Theropoda: A group of mostly carnivorous, bipedal, saurischian dinosaurs including *Tyrannosaurus* and other extremely large predators. A number of theropods evolved herbivory during the Cretaceous Period. Theropods first appeared during the Late Triassic and were the sole large terrestrial carnivores from the Early Jurassic until their extinction at the close of the Cretaceous. Birds evolved in the Late Jurassic from therapod ancestors.

Thrust fault: A low-angle reverse fault, or break in the Earth's crust with relative movement of the rocks. Thrusts push older lower strata over younger, higher strata.

Tibiotarsus: A large bone in the leg of a bird formed by the fusion of the proximal part of the tarsus with the tibia.

Tillite: A sedimentary rock formed by lithified till, which is unsorted heterogeneous sediment deposited directly by a glacier.

Tommotiids: Small, Cambrian, calcium phosphate shelly fossils of uncertain affinity to a higher taxonomic group. They possess a wide range of sclerite shapes and internal structures and are believed to be a sessile, tube-like animal made up of a spiral of overlapping plates.

Toothed baleen whales: Baleen whales (mysticetes), also called whalebone whales or great whales, are characterised by having baleen plates (whalebone) for filtering food from water. This distinguishes them from the toothed whale group, which also includes dolphins and porpoises (odontocetes). The earliest fossil baleen whales had teeth before baleen evolved.

Trace fossils: Also called ichnofossils, trace fossils include burrows, borings and etchings, tracks and trackways, coprolites, gastroliths, regurgitaliths, nests, leaf mines, and bite and gnaw structures. They also include secretions modified by organismal activity, such as cocoons, pupal cases, spiderwebs, embedment structures and plant galls.

Trackway: An animal travel route indicated by a set of footprint impressions, originally made in the soft earth and later indurated to form a fossil trackway.

Transgression: A geological process whereby flooding from the sea covers previously exposed land. The opposite event is regression, whereby areas of submerged sea floor are exposed above the sea level.

Travertine: A sedimentary rock formed by the chemical precipitation of carbonate minerals from mineral springs or streams of carbonate-saturated water.

Tree ferns: A group of relatively primitive ferns generally characterised by ascending trunk-like stems and a tree-like (arborescent) habit.

Triassic: The earliest period of the Mesozoic Era. Extends from 251.0 ± 0.4–199.6 ± 2.0 Ma.

Trichomes: Fine, hair-like outgrowths or appendages on plants and certain protists that are of diverse structure and function. Examples are hairs, glandular hairs, scales and papillae.

Trilobites: Well known group of extinct marine arthropods characterised by a generally sub-elliptical, dorsal, chitinous exoskeleton divided longitudinally into three distinct lobes. They have a distinct, relatively large head shield (cephalon), a thorax comprising articulated transverse segments, and a tail (pygidium) consisting of the fused hindmost segments. Trilobites evolved into a range of trophic niches including benthic predators, scavengers or filter feeders, and some free-swimming (nektic) plankton feeders. Trilobite fossils first appeared during the Early Cambrian and flourished throughout the Early Palaeozoic before most became extinct during the Devonian. Final extinction occurred at the end of the Permian.

Tristichopteridae: A diverse and successful family of Late Devonian tetrapodomorph fishes that share skull shape and reduced posterior fins with elpistostegalian fishes, which lie close to the origin of tetrapods. Tristichopterids range in size from a few tens of centimetres to several metres.

Tuffaceous: Of the nature or texture of tuff, a rock composed of compacted volcanic ash that varies in size from fine sand to coarse gravel.

Type locality: Typical or representative location commonly used in geology for formations, structures, rock types, minerals and fossils. Also refers to the location where biological type specimens are found.

Vascular: Relating to the circulatory system of animals or plants.

Venation: The distribution or arrangement of a system of veins in an organism structure such as a leaf blade or the wing of an insect.

Vermes: An obsolete taxonomic group (worms) for all non-arthropod invertebrate animals.

Vertebrate: An animal with a spinal chord that belongs to the subphylum Vertebrata.

Warping: Slight upward or downward deformation, folding or flexing of rocks at broad or regional scales.

Wynyardiidae: An extinct family of marsupials, originally described from a fossil of a possum-like marsupial from the early Miocene of Wynyard in Tasmania.

Xeric: Characterised by very dry conditions, neither very wet (hydric) nor intermediate (mesic).

Xeromorphic characters: Of, relating to, or characteristic of adaptations to a dry climate.

Xiphosurids: A group of marine chelicerate arthropods that includes a large number of extinct lineages and only four recent species (e.g. horseshoe crabs). Their fossil record extends from the Early Silurian to the present.

Zygomaturinae: An extinct subfamily of megafaunal diprotodontid marsupials, including species of the Pleistocene *Zygomaturus*.

References and further reading

Abele C (1979) Geology of the Anglesea area, central coastal Victoria. *Memoirs of the Geological Survey of Victoria* **31**, 1–71.

Abele C (1988a) Central coastal basins. In *Geology of Victoria*. 2nd edn. (Eds JG Douglas and JA Ferguson) pp. 303–322. Geological Society of Australia (Victoria Division), Melbourne.

Abele C (1988b) Torquay–Anglesea–Aireys Inlet. In *Victorian Geology Excursion Guide* (Eds I Clarke and B Cook) pp. 119–131. Australian Academy of Science, Canberra.

Abele C, Kenley PR, Holdgate G and Ripper D (1988) Otway Basin. In *Geology of Victoria*. 2nd edn. (Eds JG Douglas and JA Ferguson) pp. 272–303. Geological Society of Australia (Victoria Division), Melbourne.

AHDB (2004) 'Dinosaur Stampede National Monument'. Reference Number 105664. *Australian Heritage Database*. Australian Government Department of Sustainability, Environment, Water, Population and Communities. <www.environment.gov.au/cgi-bin/ahdb/search.pl>

Ahlberg P, Trinajstic K, Johanson Z and Long J (2009a) Pelvic claspers confirm chondrichthyan-like internal fertilization in arthrodires. *Nature* **460**, 888–889.

Ahlberg P, Trinajstic K and Long JA (2009b) The body musculature of arthrodire placoderms. *Journal of Vertebrate Paleontology* Supplement **29**(3), 52A.

Alexander EM and Gravestock DI (1990) Sedimentary facies in the Sellick Hill Formation, Fleurieu Peninsula, SA. In *The Evolution of a Late Pre Cambrian–Early Palaeozoic Rift Complex: the Adelaide Geosyncline* (Eds J Jago and P Moore) pp. 269–289. Special Publication 16, Geological Society of Australia, Brisbane.

Alley NF (1989) 'Preliminary palynological dating of macrofloras from Eyre Formation, Nelly Creek, Lake Eyre Basin'. Department of Mines and Energy of South Australia Report Book 89/46, Adelaide.

Alley NF, Krieg GW and Callen RA (1996) Early Tertiary Eyre Formation, lower Nelly Creek southern Lake Eyre Basin, Australia: palynological dating of macrofloras and silcrete, and palaeoclimatic interpretations. *Australian Journal of Earth Sciences* **43**(1), 71–84.

Alley NF and Pledge NS (2000) The plants, animals and environments of the last 280 million years. In *Lake Eyre Monograph Series 5* (Ed. W Slaytor) pp. 35–82. Royal Geographical Society of South Australia, Adelaide.

Anderson JM (1999) *Towards Gondwana Alive: promoting biodiversity and stemming the sixth extinction*. Gondwana Alive Society, Pretoria, South Africa.

Anderson KL, Lackie MA, Clark DA and Schmidt PW (2004) Return to Black Mountain: palaeomagnetic reassessment of the Chatsworth and Ninmaroo Formations, western Queensland, Australia. *Geophysical Journal International* **157**(1), 87–104.

Archer M, Arena DA, Bassarova M, Beck R, Black K, Boles WE, Brewer P, Cooke BN, Crosby K, Gillespie A, Godthelp H, Hand SJ, Kear B, Louys J, Morrell A, Muirhead J, Roberts KK, Scanlon JD, Travouillon KJ and Wroe S (2006) Current status of species-level representation in faunas from selected fossil localities in the Riversleigh World Heritage Area, northwestern Queensland. *Alcheringa Special Issue* **1**, 1–17.

Archer M, Flannery TF, Ritchie A and Molnar RE (1985) First Mesozoic mammal from Australia – an early Cretaceous monotreme. *Nature* **318**, 363–366.

Archer M, Hand SJ and Godthelp H (1991) *Riversleigh*. Reed Books, Sydney.

Arsenault M, Desbiens S, Janvier P and Kerr J (2004) New data on the soft tissues and external morphology of the antiarch *Bothriolepis canadensis* (Whiteaves, 1880), from the Upper Devonian of Miguasha, Quebec. In *Recent Advances in the Origin and Early Radiation of Vertebrates* (Eds G Arratia, MVH Wilson and R Cloutier) pp. 439–454. Verlag Dr Friedrich Pfeil, München.

Baillie PW, Banks MR and Rickards RB (1978) Early Silurian graptolites from Tasmania and their significance. *Search* **9**, 46–47.

Baird RF (1985) Avian fossils from Quaternary deposits in Green Waterhole Cave, South East South Australia. *Records of the Australian Museum* **37**, 353–370.

Baird RF (1991) Taphonomy of late Quaternary cave localities yielding vertebrate remains in Australia. In *Vertebrate Palaeontology of Australasia* (Eds P Vickers-Rich, JM Monaghan, RF Baird and TH Rich) pp. 267–309. Pioneer Design Studio and the Monash University Publications Committee, Melbourne.

Balme J, Merrilees D and Porter JK (1978) Late

Quaternary mammal remains, spanning about 30 000 years, from excavations in Devil's Lair, Western Australia. *Journal of the Royal Society of Western Australia* **61**, 33–65.

Balmer J, Whinam J, Kelman J, Kirkpatrick JB and Lazarus E (2004) 'A review of the floristic values of the Tasmanian Wilderness World Heritage Area'. Nature Conservation Report 2004/3, Department of Primary Industries Water and Environment, Hobart.

Banks MR (1957) The stratigraphy of Tasmanian limestones. In *Limestones in Tasmania* (Ed. TD Hughes) pp. 39–85. Tasmanian Department of Mines Geological Survey Mineral Resources 10, Hobart.

Banks MR (1962) Cambrian System. *Journal of the Geological Society of Australia* **9**, 127–145.

Banks MR (1988) The base of the Silurian System in Tasmania. *Bulletin of the British Museum of Natural History* (Geology) **43**, 191–194.

Banks MR and Baillie PW (1989) Late Cambrian to Devonian. In *Geology and Mineral Resources of Tasmania* (Eds CF Burrett, EL Martin) pp. 182–237. Geological Society of Australia, Special Publication 15. Geological Society of Australia, Sydney.

Banks MR and Burrett CF (1980) A preliminary Ordovician biostratigraphy of Tasmania. *Journal of the Geological Society Australia* **26**, 363–376.

Banks MR, Cosgriff JW and Kemp NR (1978) A Tasmanian Triassic stream community. *Australian Natural History* **19**(5), 150–157.

Banks MR and Johnson JH (1957) Maclurites and Girvanella in the Gordon River Limestone (Ordovician) of Tasmania. *Journal of Paleontology* **31**, 632–640.

Barghoorn ES and Schopf JW (1965) Microorganisms from the Late Precambrian of central Australia. *Science* **50**, 337–339.

Barnes RW and Jordan GJ (2000) Eucryphia (Cunoniaceae) reproductive and leaf macrofossils from Australian Cainozoic sediments. *Australian Systematic Botany* **13**, 373–394.

Bartholomai A (1966) Fossil footprints in Queensland. *Australian Natural History* **15**, 147–150.

Bartholomai A (1979) New lizard-like reptiles from the Early Triassic of Queensland. *Alcheringa* **3**, 225–234.

Bartholomai A and Howie A (1970) Vertebrate fauna from the Lower Trias of Australia. *Nature* **225**(5237), 1063.

Basden A, Burrow C, Hocking M, Parkes R and Young G (2000) Siluro-Devonian microvertebrates from southeastern Australia. *Courier Forschungsinstitut Senckenberg* **223**, 201–222.

Baynes A, Merrilees D and Porter JK (1976) Mammal remains from the upper levels of a late Pleistocene deposit in Devil's Lair, Western Australia. *Journal of the Royal Society of Western Australia* **58**, 97–126 [for 1975].

Bean LB (2006) The leptolepid fish *Cavenderichthys talbragarensis* (Woodward, 1895) from the Talbragar fish bed (Late Jurassic) near Gulgong, New South Wales. *Records of the Western Australian Museum* **23**, 43–76.

Beattie R (2007) The geological setting and palaeoenvironmental and palaeoecological reconstructions of the Upper Permian insect beds at Belmont, New South Wales, Australia. *African Invertebrates* **48**(1), 41–57.

Bedford R and Bedford WR (1934) New species of Archaeocyathinae and other organisms for the Lower Cambrian of Beltana, South Australia. *Memoirs of the Kyancutta Museum South Australia* **1**, 1–17.

Bedford R and Bedford WR (1936) Further notes on Cyathospongia (Archaeocyathi) and other organisms from Lower Cambrian of Beltana, South Australia. *Memoirs of the Kyancutta Museum South Australia* **3**, 21–26.

Bedford R and Bedford WR (1937) Further notes on Archaeos (Pleospongia) from the Lower Cambrian of South Australia. *Memoirs of the Kyancutta Museum South Australia* **4**, 27–38.

Bedford R and Bedford WR (1939) Development and classification of Archaeos (Pleospongia). *Memoirs of the Kyancutta Museum South Australia* **6**, 67–82.

Bender PA and Hancox PJ (2004) Newly discovered fish faunas from the Early Triassic, Karoo Basin, South Africa, and their correlative implications. *Gondwana Research* **7**(1), 185–192.

Bentley CJ and Jago JB (2004) Wuaniid trilobites of Australia. *Memoirs of the Association of Australasian Palaeontologists* **30**, 179–191.

Bigwood AJ and Hill RS (1985) Tertiary araucarian macrofossils from Tasmania. *Australian Journal of Botany* **33**, 645–656.

Black K and Archer M (1997) Silvabestius gen. nov., primitive zygomaturines (Marsupialia, Diprotodontidae) from Riversleigh, northwestern Queensland. *Memoirs of the Queensland Museum* **41**, 193–208.

Black MP, Mooney SD and Martin HA (2006) A >43,000-year vegetation and fire history from Lake Baraba, New South Wales, Australia. *Quaternary Science Reviews* **25**(21–22), 3003–3016.

Blackburn DT and Sluiter IRK (1994) The Oligo-Miocene coal floras of southeastern Australia. In *History of the Australian Vegetation: Cretaceous to recent.* (Ed. RS Hill) pp. 328–367 Cambridge University Press, Cambridge, UK.

Borsuk-Bialynicka M, Cook E, Evans SE and Maryanska T (1999) A microvertebrate assemblage from the Early Triassic of Poland. *Acta Palaeontologica Polonica* **44**(2), 167–188.

Bourne SJ and Reed EH (2009) The Australian fossil mammal sites (Riversleigh/Naracoorte). *World Heritage* **52**, 46–47.

Braddy SJ (2003) Trace fossils from the Tumblagooda Sandstone of Western Australia. *The Palaeontological Association Newsletter* **53**, 68–69.

Branagan DF and Vallance TG (2008) Some unpublished correspondence of the Rev. W.B. Clarke. *Journal and Proceedings of the Royal Society of New South Wales* **141**, 1–31.

Brasier MD, Green OR, Jephcoat AP, Kleppe AK, Van Kranendonk MJ, Lindsay JF, Steele A and Grassineau NV (2002) Questioning the evidence for Earth's oldest fossils. *Nature* **416**, 76–81.

Brasier M, Green O, Lindsey J and Steele A (2004) Earth's oldest (approximately 3.5 Ga) fossil and the 'Early Eden hypothesis': questioning the evidence. *Origins of Life and Evolution of the Biosphere* **34**, 257–269.

Brewer P, Archer M and Hand SJ (2008) Additional specimens of the oldest wombat *Rhizophascolonus crowcrofti* (Vombatidae; Marsupialia) from the Wipajiri Formation, South Australia: an intermediate morphology? *Journal of Vertebrate Paleontology* **28**, 1144–1148.

Briggs DEG and Nedin C (1997) The taphonomy and affinities of the problematic fossil Myoscolex from Lower Cambrian Emu Bay Shale. *Journal of Paleontology* **71**, 22–32.

Bromfield KE (2004) Palaeoenvironmental reconstruction of the Jurassic, using plant macrofossils from a site at Lune River, Southeast Tasmania. BSc (Hons) thesis. University of Tasmania, Australia.

Bromfield KE, Burrett CF, Leslie R A, and Meffre S (2007) Jurassic volcaniclastic–basaltic andesitic–dolerite sequence in Tasmania: new age constraints for fossil plants from Lune River. *Australian Journal of Earth Sciences* **54**, 965–974.

Brown AV (Ed.) (1989) 'Geological atlas 1:50,000 series, Sheet 21 (7916S)'. Geological Survey Explanatory Report, Tasmanian Department of Mines, Smithton.

Brown AV, Findlay RH, McClenaghan MP and Seymour DP (1991) 'Synopsis of the regional geology of the Macquarie Harbour, Point Hibbs and Montgomery 1:50 000 map sheets'. Report of Division of Mines and Mineral Resources Tasmania 1990/21, Rosny Park, Tas.

Brown IA (1948) Lower Ordovician brachiopods from Junee district, Tasmania. *Journal of Paleontology* **22**, 35–39.

Burke AC, Gaffney ES and Rich TH (1983) Miocene turtles from Lake Tarkarooloo, South Australia. *Alcheringa* **7**(2), 151–154.

Burrett CF (1978) Middle-Upper Ordovician conodonts and stratigraphy of the Gordon limestone sub-group, Tasmania. PhD thesis. University of Tasmania, Australia.

Burrett CF (1984) Early Devonian deformation and metamorphism of conodonts in western Tasmania – economic, theoretical and nomenclatural implications. In *Mineral Exploration and Tectonic Processes in Tasmania* (Eds PW Baillie and PI Collings) pp. 14–17. Geological Society of Australia (Tasmanian Division), Hobart.

Burrow CJ, Jones AS and Young GC (2005) X-ray microtomography of 410 million-year-old optic capsules from placoderm fish. *Micron* **36**, 551–557.

Burrow CJ, Long JA, Turner S (1998) Lower Devonian microvertebrates from Tasmania. *Alcheringa* **22**, 9–20.

Callen RA and Tedford RH (1976) New Late Cainozoic rock units and depositional environments, Lake Frome Area, South Australia. *Transactions of Royal Society of South Australia* **100**(3), 125–167.

Camp CL and Banks MR (1978) A proterosuchian reptile from the Early Triassic of Tasmania. *Alcheringa* **2**, 143–158.

Campbell KSW and Barwick RE (2000) The braincase, mandible and dental structures of the Early Devonian lungfish *Dipnorhynchus kurikae* from Wee Jasper, New South Wales. *Records of the Australian Museum* **52**(1), 103–128.

Campbell KSW and Bell MW (1977) A primitive amphibian from the Late Devonian of New South Wales. *Alcheringa* **1**, 369–381.

Carey SP (1989) Point Hibbs area. In *Geology and Mineral Deposits of Tasmania* (Eds CF Burrett and EL Martin) pp. 229–230. Geological Survey of Australia Special Publication 15. Geological Society of Australia, Sydney.

Carey SP and Berry RF (1988) Thrust sheets at Point Hibbs, Tasmania: palaeontology, sedimentology and structure. *Australian Journal of Earth Science* **35**, 169–180.

Carpenter RJ (1988) Early Tertiary Tmesipteris (Psilotaceae) macrofossil from Tasmania. *Australian Systematic Botany* **1**, 171–176.

Carpenter RJ (1991) Macrozamia from the Early Tertiary of Tasmania and a study of the cuticles of extant species. *Australian Systematic Botany* **4**, 433–444.

Carpenter RJ (1991) Palaeovegetation and environment at Cethana, Tasmania. PhD thesis. University of Tasmania, Australia.

Carpenter RJ (1994) Cuticular morphology and aspects of the ecology and fossil history of North Queensland rainforest Proteaceae. *Botanical Journal of the Linnean Society* **116**, 249–303.

Carpenter RJ and Buchanan AM (1993) Oligocene leaves, fruit and flowers of the Cunoniaceae from Cethana, Tasmania. *Australian Systematic Botany* **6**, 91–109.

Carpenter RJ and Hill RS (1988) Early Tertiary Lomatia (Proteaceae) macrofossils from Tasmania, Australia. *Review of Palaeobotany and Palynology* **56**, 141–150.

Carpenter RJ, Hill RS and Jordan GJ (1994) Cenozoic vegetation in Tasmania: macrofossil evidence. In *History of the Australian vegetation: Cretaceous to Recent* (Ed. RS Hill) pp. 276–298. Cambridge University Press, Cambridge, UK.

Carter JD (1987) 'Important geological localities beyond the Perth Region. Their significance and value, protection and preservation'. Report for the Geological Society of Australia, Western Australia Division, Perth, WA.

Chambers TC, Drinnan AN and McLoughlin S (1998) Some morphological features of Wollemi Pine (*Wollemia nobilis*: Araucariaceae) and their comparison to Cretaceous plant fossils. *International Journal of Plant Sciences* **159**, 160–171.

Chapman F (1922) *Sherbornina*: a new genus of the Foraminifera from Table Cape, Tasmania. *Linnean Society of London, Journal of Zoology* **34**(230), 501–532.

Chatterton BDE (1971) Taxonomy and ontogeny of Siluro-Devonian trilobites from near Yass, New South Wales. *Palaeontographica Abteilung A*, **137**(1–3), 1–108.

Chatterton BDE (1973) Brachiopods of the Murrumbidgee Group, Taemas, New South Wales. *Bulletin of the Bureau of Mineral Resources, Geology and Geophysics Australia* **137**, 1–146.

Chinsamy A, Rich TH and Vickers-Rich P (1998) Polar dinosaur bone histology. *Journal of Vertebrate Palaeontology* **18**, 385–390.

Christophel DC (1994) The early Tertiary macrofloras of continental Australia. In *History of the Australian vegetation: Cretaceous to recent*. (Ed. RS Hill) pp. 262–275. Cambridge University Press, Cambridge, UK.

Christophel DC, Harris WK and Syber AK (1987) The Eocene flora of the Anglesea Locality, Victoria. *Alcheringa* **11**, 303–323.

Christophel DC, Scriven LJ and Greenwood DR (1992) An Eocene megafossil flora from Nelly Creek, South Australia. *Transactions of the Royal Society of South Australia* **116**, 65–76.

Clack JA (1997) Devonian tetrapod trackways and trackmakers: a review of the fossils and footprints. *Palaeogeography, Palaeoclimatology, Palaeoecology* **130**, 227–250.

Clack JA (2002) *Gaining ground: Origins and evolution of tetrapods*. Indiana University Press, Bloomington.

Clarke MJ and Baillie PW (1984) 'Geological survey explanatory report: geological atlas 1:50 000 series, sheet 77 (8512 N): Maria'. Tasmanian Department of Mines, Hobart.

Clarke MJ and Banks MR (1975) The stratigraphy of the lower (Permo-Carboniferous) parts of the Parmeener Super-Group, Tasmania. In *Gondwana Geology: Papers Presented at the 3rd Gondwana Symposium*. 1973, ANU, Canberra (Ed. KSW Campbell) pp. 453–467. ANU Press, Canberra.

Clarke WB (1843) On a fossil pine forest at Kurrur-Kurran, in the inlet of Awaaba, east coast of Australia. *Proceedings of the Geological Society of London* **6**(1), 161–164.

Clarke WB (1885) On a fossil pine forest at Kurrur-Kurran, in the inlet of Awaaba, east coast of Australia. *Annual Report of the Department of Mines of NSW for the year 1884*, pp. 156–159. Government Printer, Sydney.

Clemens WA and Plane M (1974) Mid-Tertiary Thylacoleonidae (Marsupialia, Mammalia). *Journal of Paleontology* **48**(4), 652–660.

Cloutier R and Arratia G (2004) Early diversification of actinopterygians. In *Recent advances in the origin and early radiation of vertebrates* (Eds G Arratia, MVH Wilson and R Cloutier) pp. 217–270. Verlag Dr Friedrich Pfeil, München.

Cockbain AE (1984) Stromatoporoids from the Devonian reef complexes Canning Basin, Western Australia. *Bulletin of the Geological Survey of Western Australia* **129**, 1–48.

Colbert EH and Merrilees D (1967) Cretaceous dinosaur footprints from Western Australia. *Journal of the Royal Society of Western Australia* **50**, 21–25.

Colhoun EA (1989) The Quaternary. In *The Geology and Mineral Resources of Tasmania* (Eds CF Burrett and EL Martin) pp. 410–418. Geological Society of Australia, Brisbane.

Colhoun EA and Van de Geer G (1988) Darwin Crater, the King and Linda Valleys. In *Cainozoic Vegetation of Tasmania* (Ed. EA Colhoun) pp. 30–71. University of Newcastle Special Papers, Department of Geography, University of Newcastle, NSW.

Colhoun EA and Van de Geer G (1998) Pollen analysis of 0–20 m at Darwin Crater western Tasmania, Australia. In *International Project on Palaeolimnology and Late Cainozoic Climate* (Ed. S Horie). IPPCE **11**, 68–89.

Condon MA (1967) The geology of the Carnarvon Basin, Western Australia, Part 2. Permian stratigraphy. *Australia Bureau of Mineral Resources* **77**, 1–191.

Conkin JE and Conkin BM (1993) Permian Foraminifera of Tasmania. *University of Louisville Studies in Paleontology and Stratigraphy* **21**, 1–37.

Conran JG, Christophel DC and Cunningham L (2003) An Eocene monocotyledon from Nelly Creek, central Australia, with affinities to Hemerocallidaceae (Lilianae: Asparagales). *Alcheringa* **27**, 107–115.

Constantine A (2001) Sedimentology, stratigraphy and palaeoenvironment of the Upper Jurassic–Lower Cretaceous non-marine Strzelecki Group, Gippsland Basin, Southeastern Australia. PhD thesis. Monash University, Australia.

Conway-Morris S and Jenkins RJF (1985) Healed injuries in Early Cambrian trilobites from South Australia. *Alcheringa* **9**, 167–177.

Cook A (2004) 'Expert consultancy report: The assessment of the dinosaur trackways within the Lark Quarry Conservation Park for potential National Heritage values.' Queensland Museum, Hendra.

Cook A, Hocknull S and Spring K (2002) 'Report on the Mount Morgan Clay Caverns Dinosaur footprint and trackways site'. Queensland Museum, Brisbane.

Cook AG (1993a) Two bivalves from the Middle Devonian Burdekin Formation. *Memoirs of the Queensland Museum* **33**(1), 49–53.

Cook AG (1993b) *Fletcherviewia septata*: A new high-spired septate gastropod from the Devonian of north Queensland. *Journal of Paleontology* **67**(5), 816–821.

Cook AG (1995) Sedimentology and depositional environments of the Middle Devonian Big Band Arkose and Burdekin Formation, Fanning River Group. Burdekin Subprovince, north Queensland. *Memoirs of the Queensland Museum* **38**(1), 53–91.

Cook AG (1997) Gastropods from the Burdekin Formation, Middle Devonian north Queensland. *Memoirs of the Queensland Museum* **42**(1), 37–49.

Cook AG (1999) Stromatoporoid paleoecology and systematics from the Middle Devonian Fanning River Group, north Queensland. *Memoirs of the Queensland Museum* **43**(2), 463–551.

Cookson IC and Duigan SL (1950) Fossil Banksieae from Yallourn, Victoria, with notes on the morphology and anatomy of living species. *Australian Journal of Scientific Research, Series B* **3**, 133–165.

Cookson IC and Eisenack A (1967) Some Early Tertiary microplankton and pollen grains from a deposit near Strahan, western Tasmania. *Proceedings of the Royal Society of Victoria* **80**, 131–140.

Corbett KD and Banks MR (1974) Ordovician stratigraphy of the Florentine Synclinorium, southwest Tasmania. *Papers and Proceedings of the Royal Society of Tasmania* **107**, 207–238.

Corbett K and Howard K (2006) The Darwin Meteorite Crater. *Tasmania 40° South* **40**, 28–31.

Cosgriff JW (1974) Lower Triassic Temnospondyli of Tasmania. *Geological Society of America, Special Paper* **149**, 1–134.

Cotter KL (1997) Neoproterozoic microfossils from the Officer Basin, Western Australia. *Alcheringa* **21**, 247–270.

Couper J (1965) Late Silurian to Early Devonian stratigraphy of the Yea–Molesworth district, Victoria. *Proceedings of the Royal Society of Victoria* **79**, 1–8.

Coventry RJ (1976) Abandoned shorelines and the late Quaternary history of Lake George. *Journal of the Geological Society of Australia* **23**(3), 249–273.

Craig RS (2002) The palaeobiogeography of Late Cretaceous and Cenozoic brachiopods from Western Australia. *Palaeogeography, Palaeoclimatology, Palaeoecology* **179**, 267–92.

Crick RE and Stanley GD (1997) Curt Teichert May 8, 1905 – May 10, 1996. *Journal of Paleontology* **71**(4), 750–752.

Crimes TP (1970) The significance of trace fossils in sedimentology, stratigraphy and palaeoecology with examples from lower Palaeozoic strata. In *Trace Fossils* (Eds TP Crimes and JC Harper) pp. 101–126. Seed House Press, Liverpool, UK.

Crooks AF and Gatehouse CG (1980) Large trilobite track from Ellery Creek, Northern Territory. *Geological Survey of South Australia* **73**, 13–15.

Crosby K and Archer M (2000) Durudawirines, a new group of phalangeroid marsupials from the Miocene of Riversleigh, northwestern Queensland. *Journal of Paleontology* **74**, 327–335.

Crowell JC and Frakes LA (1971) Late Palaeozoic glaciation in Australia. *Journal of the Geological Society of Australia* **17**, 115–155.

Damiani RJ (1999) Parotosuchus (Amphibia, Temnospondyli) in Gondwana: biostratigraphic and palaeobiogeographic implications. *South African Journal of Science* **95**, 458–460.

Darragh TA (1970) Catalogue of Australian Tertiary Mollusca (except chitons). *Memoirs of the National Museum of Victoria* **31**, 125–212.

Darragh TA (1985) Molluscan biogeography and biostratigraphy of the Tertiary of southern Australia. *Alcheringa* **9**, 83–116.

Darragh TA and Kendrick GW (1991) Maastrichtian Bivalvia (excluding Inoceramidae) from the Miria Formation, Carnarvon Basin, north western Australia. *Records of the Western Australian Museum, Supplement* **36**, 1–102.

Darragh TA and Kendrick GW (1994) Maastrichtian Scaphopoda and Gastropoda from the Miria Formation, Carnarvon Basin, northwestern Australia. *Records of the Western Australian Museum, Supplement* **48**, 1–76.

Darrenougue N, de Deckker P, Fitzsimmons K, Norman M, Reed E, van der Kaars S and Fallon S (2009) A late Pleistocene record of aeolian sedimentation in Blanche Cave, Naracoorte, South Australia. *Quaternary Science Reviews* **28**, 2600–2615.

David TWE (1907) The geology of the Hunter River Coal Measures, New South Wales. *Memoirs of the Geological Survey of New South Wales* **4**, Sydney.

Dawson L (1982) Marsupial fossils from Wellington Caves, New South Wales. PhD thesis. University of New South Wales, Australia.

Dawson L (1985) Marsupial fossils from Wellington Caves, New South Wales: the historic and scientific significance of the collections in the Australian Museum, Sydney. *Records of the Australian Museum* **37**, 55–69.

Dawson L (1999) Wellington Caves: new life for a dormant system. *Riversleigh Notes* **43**, 2–4.

Dawson L (2004) A new Pliocene tree kangaroo species (Marsupialia, Macropodinae) from the Chinchilla Local Fauna, southeastern Queensland. *Alcheringa* **28**(1), 267–273.

Dawson L and Augee ML (1997) The Late Quaternary sediments and fossil vertebrate fauna from Cathedral Cave, Wellington Caves, New South Wales. *Proceedings of the Linnean Society of New South Wales* **117**, 51–78.

Dawson L, Muirhead J and Wroe S (1999) The Big Sink Local Fauna: a lower Pliocene mammalian fauna from the Wellington Caves complex, Wellington, New South Wales. *Records of the Western Australian Museum*, Supplement **57**, 265–290.

Debrenne F (1969) Lower Cambrian Archaeocyatha from the Ajax Mine, Beltana South Australia. *Bulletin of the British Museum (Natural History), Geology* **17**(7), 297–376.

Debrenne F and Gravestock D (1990) Archaeocyatha from the Sellick Hill Formation and the Fork Tree Limestone on Fleurieu Peninsula, SA. In 'The Evolution of a Late Pre Cambrian–Early Palaeozoic Rift Complex: The Adelaide Geosyncline' (Eds J Jago and P Moore) pp. 290–309. Geological Society of Australia, Special Publication 16. Geological Society of Australia, Sydney.

Department of Environment and Heritage (2001) 'Naracoorte Caves National Park Plan of Management'. South Australian Department for Environment and Heritage, Adelaide.

Department of Environment and Heritage (2002) 'A Review of Lake Frome and Strzelecki Regions 1991–2001'. South Australian Department of Environment and Heritage, Adelaide.

Dennant J (1889) Notes on the Muddy Creek beds, with brief remarks on other Tertiary strata of south western Victoria. *Transactions of the Royal Society of South Australia* **11**, 30–59.

Dennant J and Kitson AE (1903) Catalogue of the described species of fossils (except Bryozoa and Foraminifera) in the Cainozoic fauna of Victoria, South Australia and Tasmania. *Records of the Geological Survey of Victoria* **1**(2), 89–147.

Dennis KD and Miles RS (1979a) A second eubrachythoracid arthrodire from Gogo, Western Australia. *Zoological Journal of the Linnean Society* **67**, 1–29.

Dennis KD and Miles RS (1979b) Eubrachythoracid arthrodires with tubular rostral plates from Gogo, Western Australia. *Zoological Journal of the Linnean Society* **67**, 297–328.

Dennis KD and Miles RS (1980) New durophagous

arthrodires from Gogo, Western Australia. *Zoological Journal of the Linnean Society* **69**, 43–85.

Dennis KD and Miles RS (1981) A pachyosteomorph arthrodire from Gogo, Western Australia. *Zoological Journal of the Linnean Society* **73**, 213–258.

Dennis KD and Miles RS (1982) A eubrachythoracid arthrodire with a snub-nose from Gogo, Western Australia. *Zoological Journal of the Linnean Society* **75**, 153–166.

Dennis-Bryan K (1987) A new species of eastmanosteid arthrodire (Pisces: Placodermi) from Gogo, Western Australia. *Zoological Journal of the Linnean Society* **90**, 1–64.

Dennis-Bryan K and Miles RS (1983) Further eubrachythoracid arthrodires from Gogo, Western Australia. *Zoological Journal of the Linnean Society* **77**, 145–173.

Desmarchelier JM, Goede A, Ayliffe LK, McCulloch MT, and Moriarty KC (2000) Stable isotope record and its paleoenvironmental interpretation for a late Middle Pleistocene speleothem from Victoria Fossil Cave, Naracoorte, South Australia. *Quaternary Science Reviews* **19**, 763–774.

DEST (1993) 'Nomination of Australian fossil sites (a serial nomination of sites at Murgon, Riversleigh and Naracoorte)'. Australian Government Department of the Environment, Sport and Territories, Canberra.

Dettmann ME (1994) Cretaceous vegetation: the microfossil record. In *History of the Australian Vegetation: Cretaceous to Recent.* (Ed. RS Hill) pp. 143–170. Cambridge University Press, Cambridge, UK.

Dickinson JA, Wallace MW, Holdgate GR, Gallagher SJ and Thomas L (2002) Origin and timing of the Miocene–Pliocene unconformity in southeast Australia. *Journal of Sedimentary Research* **72**, 317–332.

Dinnick M (1985) Stratigraphy, sedimentology and palaeontology of a Cambrian molassic sequence, Cape d'Estaing to Point Marsden, north east coast of Kangaroo Island, SA. BSc (Hons) thesis. University of Adelaide, Australia.

Dixon M and Haig DW (2004) Foraminifera and their habitats within a cool-water carbonate succession following glaciation, Early Permian (Sakmarian), Western Australia. *Journal of Foraminiferal Research* **34**, 308–324.

Dodds B (1949). Mid Triassic Blattoidea from the Mount Crosby insect bed. *University of Queensland Papers, Department of Geology* **3**, 1–11.

Dortch CE and Merrilees D (1973) Human occupation of Devil's Lair, Western Australia during the Pleistocene. *Archaeology and Physical Anthropology in Oceania* **8**, 89–115.

Douglas J and Holmes F (2006) The Baragwanathia story: an update. *The Fossil Collector* **77**, 9–26.

Douglas JG (1983) *What Fossil Plant Is That? A guide to the ancient floras of Victoria.* Field Naturalists Club of Victoria, Melbourne.

Douglas JG (1994) Cretaceous vegetation: the macrofossil record. In *History of the Australian Vegetation: Cretaceous to Recent.* (Ed. RS Hill) pp. 171–188. Cambridge University Press, Cambridge, UK.

Douglas JG (2003) *Baragwanathia* – discovery of some of the world's first land plants. In *Geology of Victoria* (Ed. WD Birch) pp. 689–691. Geological Society of Australian Special Publication No. 23, Geological Society of Australia (Victoria Division), Melbourne.

Douglas, K (2010) *Pictures of Time Beneath: Science, heritage and the uses of the deep past.* CSIRO Publishing, Melbourne.

Drinnan AN and Chambers TC (1986) Flora of the Lower Cretaceous Koonwarra Fossil Bed (Korumburra Group), south Gippsland, Victoria. *Memoirs of the Association of Australasian Palaeontologists* **3**, 1–77.

Droser ML, Gehling JG and Jensen S (2006) Assemblage palaeoecology of the Ediacara biota: the unabridged edition? *Palaeogeography, Palaeoclimatology, Palaeoecology* **232**, 131–147.

Droser ML, Jensen S and Gehling JG (2004) Development of early Palaeozoic ichnofabrics: evidence from shallow marine siliciclastics. In *The application of Ichnology to palaeoenvironmental and stratigraphic analysis* (Ed. D McIlroy) pp. 383–396. The Geological Society Special Publications 228, London.

Droser ML, Jensen S and Gehling JG (2005) Ediacaran Trace Fossils: True and false. In *Evolving Form and Function: Fossils and development: Proceedings of a symposium honoring Adolf Seilacher for his contribution to palaeontology* (Ed. DEG Briggs) pp. 125–138. Peabody Museum of Natural History, Yale University, New Haven, CT.

Druce EC and Jones RL (1971) Cambro-Ordovician conodonts from the Burke River Structural Belt, Queensland. *Bulletin of the Bureau of Mineral Resources, Geology and Geophysics Australia* **110**, 1–158

Druce EC, Shergold JH and Radke BM (1982) A reassessment of the Cambrian-Ordovician boundary section at Black Mountain, western Queensland, Australia. In *The Cambrian-Ordovician Boundary: sections, fossil distributions and correlations* (Eds MG Bassett and WT Dean) pp 193–209. National Museum of Wales, Cardiff, UK.

Dulhunty JA and Eadie J (1969) Geology of the Talbragar fossil fish bed area. *Journal and Proceedings of the Royal Society of New South Wales* **102**, 1–4.

Duncan IJ, Briggs DEG and Archer M (1998) Three-dimensionally mineralised insects and millipedes from the Tertiary of Riversleigh, Queensland, Australia. *Palaeontology* **41**, 835–851.

Dziewa T (1980) Early Triassic osteichthyans from the Knocklofty Formation of Tasmania. *Papers and Proceedings of the Royal Society of Tasmania* **114**, 145–160.

Edgecombe GD (1998). Early myriapodous arthropods from Australia: Maldybulakia from the Devonian of New South Wales. *Records of the Australian Museum* **50**, 293–313.

Erwin DH (1993) *The Great Paleozoic Crisis: Life and death in the Permian.* Columbia University Press, New York.

Etheridge R Jr (1917) Descriptions of some Queensland Palaeozoic and Mesozoic fossils 3. A remarkable univalve from the Devonian Limestone of the Burdekin River, Queensland. *Geological Survey of Queensland Publication* **260**, 13–16.

Etheridge R Jr (1920) Small gasteropoda from the Lower Cretaceous of Queensland. *Geological Survey of Queensland Publication* **269**, 1–19.

Evans JW (1961) Some Upper Triassic Hemiptera from Queensland. *Memoirs of the Queensland Museum* **14**, 13–23.

Evans JW (1971) Some Upper Triassic Hemiptera from Mount Crosby. *Memoirs of the Queensland Museum* **16**, 145–152.

Everard GB (1957) Maria Island. In *Limestones in Tasmania* (Ed. TD Hughes) pp. 235–241. Mineral Resources Geological Survey No. 10, Tasmania. Government Printer, Hobart.

Fedonkin MA, Gehling JG, Grey K, Narbonne GM and Vickers-Rich P (2007) *The Rise of Animals: Evolution and diversification if the Kingdom Animalia.* The Johns Hopkins University Press, Baltimore.

Field JH, Dodson JR and Prosser IP (2002) A Late Pleistocene vegetation history from the Australian semi-arid zone. *Quaternary Science Reviews* **21**(8–9), 1023–1037.

Field JH and Fullagar R (2001) Archaeology and Australian megafauna. *Science* **294**(5540), 7a.

Fitzgerald EMG (2003) Oligocene Cetacea (Mammalia) from the southern margin of Australia. *Geological Society of Australia Abstracts* **70**, 17.

Fitzgerald EMG (2004) A review of the Tertiary fossil Cetacea (Mammalia) localities in Australia. *Memoirs of Museum Victoria* **61**, 183–208.

Fitzgerald EMG (2005a) Toothed mysticetes (Mammalia: Cetacea) from the Late Oligocene of Australia. In *Evolution of Aquatic Tetrapods: Fourth Triannual Convention Abstracts* (Ed. MD Uhen) p. 25. Cranbrook Institute of Science Miscellaneous Publications 1, Michigan.

Fitzgerald EMG (2005b) Bizarre baleen whales from Australia's ancient seas. *Australasian Science* **26**, 23–27.

Fitzgerald EMG (2005c) Diversity and novelty in the early evolution of baleen whales (Mammalia: Cetacea: Mysticeti). *Geological Society of Australia Abstracts* **80**, 30.

Fitzgerald EMG (2006) A bizarre new toothed mysticete (Cetacea) from Australia and the early evolution of baleen whales. *Proceedings of the Royal Society B: Biological Sciences* **273**, 2955–2963.

Fitzgerald EMG (2010) The morphology and systematics of *Mammalodon colliveri* (Cetacea: Mysticeti), a toothed mysticete from the Oligocene of Australia. *Zoological Journal of the Linnean Society* **158**, 367–476.

Flannery TF, Archer M, Rich TH and Jones RK (1995) A new family of monotremes from the Cretaceous of Australia. *Nature* **377**, 418–420.

Flannery TF, Rich THV, Turnbull WD and Lundelius EL (1992) The Macropodoidea of the early Pliocene Hamilton local fauna from Victoria, Australia. *Fieldiana Geology* **1443**, 1–37.

Flannery TF, Turnbull WD, Rich THV and Lundelius EL (1987) The phalangerids (Marsupialia: Phalangeridae) of the early Pliocene Hamilton Local Fauna, southwestern Victoria. In *Possums and Opposums: Studies in evolution*, volume 2 (Ed. M Archer) pp. 537–546. Surrey Beatty & Sons Pty Ltd and the Royal Zoological Society of New South Wales, Mosman, Australia.

Fleming PJG (1966) Notes on Triassic fossils from the Mount Crosby Formation in the Parish of Chuwart. *Queensland Government Mining Journal* **63**, 119–120.

Flint RB, Ambrose GJ and Campbell KSW (1980) Fossiliferous Lower Devonian boulders in

Cretaceous sediments of the Great Artesian Basin. *Transactions of the Royal Society of South Australia* **104**(3), 57–65.

Flood PG (1974) Lower Devonian brachiopods from the Point Hibbs Limestone of Western Tasmania. *Papers and Proceedings of the Royal Society of Tasmania* **108**, 113–136.

Flynn TT (1948) A description of *Prosqualodon davidis* Flynn, a fossil cetacean from Tasmania. *Transactions of the Zoological Society of London* **26**, 153–197.

Forbes MS and Bestland EA (2006) Guano-derived deposits within the sandy cave fills of Naracoorte, South Australia. *Alcheringa Special Issue* **1**, 129–146.

Forbes MS and Bestland EA (2007) Origin of the sediment deposits of the Naracoorte Caves, South Australia. *Geomorphology* **86**: 369–392.

Forbes MS, Bestland EA, Wells RT and Krull ES (2007) Palaeoenvironmental reconstruction of the Late Pleistocene to Early Holocene Robertson Cave sedimentary deposit, Naracoorte, South Australia. *Australian Journal of Earth Sciences* **54**(4): 541–559.

Fordyce RE (1982) A review of Australian fossil Cetacea. *Memoirs of the National Museum of Victoria* **43**, 43–58.

Fordyce RE (1983) Rhabdosteid dolphins (Mammalia: Cetacea) from the middle Miocene, Lake Frome area, South Australia. *Alcheringa* **7**(1), 27–40.

Fordyce RE (1984) Evolution and zoogeography of cetaceans in Australia. In *Vertebrate Zoogeography and Evolution in Australia* (Eds M Archer and G Clayton) pp. 929–948. Hesperian Press, Perth.

Fordyce RE (1991) The Australasian marine vertebrate record and its climatic and geographic implications. In *The Vertebrate Palaeontology of Australasia* (Eds P Vickers-Rich, JM Monaghan and TH Rich) pp. 1165–1190. Pioneer Design Studio and Monash University Publications Committee, Melbourne.

Forsyth SM (1989) Upper Parmeer Supergroup. In *Geology and Mineral resources of Tasmania* (Eds CF Burrett and EL Martin) pp. 309–333. Geological Society of Australia Special Publication 15, Brisbane.

Fortowski D, Crick I and Lau G (1988) 'The geological heritage of the Northern Territory. A compendium'. Conservation Commission of the Northern Territory, Geological Society of Australia, Darwin.

Foster CB, Logan GA and Summons RE (1999) The Permian–Triassic boundary in Australia: organic carbon isotopic anomalies relate to organofacies, not a biogeochemical 'event'. *Ninth Annual V M Goldschmidt Conference*. Cambridge, Massachusetts. <www.lpi.usra.edu/meetings/gold99/pdf/7301.pdf>

Frakes LA, Alley NF and Deynoux M (1995) Early Cretaceous ice rafting and climate zonation in Australia. *International Geology Review* **37**, 567–583.

Frankland G (1836) A Notice on Maria Island, on the East Coast of Van Diemens Land. *Proceedings of the Geological Society of London* **2**, 415.

Fraser R and Wells RT (2006) A palaeontological investigation of the Late Pleistocene fossil deposit in Grant Hall, Victoria Fossil Cave, Naracoorte, South Australia. *Alcheringa Special Issue* **1**, 147–161.

Fudali RF and Ford RJ (1979) Darwin glass and Darwin crater: a progress report. *Meteoritics* **14**(3), 283–296.

Gaffney ES, Archer M and White A (1992) *Warkalania*, a new meiolaniid turtle from the Tertiary Riversleigh deposits of Queensland. *The Beagle. Records of the Northern Territory Museum of Arts and Sciences* **9**, 35–47.

Gaffney ES, Kool L, Brinkman DB, Vickers-Rich P and Rich T (1998) Otwayemys, a new Cryptodiran turtle from the Early Cretaceous of Australia. *American Museum Novitates* **3233**, 1–28.

Gardiner BA, Schaeffer B and Masserie JA (2005) A review of the lower actinopterygian phylogeny. *Zoological Journal of the Linnean Society* Volume **144**(4), 511–525.

Gardiner BG (1984) Relationships of the palaeoniscoid fishes, a review based on new specimens of Mimia and Moythomasia from the Upper Devonian of Western Australia. *Bulletin of the British Museum of Natural History (Geology)* **37**, 173–428.

Gardiner BG and Miles RS (1975) Devonian fishes of the Gogo Formation, Western Australia. *Colloques internationale de C.N.R.S.* **218**, 73–79.

Gardiner BG and Miles RS (1990) A new genus of eubrachythoracid arthrodire from Gogo, Western Australia. *Zoological Journal of the Linnean Society* **99**, 159–204.

Garratt MJ (1978) New evidence for a Silurian (Ludlow) age for the earliest Baragwanathia flora. *Alcheringa* **2**, 217–224.

Garratt MJ and Rickards RB (1984) Graptolite biostratigraphy of early land plants from Victoria, Australia. *Proceedings of the Yorkshire Geological Society* **44**, 377–384.

Garratt MJ, Tims JD, Rickards RB, Chambers TC and Douglas JG (1984) The appearance of Baragwanathia (Lycophytina) in the Silurian. *Botanical Journal of the Linnean Society* **89**, 355–358.

Gee RD, Moore WR and Pike GP (1969) The geology of the Lower Gordon River particularly the Devonian sequence. *Records of the Geological Survey of Tasmania* **8**, 1–23.

Gehling JG (1987) Earliest known echinoderm – a new Ediacaran fossil from the Pound Subgroup of South Australia. *Alcheringa* **11**, 337–345.

Gehling JG (1988) A cnidarian of actinian-grade from the Ediacaran Pound Subgroup, South Australia. *Alcheringa* **12**, 299–314.

Gehling JG, Droser ML, Jensen S and Runnegar BN (2005) Ediacara organisms: relating form to function. In *Evolving Form and Function: Fossils and development. Proceedings of a symposium honoring Adolf Seilacher for his contribution to palaeontology* (Ed. DEG Briggs) pp. 43–66. Peabody Museum of Natural History, Yale University, New Haven, CT.

Gehling JG and Rigby JK (1996) Long-expected sponges from the Neoproterozoic Ediacara Fauna, Pound Subgroup, South Australia. *Journal of Paleontology* **70**, 185–195.

Gill ED (1957) The stratigraphical occurrence and palaeoecology of some Australian Tertiary marsupials. *Memoirs of the National Museum of Victoria* **21**, 135–203.

Gillette DD and Lockley MG (Eds) (1989) *Dinosaur Tracks and Traces*. Cambridge University Press, Cambridge, UK.

Glaessner MF (1979) Lower Cambrian Crustacea and Annelid worms from Kangaroo Island. *Alcheringa* **3**, 21–31.

Gorter JD and Young GC (1981) A new fish fauna of Middle Devonian age from the Taemas/Wee Jasper region of New South Wales. *Bureau of Mineral Resources Bulletin* **209**, 83–147.

Gould C (1862) Macquarie Harbour. *Tasmanian House of Assembly Paper* **26**. Tasmanian Government Printer, Hobart.

Gould C (1866) On the position of the Gordon Limestone relative to other Palaeozoic formations. *Papers and Proceedings Royal Society of Tasmania* (1866), 27–29.

Gould RE and Delevoryas T (1977) The biology of Glossopteris: evidence from petrified seed-bearing and pollen-bearing organs. *Alcheringa* **1**, 387–399.

Gradstein FM, Ogg JG and Smith AG (2004) *A Geological Time Scale, 2004*. Cambridge University Press, Cambridge, UK.

Gravestock DI (1984) Archaeocyatha from the lower parts of the Lower Cambrian carbonate sequence in South Australia. *Memoirs of the Association of Australasian Palaeontologists* **2**, 1–139.

Gravestock DI, Alexander EM, Demidenko YuE, Esakova NV, Holmer LE, Jago JB, Lin TR, Melnikova LM, Parkhaev PY, Rozanov AY, Ushatiskaya GT, Zang WL, Zhegallo EA, Zhuravlev AY (Eds) (2001) The Cambrian biostratigraphy of the Stansbury Basin, South Australia. *Transactions of the Paleontological Institute, Russian Academy of Sciences* **282**, 1–344.

Green GR (1990) Palaeozoic geology and mineral deposits of Tasmania. In *Geology and Mineral Deposits of Australia and Papua New Guinea* (Ed. FE Hughes) pp. 1207–1223. Australian Institute of Mining and Metallurgy Monograph Serial 14. Australasian Institute of Mining and Metallurgy, Melbourne.

Greenwood DR, Callen RA and Alley NF (1990) 'The correlation and depositional environment of Tertiary strata based on macrofloras in the southern Lake Eyre Basin, South Australia'. Department of Mines and Energy of South Australia Report Book 90/15, Adelaide.

Grun R, Moriarty KC and Wells RT (2001) Electron spin resonance dating of the fossil deposits in the Naracoorte Caves, South Australia. *Journal of Quaternary Science* **16**, 49–59.

Grun R, Wells R, Eggins S, Spooner N, Aubert M, Brown L and Rhodes E (2008) Electron spin resonance dating of South Australian megafauna sites. *Australian Journal of Earth Sciences* **55**, 917–935.

Guerin G and Hill RS (2003) *Gymnostoma tasmanianum* sp. nov., a fossil Casuarinaceae from the Early Oligocene of Little Rapid River, Tasmania, Australia. *International Journal of Plant Sciences* **164**, 629–634.

Gulline AB (1959) The underground water resources of the Smithton district. *Tasmanian Department of Mines, Underground Water Supply Paper* **5**, 1–72.

Haines PW (1982) Trace fossils and depositional environment of the Cambro-Ordovician Pacoota Sandstone, Amadeus Basin, Central Australia. BSc (Hons) thesis. University of Adelaide, Australia.

Haines PW (2005) Neoproterozoic to Early Cambrian stratigraphy and depositional history of the southwest margin of the Georgina Basin, central

Australia. In *Central Australian Basins Symposium; petroleum and minerals potential*, 16–18 August 2005, pp. 13–15. Northern Territory Geological Survey, Alice Springs.

Hall TS (1911) On the systematic position of the species of Squalodon and Zeuglodon described from Australia and New Zealand. *Proceedings of the Royal Society of Victoria* **23**, 257–265.

Hand SJ and Archer M (2005) A new hipposiderid genus (Microchiroptera) from an early Miocene bat community in Australia. *Palaeontology* **48**, 1–13.

Hand SJ, Archer M and Godthelp H (1990) Australian fossil hipposiderids: new evidence about the evolutionary history and radiation of Old World leaf-nosed bats. *Macroderma* **5**, 12.

Hand S, Dawson L and Augee M (1988) *Macroderma koppa*, a new Tertiary species of false vampire bat (Microchiroptera: Megadermatidae) from Wellington Caves, New South Wales. *Records of the Australian Museum* **40**, 343–351

Hand SJ, Mackness BS, Wilkinson CE and Wilkinson D (1999) First Australian Pliocene molossid bat: *Mormopterus* (Micronmus) sp. from the Chincilla Local Fauna, southeastern Queensland. *Records of the Western Australian Museum Supplement* **57**, 291–298.

Hand SJ, Murray P, Megirian D, Archer M and Godthelp H (1998) Mystacinid bats (Microchiroptera) from the Australian Tertiary. *Journal of Paleontology* **72**, 538–545.

Hanke GF, Davis SP and Wilson MVH (2001) New species of the acanthodian genus *Tetanopsyrus* from northern Canada and comments on related taxa. *Journal of Vertebrate Paleontology* **21**, 740–753.

Hanke GF and Wilson MVH (2004) New teleostome fishes and acanthodian systematics. In *Recent Advances in the Origin and Early Radiation of Vertebrates* (Eds G Arratia, MVH Wilson and R Cloutier) pp. 189–216. Verlag Dr Friedrich Pfeil, München.

Hann W (1873a) Report from Mr W. Hann, leader of the northern expedition party. *Votes and Proceedings of the Legislative Assembly of Queensland* **1**, 1031–1043.

Hann W (1873b) Copy of the diary of the Northern Expedition under the leadership of Mr William Hann. *Votes and Proceedings of the Legislative Assembly of Queensland* **1**, 1045–1070.

Harris WJ (1916) The palaeontological sequence of the Lower Ordovician rocks in the Castlemaine district. Part. 1. *Proceedings of the Royal Society of Victoria* **29**, 50–74.

Harris WJ (1933) *Isograptus caduceus* and its allies in Victoria. *Proceedings of the Royal Society of Victoria* **46**, 79–115.

Harris WJ and Thomas DE (1938a) A revised classification and correlation of the Ordovician graptolite beds of Victoria. *Mining and Geological Journal* **1**(3), 62–72.

Harris WJ and Thomas DE (1938b) Victorian graptolites, part 5. *Mining and Geological Journal* **1**(2), 70–81.

Harris WJ and Thomas DE (1939) Victorian graptolites, part 6, some multiramous forms. *Mining and Geological Journal* **2**(1), 55–60.

Harris WJ and Thomas DE (1940) Victorian graptolites (new series), part 7. *Mining and Geological Journal* **2**(2), 128–136.

Harris WJ and Thomas DE (1954) Notes on the geology of the Wellington–Macalister area. *Mining and Geological Journal* **5**(3), 34–49.

Harris WJ and Thomas DE (1955) Victorian graptolites, part 13. Graptolites from the Wellington River, part 1. *Mining and Geological Journal* **5**(6), 35–45.

Heidecker E (1959) Middle Devonian molluscs from the Burdekin Formation of north Queensland. *Department of Geology, University of Queensland Papers* **5**(2), 1–11.

Henderson RA, Kennedy WJ and McNamara KJ (1992) Maastrichtian heteromorphy ammonites from the Carnarvon Basin, Western Australia. *Alcheringa* **16**, 133–170.

Henderson RA and McNamara KJ (1985) Maastrichtian non-heteromorph ammonites from the Miria Formation, Western Australia. *Palaeontology* **28**, 35–88.

Heusser LE and Van de Geer G (1994) Direct correlation of terrestrial and marine palaeoclimatic records from four glacial-interglacial cycles-DSDP Site 594, Southwest Pacific. *Quaternary Science Review* **13**, 273–282.

Hill D (1942a) Some Tasmanian Palaeozoic Corals. *Papers and Proceedings of the Royal Society of Tasmania* (1943), 3–12.

Hill D (1942b) The Middle Devonian corals of Queensland 3. Burdekin Downs, Fanning River and Reid's Gap, north Queensland. *Proceedings of the Royal Society of Queensland* **53**, 229–268.

Hill D (1943) A reinterpretation of the Australian Palaeozoic record, based on a study of the rugose corals. *Proceedings of the Royal Society of Queensland* **54**, 53–66.

Hill D and Jell J (1970) Devonian corals from the

Canning Basin, Western Australia. *Bulletin of the Geological Survey of Western Australia* **121**, 1–158.

Hill RS (1984) Tertiary *Nothofagus* macrofossils from Cethana, Tasmania. *Alcheringa* **8**, 81–86.

Hill RS (1987) Discovery of *Nothofagus* fruits corresponding to an important Tertiary pollen type. *Nature* **327**, 56–58.

Hill RS (1989) New species of *Phyllocladus* (Podocarpaceae) macrofossils from south eastern Australia. *Alcheringa* **13**, 193–208.

Hill RS (1990) *Araucaria* (Araucariaceae) species from Australian Tertiary sediments – a micromorphological study. *Australian Systematic Botany* **3**, 203–220.

Hill RS (1991a) Tertiary *Nothofagus* (Fagaceae) macrofossils from Tasmania and Antarctica and their bearing on the evolution of the genus. *Botanical Journal of the Linnean Society* **105**, 73–112.

Hill RS (1991b) History of rainforest – evidence from plant macrofossils. In *The Rainforest Legacy: Australian National Rainforest Study*, volume 3 (Eds G Werren and P Kershaw) pp. 17–29. Special Australian Heritage Publication Series Number 7, Canberra.

Hill RS (1992a) Australian vegetation during the Tertiary: macrofossil evidence. *The Beagle. Records of the Northern Territory Museum of Arts and Sciences* **9**, 1–10.

Hill RS (1992b) *Nothofagus*: evolution from a southern perspective. *Trends in Ecology and Evolution* **7**, 190–194.

Hill RS (1994) *Nothofagus smithtonensis* (Nothofagaceae), a new species based on a reproductive structure from Early Oligocene sediments in northwest Tasmania, Australia. *Review of Palaeobotany and Palynology* **80**, 115–121.

Hill RS (1994) The history of selected Australian taxa. In *History of the Australian Vegetation: Cretaceous to Recent* (Ed. RS Hill) pp. 390–419. Cambridge University Press, Cambridge, UK.

Hill RS (1995a) Conifer origin, evolution and diversification in the southern hemisphere. In *Ecology of the Southern Conifers* (Eds NJ Enright and RS Hill) pp. 10–29. Melbourne University Press, Melbourne.

Hill RS (1995b) 'Tertiary fossil deposits of north western and north eastern Tasmania'. National Estate Grants Program, project no. 9427. Department of Plant Science, University of Tasmania, Hobart.

Hill RS (1997) The riddle of unique southern hemisphere *Nothofagus* on southwest Pacific Islands: its challenge to biogeographers. In *Processes and Patterns of Island Evolution: New Guinea to the central Pacific* (Eds A Keast and SE Miller) pp. 247–260. SPB Academic Publishers, Amsterdam.

Hill RS (2001) Biogeography, evolution and palaeoecology of *Nothofagus* (Nothofagaceae): the contribution of the fossil record. *Australian Journal of Botany* **49**, pp. 321–332.

Hill RS and Bigwood AJ (1987) Tertiary gymnosperms from Tasmania: Araucariaceae. *Alcheringa* **11**, 325–335.

Hill RS and Brodribb TJ (1999) Turner Review No. 2 – Southern conifers in time and space. *Australian Journal of Botany* **47**, 639–696.

Hill RS and Carpenter RJ (1989) Tertiary gymnosperms from Tasmania: Cupressaceae. *Alcheringa* **13**, 89–102.

Hill RS and Carpenter RJ (1991a) Evolution of *Acmopyle* and *Dacrycarpus* (Podocarpaceae) foliage as inferred from macrofossils in south-eastern Australia. *Australian Systematic Botany* **4**, 449–479.

Hill RS and Carpenter RJ (1991b) Extensive past distributions for major Gondwanic floral elements: macrofossil evidence. *Papers and Proceedings of the Royal Society of Tasmania* **124**, 239–247.

Hill RS and Christophel DC (2001) Two new species of *Dacrydium* (Podocarpaceae) based on vegetative fossils from middle Eocene sediments at Nelly Creek, South Australia. *Australian Systematic Botany* **14**, 193–205.

Hill RS and Dettmann ME (1996) Origin and diversification of the genus *Nothofagus*. In *The Ecology and Biogeography of Nothofagus Forests* (Eds TT Veblen, RS Hill and J Read) pp. 11–24. Yale University Press, New Haven, CT.

Hill RS and Jordan GJ (1995) Plant macrofossils as bioclimatic indicators during the Pliocene–Pleistocene. In *Climate Succession and Glacial History of the Southern Hemisphere over the Past Five Million Years* (Ed. PG Quilty) pp. 21–22. Australian National Antarctic Research Expeditions (ANARE) Research Notes 94. Australian Antarctic Division, Kingston, Tas.

Hill RS and Jordan GJ (1996) Macrofossils as indicators of Pliocene–Pleistocene climates in Tasmania and Antarctica. *Papers and Proceedings of the Royal Society of Tasmania* **130**, 9–15.

Hill RS, Jordan GJ and Carpenter RJ (1993) Taxodiaceous macrofossils from Tertiary and Quaternary sediments in Tasmania. *Australian*

Systematic Botany **6**, 237–249.

Hill RS, Jordan GJ and Macphail MK (1996) History and paleoecology of Australian *Nothofagus* forests. In *The ecology and biogeography of Nothofagus forests* (Eds TT Veblen, RS Hill and J Read) pp. 182–199. Yale University Press, New Haven, CT.

Hill RS, Jordan GJ, Carpenter RJ and Smith S (1993) Past environments and the Gondwana connection. In *Tasmanian Wilderness: World Heritage values.* (Eds SJ Smith and MR Banks) pp. 38–48. Royal Society of Tasmania, Hobart.

Hill RS and Macphail MK (1985) A fossil flora from rafted Pliocene–Pleistocene mudstones at Regatta Point, western Tasmania. *Australian Journal of Botany* **33**, 497–517.

Hill RS, Macphail MK and Jordan GJ (1999) Tertiary history and origins of the flora vegetation. In *Vegetation of Tasmania* (Eds JB Reid, RS Hill, MJ Brown and MJ Hovenden) pp. 39–63. Flora of Australia Supplementary Series Number 8, Australian Biological Resources Study, Canberra.

Hill RS and Merrifield HE (1993) An Early Tertiary macroflora from West Dale, southwestern Australia. *Alcheringa* **17**, 285–326.

Hill RS and Paull R (2003) *Fitzroya* (Cupressaceae) macrofossils from Cenozoic sediments in Tasmania, Australia. *Review of Palaeobotany and Palynology* **126**, 145–152.

Hill RS and Pole MS (1992) Leaf and shoot morphology of extant *Afrocarpus, Nageia* and *Retrophyllum* (Podocarpaceae) species, and species with similar leaf arrangement, from Tertiary sediments in Australasia. *Australian Systematic Botany* **5**, 337–358.

Hill RS and Pole MS (1994) Two new species of Pterostoma R.S. Hill from Cenozoic sediments in Australasia. *Review of Palaeobotany and Palynology* **80**, 123–130.

Hill RS and Scriven LJ (1997) Palaeoclimate across an altitudinal gradient in the Oligo-Miocene of northern Tasmania: an investigation of nearest living relative analysis. *Australian Journal of Botany* **45**, 493–505.

Hill RS and Scriven LJ (1999) *Falcatifolium* (Podocarpaceae) macrofossils from Palaeogene sediments in south-eastern Australia. *Australian Systematic Botany* **11**, 711–720.

Hill RS and Whang SS (1997) A new species of *Fitzroya* (Cupressaceae) from Oligocene sediments in north-western Tasmania. *Australian Systematic Botany* **9**, 867–875.

Hills CL (1914) Geological reconnaissance of the country between Cape Sorell and Point Hibbs. *Bulletin of the Geological Survey of Tasmania* **18**, 1–23.

Hills ES (1934) Tertiary freshwater fishes from southern Queensland. *Memoirs of the Queensland Museum* **10**, 157–174.

Hocking RM (1991) 'The Silurian Tumblagooda Sandstone, Western Australia'. Geological Survey of Western Australia Report 27. Geological Survey of Western Australia, Perth, WA.

Hocknull SA (2005) Additional specimens of *Bohra* (Marsupialia, Macropodidiae) from the Pliocene of Queensland. *Memoirs of the Queensland Museum* **51**(1), 26.

Hofmann HJ, Grey K, Hickman AH and Thorpe RI (1999) Origin of 3.45 Ga coniform stromatolites in Warrawoona Group, Western Australia. *Geological Society of America Bulletin* **111**(8), 1256–1262.

Hofmann HJ, Grey K, Hickman AH and Thorpe RI (1999) Origin of 3.45 Ga coniform stromatolites in Warrawoona Group, Western Australia. *Geological Society of America, Abstracts with Programs* **31**(7), 361.

Holdgate GR, Cartwright I, Blackburn DT, Wallace MW, Gallagher SJ, Wagstaff BE and Chung L (2007) The Middle Miocene Yallourn coal seam – the last coal in Australia. *International Journal of Coal Geology* **70**, 95–115.

Holdgate GR and Gallagher SJ (2003) Tertiary. In *Geology of Victoria* (Ed. WD Birch) pp. 289–335. Geological Society of Australian Special Publication No. 23, Geological Society of Australia (Victoria Division), Melbourne.

Holland T and Long JA (2009) On the phylogenetic position of *Gogonasus andrewsae* Long 1985, within the Tetrapodomorpha. *Acta Zoologica* (Stockholm) **90**(Supp 1), 285–296.

Holloway D (1994) Early Silurian trilobites from the Broken River area, north Queensland. *Memoirs of Museum Victoria* **54**(2), 243–269.

Holmes F (1997) Common Tertiary macrofossils from the coastal cliffs southwest of Torquay, Victoria. *The Fossil Collector* **50**, 5–20.

Holmes WBK and Anderson HM (2005) The Middle Triassic megafossil flora of the Basin Creek Formation, Nymboida Coal Measures, New South Wales, Australia. Part 5. The genera *Lepidopteris, Kurtziana, Rochipteris* and *Walkomiopteris. Proceedings of the Linnean Society of New South Wales* **126**, 79–139.

Howard KT (2004) Origin of Darwin Glass. PhD thesis. University of Tasmania, Australia.

Hughes PJ, Huxley JK, Judd CD, Walker-Smith GK, Hammond CA and Rozefelds AC (Eds) (2007) *Collection*. Tasmanian Museum and Art Gallery, Hobart.

Hutchinson MN and Mackness BS (2002) Fossil lizards from the Pliocene Chinchilla Local Fauna, Queensland, with description of a new species. *Records of the South Australian Museum* **35**, 169–184.

Inan K, Summons TS and King RL (1992) 'Limestone resources of Victoria'. Geological Survey of Victoria Report 97. Department of Primary Industries, Melbourne.

International Commission on Stratigraphy (2009) International Stratigraphic Chart, http://www.stratigraphy.org/column.php?id=Chart/Time%20 Scale.

Jack RL and Etheridge R Jr (1892) Geology and palaeontology of Queensland and New Guinea. *Geological Survey of Queensland Publication* **92**.

Jackson JH, Pedder AEH and Philip GM (1970) Lower Devonian biostratigraphy in the Wee Jasper region of New South Wales. *Journal of Paleontology* **44**(2), 206–251.

Jackson WD (1999a) The Tasmanian legacy of man and fire. *Papers and Proceedings of the Royal Society of Tasmania* **133**(1), 1–14.

Jackson WD (1999b) Palaeohistory of vegetation change: the last two million years. In *Vegetation of Tasmania* (Eds JB Reid, RS Hill, MJ Brown and MJ Hovenden) pp. 64–88. *Flora of Australia* Supplementary Series Number 8, Australian Biological Resources Study, Canberra.

Jago JB (1973) Cambrian agnostid communities in Tasmania. *Lethaia* **6**, 405–421.

Jago JB (1976) Late Middle Cambrian agnostid trilobites from northwestern Tasmania. *Palaeontology* **19**, 133–172.

Jago JB and Anderson A (2004) A late Middle Cambrian bivalved arthropod with antennae from Tasmania. *Memoirs of the Australasian Association of Palaeontologists* **30**, 79–81.

Jago JB and Bentley CJ (2007) Late Middle Cambrian trilobites from Christmas Hills, northwestern Tasmania. *Memoirs of the Association of Australasian Palaeontologists* **34**, 283–309.

Jago JB and Buckley JH (1971) An abrupt Upper Middle Cambrian faunal change, Christmas Hills, Tasmania. *Papers and Proceedings of the Royal Society of Tasmania* **105**, 83–85.

Jago JB and Daily B (1974) The trilobite genus *Clavagnostus* Howell from the Cambrian of Tasmania. *Palaeontology* **17**, 95–109.

James NP, Bone Y (2007) A late Pliocene–early Pleistocene inner-shelf subtropical seagrass-dominated carbonate: Roe Calcarenite Great Australian Bight Western Australia. *Palaios* **22**(4), 343–359.

James NP, Bone Y, Carter RM and Murray-Wallace CV (2006) Origin of the Late Neogene Roe Plains and their calcarenite veneer: implications for sedimentology and tectonics in the Great Australian Bight. *Australian Journal of Earth Sciences* **53**, 407–419.

Jell JS and Hill D (1970) The Devonian coral fauna of the Point Hibbs Limestone. *Papers and Proceedings of the Royal Society of Tasmania* **104**, 1–16.

Jell JS and Pedder AEH (1969) *Martinophyllum*, a new genus of Devonian rugose corals. *Journal of Paleontology* **43**, 735–740.

Jell JS, Simpson A and Mawson R and Talent JA (1993) Biostratigraphic summary. In *Geology of the Broken River Province, north Queensland* (Eds IW Withnall and SC Lang), 239–245. Queensland Geology 4, Department of Minerals and Energy, Queensland, Brisbane.

Jell PA (2004) Fossil insects of Australia. *Memoirs of the Queensland Museum* **50**(1), 1–124.

Jell PA, Burrett CF and Banks MR (1985) Cambrian and Ordovician echinoderms from eastern Australia. *Alcheringa* **9**, 183–208.

Jell PA and Duncan PN (1986) Invertebrates, mainly insects, from the freshwater, Lower Cretaceous Koonwarra Fossil Bed (Korumburra Group), South Gippsland, Victoria. *Memoirs of the Association of Australasian Palaeontologists* **3**, 111–205.

Jell PA, Jell JS, Johnson BD, Mawson R, and Talent JA (1988) Crinoids from Devonian limestones of eastern Australia. *Memoirs of the Queensland Muesum* **25**, 355–402.

Jell PA and Stait BA (1985) Tremadoc trilobites from the Florentine Valley Formation, Tim Shea area, Tasmania. *Memoirs of the Museum of Victoria* **46**, 1–34.

Jenkins RJF (1995) The problems and potential of using animal fossils and trace fossils in terminal Proterozoic biostratigraphy. *Precambrian Research* **73**, 51–69.

Jenkins RJF and Gehling JG (1978) A review of frond-like fossils of the Ediacara assemblage. *Records of* the South Australian Museum **17**, 347–359.

Jensen S, Droser ML and Gehling JG (2005) Trace fossil preservation and the early evolution of animals. *Palaeogeography, Palaeoclimatology, Palaeoecology* **220**, 19–29.

Johanson Z (1997a) New *Bothriolepis* and *Remigolepis* (Placodermi; Antiarchi) from the Hunter Siltstone (Upper Famennian) near Grenfell, NSW. *Alcheringa* **21**, 197–217.

Johanson Z (1997b) New *Remigolepis* (Placodermi; Antiarchi), from Canowindra, NSW, Australia. *Geological Magazine* **134**, 813–846.

Johanson Z (1997c) Devonian fish faunas from New South Wales, with emphasis on the Canowindra fauna. PhD thesis. Macquarie University, Australia.

Johanson Z (1998) The Upper Devonian fish *Bothriolepis* (Placodermi: Antiarchi) from near Canowindra, New South Wales, Australia. *Records of the Australian Museum* **50**, 315–348.

Johanson Z (2004) Late Devonian sarcopterygian fishes from eastern Gondwana (Australia and Antarctica) and their importance in phylogeny and biogeography. In *Recent Advances in the Origin and Early Radiation of Vertebrates* (Eds G Arratia, MVH Wilson and R Cloutier) pp. 285–306. Verlag Dr Friedrich Pfeil, München.

Johanson Z and Ahlberg PE (1997) New tristichopterid (Osteolepiformes; Sarcopterygii) from the Mandagery Sandstone (Famennian) near Canowindra, NSW, Australia. *Transactions of the Royal Society of Edinburgh* **88**, 39–53.

Johnson ME and Webb GE (2007) Outer rocky shores of the Mowanbini archipelago, Devonian Reef Complex, Canning Basin, Western Australia. *Journal of Geology* **115**, 583–600.

Johnston RM (1880) Third contribution to the natural history of the Tertiary marine beds of Table Cape with a description of 30 new species of mollusca. *Papers and Proceedings of the Royal Society of Tasmania* (1879), 29–41.

Johnston RM (1885a) Additions to the list of Table Cape fossils, together with further remarks upon certain fossil shells, supposed to be identical with living species. *Papers and Proceedings of the Royal Society of Tasmania* (1884), 220–224.

Johnston RM (1885b) Description of a new fossil shell from the Eocene beds, Table Cape. *Papers and Proceedings of the Royal Society of Tasmania* (1884), 232.

Johnston RM (1887) Contribution to the palaeontology of the Upper Palaeozoic rocks of Tasmania. *Papers and Proceedings of the Royal Society of Tasmania* (1886), 4–18.

Johnston RM (1888) *Systematic Account of the Geology of Tasmania*. Government Printer, Hobart.

Johnston RM (1902) Further notes on the 'Permo-Carboniferous Fossil Cliffs at Darlington, Maria Island'. *Papers and Proceedings of the Royal Society of Tasmania* (1900–1901), 11–20.

Jones OA and de Jersey NJ (1947) The flora of the Ipswich Coal Measures: morphology and floral succession. *University of Queensland Papers, Department of Geology* **3**, 1–88.

Jones PJ and Chen PJ (2000) Carboniferous and Permian *Leaioidea* (Branchiopoda: Conchostraca) from Australia: taxonomic revision and biostratigraphic implications. *Records of the Australian Museum* **52**, 223–244.

Jordan GJ (1992) Macrofossil evidence for Quaternary plant extinction and vegetation change in western Tasmania. PhD thesis. University of Tasmania, Australia.

Jordan GJ (1995a) Early–Middle Pleistocene leaves of extinct and extant Proteaceae from western Tasmania, Australia. *Botanical Journal of the Linnean Society* **118**, 19–35.

Jordan GJ (1995b) Extinct conifers and conifer diversity in the Early Pleistocene of western Tasmania. *Review of Palaeobotany and Palynology* **84**, 375–387.

Jordan GJ (1997a) Contrasts between the climatic ranges of fossil and extant taxa: causes and consequences for palaeoclimatic estimates. *Australian Journal of Botany* **45**, 465–474.

Jordan GJ (1997b) Evidence of Pleistocene plant extinction and diversity from Regatta Point, western Tasmania, Australia. *Botanical Journal of the Linnean Society* **123**, 45–71.

Jordan GJ (1999) A new Early Pleistocene species of *Nothofagus* and the climatic implications of co-occurring *Nothofagus* fossils. *Australian Systematic Botany* **12**, 757–768.

Jordan GJ, Carpenter RJ and Hill RS (1998) Macrofossil evidence of past diversity of Proteaceae in Tasmania, including nine new species. *Australian Systematic Botany* **11**, 465–501.

Jordan GJ and Dalton PJ (1995) Mosses from Early–Middle Pleistocene sediments in western Tasmania. *Alcheringa* **19**, 291–296.

Jordan GJ and Hill RS (1991) Two new *Banksia* species from Pleistocene sediments in western Tasmania. *Australian Systematic Botany* **4**, 499–511.

Jordan GJ and Hill RS (1994) Past and present

variability in leaf length of evergreen members of *Nothofagus* subgenus *Lophozonia* related to ecology and population dynamics. *New Phytologist* **127**, 377–90.

Jordan GJ and Hill RS (1995) Oligocene leaves of Epacridaceae from Little Rapid River, Tasmania and the identification of fossil Epacridaceae leaves. *Australian Systematic Botany* **8**, 71–83.

Jordan GJ and Hill RS (1996) The fossil record of the Epacridaceae. *Annals of Botany* **77**, 341–346.

Jordan GJ and Hill RS (1998) 'Cenozoic fossil plant locations in Tasmania'. Report for National Estate Grant 9610. Department of Plant Science, University of Tasmania, Hobart.

Jordan GJ and Hill RS (1999) The phylogenetic affinities of *Nothofagus* leaf fossils based on combined molecular and morphological data. *International Journal of Plant Science* **160**, 1177–1188.

Jordan GJ, Macphail MK, Barnes R and Hill RS (1995) An Early–Middle Pleistocene flora of subalpine affinities in lowland western Tasmania. *Australian Journal of Botany* **43**, 231–242.

Jordan GJ, Macphail MS and Hill RS (1996) A fertile pinnule fragment with spores of *Dicksonia* from Oligocene sediments in Tasmania. *Review of Palaeobotany and Palynology* **92**, 245–252.

Kakuwa Y and Webb J (2007) Trace fossils of a Middle to Upper Ordovician pelagic deep-ocean bedded chert in southeastern Australia. Sediment–organism interactions: A multifaceted ichnology. *SEPM Special Publication* **88**, 267–276.

Kear BP (2003) Cretaceous marine reptiles of Australia: a review of taxonomy and distribution. *Cretaceous Research* **24**, 277–303.

Kear BP (2004) Biogeographic and biostratigraphic implications of Australian Mesozoic marine reptiles. *Australian Biologist* **17**, 4–22.

Kear BP (2005) Marine reptiles from the Lower Cretaceous (Aptian) deposits of White Cliffs, southeastern Australia: implications of a high latitude, cold water assemblage. *Cretaceous Research* **26**, 769–782.

Kear BP (2006) Marine reptiles from the Lower Cretaceous of South Australia: of a high-latitude cold water assemblage. *Palaeontology* **49**(4), 837–856.

Kear BP (2007) Juvenile pliosauroid plesiosaur from the Lower Cretaceous of South Australia. *Journal of Paleontology* **81**(1), 154–162.

Kemp A (1991) Australian Mesozoic and Cainozoic lungfish. In *Vertebrate Palaeontology of Australasia* (Eds P Vickers-Rich, JM Monaghan, RF Baird and TH Rich) pp. 466–496. Pioneer Design Studio and the Monash University Publications Committee, Melbourne.

Kemp A (1993) Problematic Triassic dipnoans from Australia. In *The Non-marine Triassic* (Eds SG Lucas and M Morales) pp. 223–227. New Mexico Museum of Natural History and Science Bulletin 3, Albuquerque.

Kemp A (1996) Triassic Lungfish from Gondwana. In *Mesozoic Fishes: Systematics and paleoecology* (Eds G Arratia and G Viohl) pp. 409–416. Friedrich Pfeil Verlag, München.

Kemp N (1991) Chondrichthyans in the Cretaceous and Tertiary of Australia. In *Vertebrate Palaeontology of Australia* (Eds P Vickers-Rich, JM Monaghan, RF Baird and TH Rich) pp. 497–568. Pioneer Design Studio with the Monash University Publications Committee, Melbourne.

Kendrick KJ, McNamara KJ and Brimmel K (1997) *A Guide to the Fossils of the Roe Plains*. Department of Earth and Planetary Sciences, Western Australian Museum, Perth.

Kershaw AP, Bretherton SC and van der Kaars S (2007) A complete pollen record of the last 230 ka from Lynch's Crater, north-eastern Australia. *Palaeogeography, Palaeoclimatology, Palaeoecology*, **251**(1), 23–45.

Knight OL (1950) Fossil Insect Beds of Belmont, NSW. *Records of the Australian Museum* **22**, 251–253.

Kobayashi T (1940) Lower Ordovician fossils from Junee, Tasmania. *Papers and Proceedingsof the Royal Society of Tasmania* (1939), 61–66.

Krause J and AhYee C (1995) Victoria's Muddy Creek: a real seafood cocktail. *The Fossil Collector* **44**, 11–20.

Kruse PD (1990) Cyanobacterial-archaeocyathan-radiocyathan bioherms in the Wirrealpa Limestone of South Australia. *Canadian Journal of Earth Science* **28**, 601–615.

Lang WH and Cookson IC (1935) On a flora, including vascular plants, associated with Monograptus, in rocks of Silurian age, from Victoria, Australia. *Philosophical Transactions of the Royal Society of London, Series B, Biological Sciences* **224**(517), 421–449.

Langloh Parker K (1905) *The Euahlayi Tribe: a study of Aboriginal life in Australia*. Archibald Constable and Company Ltd, London.

de Lapparent DB and Molnar RE (2001) Eocene chelid turtles from Redbank Plains, southeast Queensland, Australia. *Geodiversitas* **23**(1), 41–79.

Laurie JR (1980) Early Ordovician orthide brachiopods from southern Tasmania. *Alcheringa* **4**, 11–23.

Laurie JR (1982) The taxonomy and biostratigraphy of the Ordovician and Early Silurian articulate brachiopods of Tasmania. PhD thesis. University of Tasmania, Australia.

Laurie JR (1991) Articulate brachiopods from the Ordovician and Lower Silurian of Tasmania. *Memoirs of the Association of Australasian Palaeontologists* **11**, 1–106.

Leichhardt L (1847) *Journal of an Overland Expedition in Australia from Moreton Bay to Port Essington during the Years 1844–5.* Boone, London.

Li G-Q (1994) Systematic position of the Australian fossil osteoglossid fish *Phareodus* (=Phareoides) *queenslandicus* Hills. *Memoirs of the Queensland Museum* **37**(1), 287–300.

Li Q, Davies PJ and McGowran B (1999) Foraminiferal sequence biostratigraphy of the Oligo-Miocene Janjukian strata from Torquay, southeastern Australia. *Australian Journal of Earth Sciences* **46**, 261–273.

Lin J-P, Gon SM, Gehling JG, Babcock LE, Zhao Y-L, Zhang X-L, Hu S-X, Yuan J-L, Yu M-Y and Peng J (2006) A Parvancorina-like arthropod from the Cambrian of South China. *Historical Biology* **18**(1), 33–45.

Lindley ID (2001) Lower Devonian ischnacanthid fish (Gnathostomata: Acanthodii) from the Taemas Limestone, Lake Burrinjuck, New South Wales. *Alcheringa* **25**(3), 269–291.

Lindley ID (2002) Acanthodian, onychodontid and osteolepidid fish from the middle-upper Taemas Limestone (Early Devonian), Lake Burrinjuck, New South Wales. *Alcheringa* **26**(1), 103–126.

Long J (2004) 'Expert consultancy report: The assessment of the dinosaur trackways within the Lark Quarry Conservation Park for potential National Heritage values'. Gogo Press, Perth.

Long J and Mackness B (1994) Studies of the late Cainozoic diprotodontid marsupials of Australia. 4. The Bacchus Marsh Diprotodons: geology, sedimentology and taphonomy. *Records of the South Australian Museum* **27**(2), 95–110.

Long JA (1983a) New bothriolepid fishes from the Late Devonian of Victoria, Australia. *Palaeontology* **26**, 295–320.

Long JA (1983b) A new diplacanthoid acanthodian from the Late Devonian of Victoria, Australia. *Memoirs of the Association of Australasian Palaeontologists* **1**, 51–65.

Long JA (1984) New phyllolepids from Victoria and the relationships of the group. *Proceedings of the Linnean Society of New South Wales* **107**, 263–308.

Long JA (1985) A new osteolepidid fish from the Upper Devonian Gogo Formation, *Western Australia. Records of the Western Australian Museum* **12**, 361–377.

Long JA (1985) The structure and relationships of a new osteolepiform fish from the Late Devonian of Victoria, Australia. *Alcheringa* **9**, 1–22.

Long JA (1986) A new Late Devonian acanthodian fish from Mt Howitt, Victoria, Australia, with remarks on acanthodian biogeography. *Proceedings of the Royal Society of Victoria* **98**, 1–17.

Long JA (1987) An unusual osteolepiform fish from the Late Devonian of Victoria, Australia. *Palaeontology* **30**(4), 839–852.

Long JA (1987) Late Devonian fishes from the Gogo Formation, Western Australia – new discoveries. *Search* **18**, 203–205.

Long JA (1988) New Palaeoniscoid fishes from the Late Devonian – Early Carboniferous of Victoria, Australia. *Memoirs of the Association of Australasian Palaeontologists* **7**, 1–64.

Long JA (1988a) New information on the Late Devonian arthrodire *Tubonasus* from Gogo, Western Australia. *Memoirs of the Association of Australasian Palaeontologists* **7**, 81–85.

Long JA (1988b) A new camuropiscid arthrodire (Pisces: Placodermi) from Gogo, Western Australia. *Zoological Journal of the Linnean Society* **94**, 233–258.

Long JA (1988c) Late Devonian fishes from the Gogo Formation, Western Australia. *National Geographic Research* **4**, 436–450.

Long JA (1988d) The extraordinary fishes of Gogo. *New Scientist* **1639**, 41–45.

Long JA (1988e) 360 million-year-old Gogo fishes. *Geo: Australasia's Geographical Magazine* **11**(3), 102–113.

Long JA (1990) Two new arthrodires (placoderm fishes) from the Upper Devonian Gogo Formation, Western Australia. *Memoirs of the Queensland Museum* **28**, 51–63.

Long JA (1991a) Arthrodire predation by Onychodus (Pisces, Crossopterygii) from the Late Devonian Gogo Formation, Western Australia. *Records of the Western Australia Museum* **15**(2), 479–482.

Long JA (1991b) The long history of Australian fossil fishes. In *Vertebrate Palaeontology of Australasia.* (Eds P Vickers-Rich, JM Monaghan, RF Baird and TH Rich) pp. 336–428. Pioneer Design Studio and Monash University Publications Committee, Melbourne.

Long JA (1992a) Cranial anatomy of two new Late Devonian lungfishes (Pisces, Dipnoi) from Mt Howitt, Victoria. *Records of the Australian Museum* **44**, 299–318.

Long JA (1992b) *Gogodipterus paddyensis* gen. nov., a new chirodipterid lungfish from the Late Devonian Gogo Formation, Western Australia. *The Beagle. Records of the Northern Territory Museum of Arts and Sciences* **9**, 11–20.

Long JA (1993a) Cranial ribs in Devonian lungfishes and the origin of air-breathing. *Memoirs of the. Australasian Association of Palaeontologists* **15**, 199–210.

Long JA (1993b) *Palaeozoic Vertebrate Biostratigraphy and Biogeography*. Belhaven Press, London.

Long JA (1993c) Window into the past – The Gogo fishes. *Western Fisheries* Autumn, 28–29.

Long JA (1994) A second incisoscutid arthrodire from Gogo, Western Australia. *Alcheringa* **18**, 59–69.

Long JA (1995) *The Rise of Fishes: 500 million years of evolution*. University of New South Wales Press, Sydney; Johns Hopkins University Press, Baltimore.

Long JA (1996). The long history of Australian fossil fish. In *Vertebrate Palaeontology of Australasia*. (Eds P Vickers–Rich, JM Monaghan, RF Baird and TH Rich). Pioneer Design Studio, Melbourne, 336–428.

Long JA (1997) Ptyctodontid fishes (Vertebrata, Placodermi) from the Late Devonian Gogo Formation, Western Australia, with a revision of the European genus Ctenurella Orvig, 1960. *Geodiversitas* **19**, 515–555.

Long JA (1999) A new genus of fossil coelacanth (Osteichthyes; Coelacanthiformes) from the Middle Devonian of southeastern Australia. *Records of the Western Australian Museum Supplement* **57**, 37–54.

Long JA (2006) *Swimming in Stone: The amazing Gogo fossils of the Kimberley*. Fremantle Arts Centre Press, Fremantle.

Long JA, Barwick RE and Campbell KSW (1997) Osteology and functional morphology of the osteolepiform fish *Gogonasus andrewsae* Long, 1985, from the Upper Devonian Gogo Formation, Western Australia. *Records of the Western Australian Museum Supplement* **53**, 1–89.

Long JA, Trinajstic K, Johanson Z (2009) Devonian arthrodire embryos and the origin of internal fertilization in vertebrates. *Nature* **457**, 1124–1127.

Long JA, Trinajstic K, Young GC and Senden T (2008) Live birth in the Devonian period. *Nature* **453**, 650–652.

Long JA and Turner S (1984) A checklist and bibliography of Australian fossil fish. In *Vertebrate Zoogeography and Evolution in Australia* (Eds M Archer and G Clayton) pp. 235–254. Hesperian Press, Perth.

Long JA and Werdelin L (1986) A new Late Devonian bothriolepid (Placodermi, Antiarcha) from Victoria, with descriptions of others from the state. *Alcheringa* **10**, 355–399.

Long JA, Young GC, Holland T, Senden TJ and Fitzgerald EMG (2006) An exceptional Devonian fish from Australia sheds light on tetrapod origins. *Nature* **444**, 199–202.

Louys J, Aplin K, Beck RMD and Archer M (2009) Cranial anatomy of Oligo-Miocene koalas (Diprotodontia: Phascolarctidae): stages in the evolution of an extreme leaf-eating specialisation. *Journal of Vertebrate Paleontology* **29**, 981–992.

Ludbrook NH (1973) Distribution and stratigraphic utility of Cenozoic molluscan faunas in southern Australia. *Science Reports of Tohoku University, Series 2: Geology, Special Volume* **6**, 241–261.

Ludbrook NH (1978) Quaternary molluscs of the western part of the Eucla Basin. *Western Australian Geological Survey Bulletin* **125**, 1–286.

Luo ZX, Kielan-Jaworowska Z and Cifelli RL (2002) In quest for a phylogeny of Mesozoic mammals. *Acta Palaeontologica Polonica* **47**, 1–78.

Mackaness G (1947) Some private Correspondence of Sir John and Lady Jane Franklin. *Australian Historical Monographs* **15**. Author, Sydney.

Macphail MK, Alley NF, Truswell EM and Sluiter IRK (1994) Early Tertiary vegetation: evidence from spores and pollen. In *History of Australian Vegetation: Cretaceous to Recent* (Ed. RS Hill) pp. 189–261. Cambridge University Press, Cambridge, UK.

Macphail MK, Jordan GJ and Hill RS (1993) Key periods in the evolution of the flora and vegetation in western Tasmania. 1. The Early–Middle Pleistocene. *Australian Journal of Botany* **41**, 673–70.

Mahoney JA and Ride WDL (1975) Index to the genera and species of fossil Mammalia described from Australia and New Guinea between 1838 and 1968. *Western Australian Museum Special Publication* **6**, 1–250.

McClenaghan MP and Findlay RH (1993) 'Geological

Atlas 1:50 000 Series. Sheet 64 (7913S) Macquarie Harbour. Explanatory Report.' Geological Survey Tasmania, Rosny Park Tas.

McClenaghan MP, Seymour DB, Villa IM (1994) Lamprophyre dyke suites from western Tasmania, their radiometric dating and the age of thrust faulting in the Point Hibbs area. *Australian Journal of Earth Sciences* **41**, 47–54.

McHenry CR, Cook AG and Wroe S (2005) Bottom-feeding plesiosaurs. *Science* **310**, 75.

McHenry B and Yates A (1993) First report of the enigmatic metazoan Anomalocaris from the southern hemisphere and a trilobite with preserved appendages from the early Cambrian of Kangaroo Island, SA. *Records of the South Australian Museum* **26**, 77–86.

McLoughlin S (1992) Late Permian megafossils from the Bowen Basin, Queensland, Australia: Part 1. *Palaeontographica Abteiling B* **228**, 105–149.

McLoughlin S (1996) Early Cretaceous macrofloras of Western Australia. *Records of the Western Australian Museum* **18**, 19–65.

McLoughlin S and Hill RS (1996) The succession of Western Australian Phanerozoic terrestrial floras. In *Gondwanan Heritage: Past, present and future of the Western Australian biota* (Eds SD Hopper, JA Chappill, MS Harvey and AS George) pp. 61–80. Surrey Beatty & Sons, Chipping Norton, NSW.

McLoughlin S and McNamara K (2001) *Ancient Floras of Western Australia*. Western Australian Museum, Quality Press, Osborn Park.

McLoughlin S, Drinnan AN and Rozefelds AC (1995) A Cenomanian Flora from the Winton Formation, Eromanga Basin, Queensland, Australia. *Memoirs of the Queensland Museum* **38**(1), 273–313.

McLoughlin S, Tosolini A-M, Nagalingum N and Drinnan AN (2002) The Early Cretaceous (Neocomian) flora and fauna of the lower Strzelecki Group, Gippsland Basin, Victoria, Australia. *Memoirs of the Association of Australasian Palaeontologists* **26**, 1–144.

McNamara KJ (1997) *Shapes of Time: The evolution of growth and development*. Johns Hopkins University Press, Baltimore.

McNamara KJ and Trewin NH (1993) A euthycarcinoid arthropod from the Silurian of Western Australia. *Palaeontology* **36**, 319–335.

Megirian D (1994) Approaches to marsupial biochronology in Australia and New Guinea. *Alcheringa* **18**, 259–274.

Megirian D, Prideaux GJ, Murray PF and Smit N (2010) An Australian land mammal age biochronological scheme. *Paleobiology* **36**(4), 658–671.

Miles RS (1971) The Holonematidae (placoderm fishes): a review based on new specimens of Holonema from the Upper Devonian of Western Australia. *Philosophical Transactions of the Royal Society of London B* **263**, 101–234.

Miles RS (1977) Dipnoan (lungfish) skulls from the Upper Devonian of Western Australia. *Zoological Journal of the Linnean Society* **61**, 1–328.

Miles RS and Dennis K (1979) A primitive eubrachythoracid arthrodire from Gogo, Western Australia. *Zoological Journal of the Linnean Society* **66**, 31–62.

Miles RS and Young GC (1977) Placoderm interrelationships reconsidered in the light of new ptyctodontids from Gogo Western Australia. *Linnean Society Symposium Series* **4**, 123–198.

Mill RR and Hill RS (2004) Validations of the names of seven Podocarpaceae macrofossils. *Taxon* **53**, 1043–1046.

Mitchell J (1925) Description of new species of Leaia. *Proceedings of the Linnean Society of New South Wales* **50**, 438–447.

Mitchell, TL (1838) *Three Expeditions into the Interior of Eastern Australia with Descriptions of the Recently Explored Region of Australia Felix and of the Present Colony of New South Wales*, 2 vols. T&W Boone, London.

Molnar RE (1991) Fossil reptiles in Australia. In *Vertebrate Palaeontology of Australia* (Eds P Vickers-Rich, JM Monaghan, RF Baird and TH Rich) pp. 605–702. Pioneer Design Studio with the Monash University Publications Committee, Melbourne.

Montgomery A (1891) Report on the hydraulic limestones on Maria Island. *Report of the Secretary for Mines, Tasmania* (1890–91), 22–25.

Moriarty KC, McCulloch MT, Wells RT and McDowell MC (2000) Mid-Pleistocene cave fills, megafaunal remains and climate change at Naracoorte, South Australia: towards a predictive model using U/Th dating of speleothems. *Palaeogeography, Palaeoclimatology, Palaeoecology* **159**,113–143.

Munson TJ (1979) Geology and Silurian rugose corals of the Jack Hills Gorge area, Broken River Province, north Queensland. BSc (Hons) thesis. University of Queensland, Australia.

Munson TJ (1987) Silurian stratigraphy and coral faunas of the Graveyard Creek subprovince, northeast Queensland. PhD thesis. University of Queensland, Australia.

Murray PF and Megirian D (1992). Continuity and contrast in middle and late Miocene vertebrate communities from the Northern Territory. *The Beagle. Records of the Northern Territory Museum of Arts and Sciences* **9**(1), 195–218.

Murray PF, Megirian D, Rich T, Plane M, Black K, Archer M, Hand S and Vickers-Rich P (2000) Morphology, systematics, and evolution of the marsupial genus *Neohelos* Stirton (Diprotodontidae, Zygomaturinae). *Museums and Art Galleries of the Northern Territory Research Report* **6**, 1–141.

Murray PF, Megirian D and Wells R (1993) *Kolopsis yperus* sp. nov. (Zygomaturinae, Marsupialia) from the Ongeva Local Fauna: new evidence for the age of the Alcoota fossil beds of central Australia. *The Beagle. Records of the Northern Territory Museum of Arts and Sciences* **10**, 155–172.

Nedin C (1995a) The palaeontology and palaeoenvironment of the early Cambrian Emu Bay Shale, Kangaroo Island, SA. PhD thesis. University of Adelaide, Australia.

Nedin C (1995b) The Emu Bay Shale, a lower Cambrian fossil Lagerstatten, Kangaroo Island SA. *Memoirs of the Association of Australasian Palaeontologists* **18**, 31–40.

Nedin C (1999) Anomalocaris predation on nonmineralised and mineralised trilobites. *Geology* **27**, 987–990.

Neudert M (1986) 'Palaeocurrents and facies in the Jurassic Razorback Beds, Mount Morgan, Queensland'. Sedcon Professional Report Q18/86. Sedcon, Paddington Qld.

Neveling J (2004) Stratigraphic and sedimentological investigation of the contact between the Lystrosaurus and the Cynognathus Assemblage Zones (Beaufort Group: Karoo Supergroup) *Council for Geoscience Bulletin* **137**, 1–164.

Newsome AE and Rochow KA (1964) Vertebrate fossils from Tertiary sediments in Central Australia. *Australian Journal of Science* **26**, 352.

Nicoll RS (1991) Differentiation of Late Cambrian–Early Ordovician species of Cordylodus (Conodonta) with biapical basal cavities. *BMR Journal of Australian Geology and Geophysics* **12**(3), 223–244.

Nicoll RS and Shergold JH (1991) Revised Late Cambrian (pre-Payntonian-Datsonian) conodont biostratigraphy at Black Mountain, Georgina Basin, western Queensland, Australia. *BMR Journal of Australian Geology and Geophysics* **12**, 93–118.

Nishida H, Pigg KB and Rigby JF (2003) Swimming sperm in an extinct Gondwanan plant. *Nature* **422**, 396–397.

Noble PJ and Webby BD (2009) Katian (Ordovician) radiolarians from the Malongulli Formation, New South Wales, Australia, a reexamination. *Journal of Paleontology* **83**(4), 548–561.

Northwood C (1999) Actinopterygians from the Early Triassic Arcadia Formation, Queensland, Australia. *Memoirs of the Queensland Museum* **43**(2), 787–796.

Northwood C (2005) Early Triassic coprolites from Australia and their palaeobiological significance. *Palaeontology* **48**(1), 49–68.

O'Dowd DJ, Brew CR, Christophel DC and Norton RA (1991) Mite–plant associations from the Eocene of southern Australia. *Science* **252**, 99–101.

Oehler DZ (1976) Transmission electron microscopy of organic microfossils from the Late Precambrian Bitter Springs Formation of Australia. *Journal of Palaeontology* **50**, 90–106.

Ogg JG, Ogg G and Gradstein FM (2008) *The Concise Geological Time Scale*. Cambridge University Press, Cambridge, UK.

Oosting AM (2004) Palaeoenvironmental and climatic changes in Australia during the Early Createceous. PhD thesis. Utrecht University, Netherlands and James Cook University, Australia.

Oxley J (1820) *Journals of Two Expeditions into the Interior of New South Wales, Undertaken by Order of the British Government in the Years 1817–1818.* John Murray, London.

Pate FD, McDowell MC, Wells RT and Smith AM (2002) Last recorded evidence for megafauna at Wet Cave, Naracoorte, South Australia 45,000 years ago. *Australian Archaeology* **54**, 53–55.

Pate DF, McDowell MC, Wells RT and Smith AM (2006) AMS radiocarbon dating of charcoal and the timing of megafaunal extinctions in the Naracoorte region of South Australia. *Alcheringa Special Issue* **1**: 277–279.

Paterson JR and Jago JB (2006) New trilobites from the lower Cambrian Emu Bay Shale Largerstatten at Big Gully, Kangaroo Island, SA. *Memoirs of the Association of Australasian Palaeontologists* **32**, 43–57.

Pattemore GA and Rigby JF (2005) Fructifications and foliage from the Mesozoic of southeast Queensland. *Memoirs of the Queensland Museum* **50**, 329–345.

Pedder AEH (1998) New and revised Lower Devonian Rugosa from western North America and Tasmania. *Journal of Paleontology* **72**(2), 224–245.

Percival IG (1985) *The Geological Heritage of New South Wales*, volume 1. NSW National Parks and Wildlife Service, Sydney.

Percival IG (1991) Late Ordovician articulate brachiopods from central New South Wales. *Memoirs of the Association of Australasian Palaeontologists* **11**, 107–177.

Peters MD and Christophel DC (1978) *Austrosequoia wintonensis*, a new taxodiaceous cone from Queensland, Australia. *Canadian Journal of Botany* **56**, 3119–3128.

Pigg KB and Nishida H (2006) The significance of silicified plant remains to the understanding of Glossopteris-bearing plants: an historical review. *Journal of the Torrey Botanical Society* **133**, 46–61.

Piper KJ, Fitzgerald EMG and Rich TH (2006) Mesozoic to early Quaternary mammal faunas of Victoria, south-east Australia. *Palaeontology* **49**, 1237–1262.

Plane M and Gatehouse CG (1968) A new vertebrate fauna from the Tertiary of northern Australia. *Australian Journal of Science* **30**(7), 272–273.

Playford G and Cornelius KD (1967) Palynological and lithostratigraphic features of the Razorback Beds, Mount Morgan district, Queensland. *Papers of the Department of Geology, University of Queensland* **6**, 81–96.

Playford PE (1980) Devonian 'Great Barrier Reef' of Canning Basin, Western Australia. *The American Association of Petroleum Geologists Bulletin* **64**(6), 814–840.

Playford PE, Cope RN, Cockbain AE, Low GH and Lowry DC (1975) Phanerozoic. In *The Geology of Western Australia*. Geological Survey of Western Australia Memoir 2, 1–541.

Playford PE, Hocking RM and Cockbain AE (2009) Devonian reef complexes of the Canning Basin, Western Australia. *Geological Survey of Western Australia*, Bulletin 145, Perth.

Playford PE and Lowry DC (1966) Devonian Reef complexes of the Canning Basin, Western Australia. *Bulletin of the Geological Survey of Western Australia* **116**, 1–150.

Pojeta J, Gilbert-Tomlinson J and Shergold JH (1977) Cambrian and Ordovician rostroconch molluscs from northern Australia. *Bulletin of the Bureau of Mineral Resources, Geology and Geophysics Australia* **171**, 1–54.

Pole MS (1998a) Early Eocene estuary at Strahan, Tasmania. *Australian Journal of Earth Sciences* **45**, 979–985.

Pole MS (1998b). The fossil flora of Melville Island, Northern Australia. *The Beagle. Records of the Northern Territory Museum of Arts and Sciences* **14**, 1–28.

Pole MS (2000) Dicotyledonous leaf macrofossils from the latest Albian–earliest Cenomanian of the Eromanga Basin, Queensland, Australia. *Paleontological Research* **4**, 39–52.

Pole MS and Bowman D (1996) Tertiary plant fossils from Australia's 'Top End'. *Australian Systematic Botany* **9**, 113–126.

Pole MS and Macphail MK (1996) Eocene Nypa from Regatta Point, Tasmania. *Review of Palaeobotany and Palynology* **92**, 55–67.

Pretyman ER (1970) *Some Notes on Maria Island and its Penal Settlements*. Tasmanian Museum and Art Gallery, Hobart.

Prideaux GJ, Roberts RG, Megirian D, Westaway KE, Hellstrom, JC and Olley JM (2007) Mammalian responses to Pleistocene climate change in southeastern Australia. *Geology* **35**, 33–36.

Pridmore P, Rich TH, Vickers-Rich P and Gambaryan PP (2005) A Tachyglossid-like humerus from the Early Cretaceous of south-eastern Australia. *Journal of Mammalian Evolution* **12**, 359–378.

Pritchard GB (1896) A revision of the fossil fauna of the Table Cape beds, Tasmania, with descriptions of the new species. *Proceedings of the Royal Society of Victoria* **8**, 74–150.

Pritchard GB (1903) On some Australian Tertiary pleurotomarias. *Proceedings of the Royal Society of Victoria* **16**, 83–91.

Pritchard GB (1913) A revision of the fossil volutes of the Table Cape beds, Tasmania, with descriptions of new species and varieties. *Victorian Naturalist* **16**, 192–201.

Pritchard GB (1939) On the discovery of a fossil whale in the older Tertiaries of Torquay, Victoria. *Victorian Naturalist* **55**, 151–159.

Quilty PG (1971) Cambrian and Ordovician dendroids and hydroids of Tasmania. *Journal of the Geological Society of Australia* **17**, 171–189.

Quilty PG (1972) The biostratigraphy of the Tasmanian marine Tertiary. *Papers and Proceedings of the Royal Society of Tasmania* **106**, 25–44.

Quilty PG (1974) Tasmanian Tertiary Foraminifera. Part 1. Textulariina, Miliolina, Nodosariacea. *Papers and Proceedings of the Royal Society of Tasmania* **108**, 31–106.

Quilty PG (1977) Tasmanian Tertiary Foraminifera. Part 2. chiefly Spirillinacea to Glabratellidae. *Papers and Proceedings of the Royal Society of*

Tasmania **111**, 69–109.

Quilty PG 1(982) Tasmanian Tertiary Foraminifera. Part 3. Discorbacea (Eponididae) to Nonionacea. *Papers and Proceedings of the Royal Society of Tasmania* **116**, 5–51.

Raggatt HG and Crespin I (1952) Geology of the Tertiary rocks between Torquay and Eastern View, Victoria. *Australian Journal of Science* **14**, 143–147.

Raggatt HG and Crespin I (1955) Stratigraphy of Tertiary rocks between Torquay and Eastern View, Victoria. *Proceedings of the Royal Society of Victoria* **67**, 75–142.

Rao CP and Green DC (1982) Oxygen- and carbon-isotopes of shallow-marine carbonates of Tasmania. *Marine Geology* **53**, 117–129.

Rasnitsyn AP and Aristov DS (2002) Two new insects from the upper Permian (Tatarian) of Belmont, New South Wales, Australia (Insecta: Hypoperlida: Anthracoptilidae = Permarraphidae; Grylloblattida: Sylvaphlebiidae). *Palaeontological Journal* **38**, 158–163.

Reed EH (2006) *In situ* taphonomic investigation of Pleistocene large mammal bone deposits from the ossuaries Victoria Fossil Cave, Naracoorte South Australia. *Helictite* **39**, 5–15.

Reed EH (2008) Pinning down the pitfall: entry points for Pleistocene vertebrate remains and sediments in the Fossil Chamber, Victoria Fossil Cave, Naracoorte, South Australia. *Quaternary Australasia* **25**(2): 2–8.

Reed EH and Bourne SJ (2000) Pleistocene fossil vertebrate sites of the south east region of South Australia. *Transactions of the Royal Society of South Australia* **124**, 61–90.

Reed EH and Bourne SJ (2009) Pleistocene fossil vertebrate sites of the south east region of South Australia II. *Transactions of the Royal Society of South Australia* **133**(1), 30–40.

Reed L and Gillieson D (2003) Mud and bones: cave deposits and environmental history in Australia. In *Beneath the Surface: A natural history of Australian Caves* (Eds B Finlayson and E Hamilton-Smith) pp. 89–110. UNSW Press, Sydney.

Reid CM (2003) Permian bryozoa of Tasmania and New South Wales: systematics and their use in Tasmanian biostratigraphy. *Memoirs of the Association of Australasian Palaeontologists* **28**, 1–133.

Resource Planning and Development Commission (1998) 'Inquiry into certain Crown lands in the Central Highlands of Tasmania'. Resource Planning and Development Commission, Hobart.

Retallack GJ (1975) The life and times of a Triassic lycopod. *Alcheringa* **1**, 3–29.

Retallack GJ (1977) Reconstructing Triassic vegetation of eastern Australasia: a new approach to the biostratigraphy of Gondwanaland. *Alcheringa* **1**(3), 247–277.

Rich L (1995) My little dino: a girl and her paleontologist parents dig for dinosaur bones in Dinosaur Cove, Australia. *Ranger Rick*, February.

Rich TH and Rich PV (1989) Polar dinosaurs and biotas of the early Cretaceous of southeastern Australia. *National Geographic Research* **5**, 15–53.

Rich TH and Vickers-Rich P (2000) *The Dinosaurs of Darkness*. University of Indiana Press, Bloomington; Allen & Unwin, Sydney.

Rich TH and Vickers-Rich P (2003) *A Century of Australian Dinosaurs*. Queen Victoria Museum, Launceston, Tasmania and Monash Science Centre, Clayton, Vic.

Rich TH and Vickers-Rich P (2004) Diversity of Early Cretaceous mammals from Victoria, Australia. *Bulletin of the American Museum of Natural History* **285**, 36–53.

Rich TH, Archer M, Hand SJ, Godthelp H, Muirhead J, Pledge NS, Flannery TF, Woodburne MO, Case JA, Tedford RH, Turnbull WD, Lundelius Jr EL, Vickers-Rich LS, Whitelaw MJ, Kemp A, Vickers-Rich P (1993) Australian Mesozoic and Tertiary terrestrial mammal localities. In *Vertebrate Palaeontology of Australasia* (Eds P Vickers-Rich, JM Monaghan, RF Baird and TH Rich) pp.1005–1069. Pioneer Design Studio and the Monash University Publications Committee, Melbourne.

Rich TH, Flannery TF, Trusler P and Vickers-Rich PV (2001b) Corroboration of the Garden of Eden Hypothesis. In *Faunal and Floral Migrations and evolution in SE Asia-Australia* (Eds I Metcalfe, LMB Smith, M Morwood, I Davidson and K Hewison) pp. 315–324. Balkema, Lisse.

Rich TH, Flannery TF, Trusler P, Kool L, van Klaveren NA and Vickers-Rich P (2001a) A second tribosphenic mammal from the Mesozoic of Australia. *Records of the Queen Victoria Museum* **110**, 1–9.

Rich TH, Flannery TF, Trusler P, Kool L, van Klaveren NA and Vickers-Rich P (2002a) Evidence that monotremes and ausktribosphenids are not sister groups. *Journal of Vertebrate Paleontology* **22**, 466–469.

Rich TH, Hopson JA, Musser AM, Flannery TF and

Vickers-Rich P (2005) Independent origins of middle ear bones in monotremes and therians. *Science* **307**, 910–914.

Rich TH, Vickers-Rich P, Constantine A, Flannery TF, Kool L and van Klaveren N (1997) A Tribosphenic Mammal from the Mesozoic of Australia. *Science* **278**, 1438–1442.

Rich TH, Vickers-Rich P, Constantine A, Flannery TF, Kool L and van Klaveren N (1999) Early Cretaceous mammals from Flat Rocks, Victoria, Australia. *Records of the Queen Victoria Museum* **106**, 1–34.

Rich TH, Vickers-Rich P and Gangloff RA (2002b) Polar dinosaurs. *Science* **295**, 979–980.

Rich THV (1976) Recent fossil discoveries in Victoria. Five late Cenozoic fossil marsupial sites in Victoria; a progress report. *Victorian Naturalist* **93**(5), 198–206.

Rich THV (1986) *Darcius duggani*, a new ektopodontid (Marsupialia, Phalangeroidea) from the early Pliocene Hamilton Local Fauna, Australia. In *Revision of the Ektopodontidae (Mammalia: Marsupialia; Phalangeroidea) of the Australian Neogene* (Eds MO Woodburne and WA Clemens) pp. 68–74. University of California Publications in Geological Science 131, Berkeley.

Rickards RB (2000) The age of the earliest club mosses: the Silurian *Baragwanathia* flora in Victoria, Australia. *Geological Magazine* **137**(2), 207–209

Rickards RB, Partridge PL and Banks MR (1991) *Psigraptus jacksoni* Rickards and Stait: systematics, reconstruction, distribution and preservation. *Alcheringa* **15**, 243–254.

Rickards RB and Stait BA (1984) *Psigraptus*, its classification, evolution and zooid. *Alcheringa* **8**, 101–111.

Ride WDL (1993) *Jackmahoneya* gen. nov. and the genesis of the macropodiform molar. *Memoirs of the Association of Australasian Palaeontologists* **15**, 441–459.

Riek EF (1952) The fossil insects of the Tertiary Redbank Plains series. Part 1. An outline of the fossil assemblage with descriptions of the fossil insects of the orders Mecoptera and Neuroptera. *University of Queensland Papers, Department of Geology* **4**, 3–14.

Riek EF (1953) Fossil mecopteroid insects from the upper Permian of New South Wales. *Records of the Australian Museum* **23**, 55–87.

Riek EF (1954) The fossil Diptera of the Tertiary Redbank Plains series, Queensland. *Proceedings of the Linnean Society of New South Wales* **79**, 58–60.

Riek EF (1955) Fossil insects from the Triassic beds at Mount Crosby, Queensland. *Australian Journal of Zoology* **3**, 654–691.

Riek EF (1968) Undescribed fossil insects from the upper Permian of Belmont, New South Wales (with an appendix listing the described species). *Records of the Australian Museum* **27**, 303–310.

Riek EF (1970a) Fossil history. In *The Insects of Australia*. CSIRO Publishing and Melbourne University Press, Melbourne, Victoria.

Riek EF (1970b) Origin of the Australian insect fauna. In *Proceedings of the Second Gondwana Symposium* volume 1, pp. 593–598. Council for Scientific and Industrial Research, Scientia, South Africa.

Riek EF (1973) A Carboniferous insect from Tasmania. *Nature* **244**, 455–456.

Riek EF (1976) Neosecoptera, a new insect suborder based on specimen discovered in the late Carboniferous of Tasmania. *Alcheringa* **1**, 227–234.

Rigby JK and Webby BD (1988) Late Ordovician sponges from the Malongulli Formation of central New South Wales, Australia. *Palaeontographica Americana* **56**, 1–147.

Ripperdan RL (1992) A polarity constraint on Cambrian–Ordovician segments of the Australian and North China apparent polar wander paths. *Eos, Transactions, American Geophysical Union* **73**(43) *Supplement*, 1–150.

Ripperdan RL and Kirschvink JL (1992) Palaeomagnetic results from the Cambrian–Ordovician boundary section at Black Mountain, Georgina Basin, western Queensland, Australia. In *Global Perspectives on Ordovician Geology* (Eds BD Webby and JR Laurie) pp. 93–103. Balkema, Rotterdam.

Ripperdan RL, Magaritz M, Nicoll RS and Shergold JH (1992) Simultaneous changes in carbon isotopes, sea level, and conodont biozones within the Cambrian–Ordovician boundary interval at Black Mountain, Australia. *Geology* **20**(11), 1039–1042.

Ritchie A (1985) *Arandaspis prionotolepis*. The southern four-eyed fish. In *Kadimakara* (Eds P Vickers-Rich and G van Tets) pp. 95–106. Pioneer Design Studios, Lilydale, Victoria.

Ritchie A (2005) *Cowralepis*, a new genus of phyllolepid fish (Pisces, Placodermi) from the Late Middle Devonian of New South Wales, Australia. *Proceedings of the Linnean Society of New South Wales* **126**, 215–259.

Ritchie A and Gilbert-Tomlinson J (1977) First Ordovician vertebrates from the southern hemisphere. *Alcheringa* **1**(4), 351–368.

Ritchie A and Johanson Z (2006) Buried treasures from the age of fishes. *Australian Age of Dinosaurs* **4**.

Rohde RA (2005) GeoWhen Database. University of California, Berkeley, <http://msee140lnx10.ecn.purdue.edu/~astacy/projects/stratigraphy/site/resources/geowhen/index.html>

Royer DL, Osborne CP and Beerling DJ (2005) Contrasting seasonal patterns of carbon gain in evergreen and deciduous trees of ancient polar forests. *Paleobiology* **31**, 141–150.

Runnegar B (1970) *Eurydesma* and *Glendella* gen nov. (Bivalves) in the Permian of eastern Australia. *Bulletin of the Bureau of Mineral Resources, Geology and Geophysics* **116**, 83–118.

Runnegar B (1979) Ecology of Eurydesma and the Eurydesma fauna, Permian of Eastern Australia. *Alcheringa* **3**, 261–285.

Runnegar B, Pojeta J, Taylor ME, Collins D (1979) New species of the Cambrian and Ordovician chitons Matthevia and Chelodes from Wisconsin and Queensland: evidence for the early history of the polyplacophoran molluscs. *Journal of Paleontology* **56**, 1374–1394.

Saini N (2005) Geology and palaeoichnology of the Razorback Beds, Mount Morgan, Queensland. BSc (Hons) thesis. University of Wollongong, Australia.

Scanlon JD, Lee MSY and Archer M (2003) Mid-tertiary elapid snakes (Squamata, Colubroidea) from Riversleigh, northern Australia: early steps in a continent-wide adaptive radiation. *Geobios* **36**, 573–601.

Schopf JW (1968a) Microflora of the Bitter Springs Formation, late Precambrian, central Australia. *Journal of Paleontology* **42**, 651–688.

Schopf JW (1968b) Precambrian palaeobiology: problems and perspectives. *Annual Reviews of Earth and Planetary Sciences* **2**, 213–249.

Schopf JW (1992) Paleobiology of the Archean In *The Proterozoic biosphere*. (Eds JW Schopf and C Klein) pp. 25–39. Cambridge University Press, New York.

Schopf JW and Blacic JM (1971) New microroganisms from the Bitter Springs Formation (late Precambrian) of north-central Amadeus Basin, Australia. *Journal of Paleontology* **42**(3), 651–688.

Schopf JW and Packer BM (1987) Early Archaean (3.3 billion to 3.5 billion-year-old) microfossils from Warrawoona Group, Australia. *Science* **237**, 70–73.

Schultze, H-P (2001) *Melanognathus*, a primitive dipnoan from the Lower Devonian of the Canadian Arctic and the interrelationships of Devonian dipnoans. *Journal of Vertebrate Paleontology* **21**(4), 781–794.

Scriven LJ and Hill RS (1997) Relationships among Tasmanian Tertiary *Nothofagus* (Nothofagaceae) populations. *Botanical Journal of the Linnean Society* **121**, 345–364.

Sheehan PM and Baillie PW (1981) A new species of Eospirifer from Tasmania. *Journal of Paleontology* **55**, 248–256.

Shergold JH (1972) Late Upper Cambrian trilobites from the Gola Beds, western Queensland. *Bulletin of the Bureau of Mineral Resources, Geology and Geophysics* **112**, 1–19.

Shergold JH (1975) Late Cambrian and Early Ordovician trilobites from the Burke River structural belt, western Queensland. *Bulletin of the Bureau of Mineral Resources,Geology and Geophysics Australia* **153**, 1–251.

Shergold JH (1980) Late Cambrian trilobites from the Chatsworth Limestone, western Queensland. *Bulletin of the Bureau of Mineral Resources, Geology and Geophysics Australia* **186**, 1–111.

Shergold JH (1982) Idamean (Late Cambrian) trilobites, Burke River structural belt, Western Queensland. *Bulletin of the Bureau of Mineral Resources,Geology and Geophysics Australia* **187**, 1–69.

Shergold JH and Nicoll RS (1992) Revised Cambrian–Ordovician boundary biostratigraphy, Black Mountain, western Queensland. In *Global Perspectives on Ordovician Geology* (Eds BD Webby and JR Laurie) pp. 81–93. Balkema, Rotterdam.

Shergold JH, Nicoll RS, Laurie JL and Radke BM (1991) 'The Cambro-Ordovician boundary at Black Mountain, western Queensland'. Sixth International Symposium on the Ordovician System, Guidebook for Field Excursion 1. Record 48, Bureau of Mineral Resources, Canberra.

Simpson A (1999) Early Silurian conodonts from the Quinton Formation of the Broken River Region (north-eastern Australia). *Abhandlungen der Geologischen Bundesanstalt (Vienna)* **54**, 181–199.

Simpson A (2000) Silurian to basal Devonian conodonts from the Broken River Crossing, northern Australia. *Records of the Western Australian Museum Supplement* **58**, 145–162.

Singh G and Geissler E (1985) Late Cainozoic history of vegetation, fire, lake levels and climate, at Lake George, New South Wales, Australia. *Philosophical Transactions of the Royal Society of London B* **311**, 379–447.

Singh G, Kershaw AP and Clark R (1981a) Quaternary

vegetation and fire history in Australia. In *Fire and the Australian Biota* (Eds AM Gill, RA Groves and IR Noble) pp. 23–54. Australian Academy of Science, Canberra.

Singh G, Opdyke ND and Bowler JM (1981b) Late Cainozoic stratigraphy, palaeomagnetic chronology and vegetational history from Lake George, New South Wales. *Journal of the Geological Society of Australia* **28**, 435–452.

Skwarko SK (1966) Cretaceous stratigraphy and palaeontology of the Northern Territory. *Bulletin of the Bureau of Mineral Resources, Geology and Geophysics, Australia* **73**, 1–135.

Skwarko SK (Ed.) (1993) Palaeontology of the Permian of Western Australia. *Bulletin of the Geological Survey of Western Australia* **136**, 1–417.

Sloan LC, Walker JCG and Moore Jr TC (1995) The role of oceanic heat transport in early Eocene climate. *Paleoceanography* **10**, 347–356.

Smith AB and Jell PA (1990) Cambrian edrioasteroids from Australia and the origin of starfishes. *Memoirs of the Queensland Museum* **28**(2), 715–778.

Smith AB and Jell PA (1999) A new cornute carpoid from the Upper Cambrian (Idamean) of Queensland. *Memoirs of the Queensland Museum* **43**, 341–350.

Smith E and Smith R (1999) *Black Opal Fossils of Lightning Ridge*. Kangaroo Press, East Roseville, NSW.

Spencer B (1901) A description of *Wynyardia bassiana*, a fossil marsupial from the Tertiary beds at Table Cape, Tasmania. *Proceedings of the Royal Society of London (1900)*, 776–794.

Spencer-Jones D (1971) Marginal Tertiary deposits of the Tyrendarra Embayment–Grassdale and Hamilton district. In *The Otway Basin of Southeastern Australia*. (Eds H Wopfner and JG Douglas) pp. 241–249. Special Bulletin, Geological Surveys of South Australia and Victoria, Adelaide.

Staines HRE (1954) Dinosaur footprints at Mount Morgan. *Queensland Government Mining Journal* **55**, 483–485.

Stait BA (1981) Ordovician nautiloids of Tasmania. PhD thesis. University of Tasmania, Australia.

Stait BA (1988) Tasmanian nautiloid faunas: biostratigraphy, biogeography and morphology. *Senckenbergiana lethaea* **69**, 87–107.

Stait BA and Laurie J (1980) Lithostratigraphy and biostratigraphy of the Florentine Valley Formation in the Tim Shea area, southwest Tasmania. *Papers and Proceedings of the Royal Society of Tasmania* **114**, 201–208.

Steiger RH and Jäger E (1978) *Subcommission on Geochronology: Convention on the use of decay constants in geochronology and cosmochronology*. AAPG Special Volume, SG 6. *Earth and Planetary Science Letters* **36**(3), 359–362.

Stevens NC (1950) Note on the occurrence of a shelly facies in the Ordovician at Cliefden Caves, Near Mandurama, NSW. *Australian Journal of Science* **13**, 83.

Stinton FC (1958) Fish otoliths from the Tertiary strata of Victoria, Australia. *Proceedings of the Royal Society of Victoria* **70**, 81–93.

Stinton FC (1963) Further studies of the Tertiary otoliths of Victoria, Australia. *Proceedings of the Royal Society of Victoria* **76**, 13–22.

Stirton RA (1955) Late Tertiary marsupials from South Australia. *Records of the South Australian Museum* **9**(3), 247–268.

Stirton RA (1957) Tertiary marsupials from Victoria. *Memoirs of the National Museum of Victoria* **21**, 121–134.

Stirton RA (1967) A diprotodontid from the Miocene Kutjamarpu Fauna, South Australia. *Bulletin of the Bureau of Mineral Resources, Geology and Geophysics Australia* **85**, 45–51.

Stirton RA, Tedford RH and Woodburne MO (1967a) A new Tertiary formation and fauna from the Tirari Desert, South Australia. *Records of the South Australian Museum* **15**(3), 428–462.

Stirton RA, Tedford RH and Woodburne MO (1967b) A phylogeny of the Tertiary Diprotodontidae and its significance in correlation. *Bulletin of the Bureau of Mineral Resources, Geology and Geophysics Australia* **85**, 149–160.

Stirton RA, Tedford RH and Woodburne MO (1968) Australian Tertiary deposits containing terrestrial mammals. *University of California Special Publications in Geological Sciences* **77**, 1–30.

Stover LE and Partridge AD (1973) Tertiary and Late Cretaceous spores and pollen from the Gippsland Basin, southeastern Australia. *Proceedings of the Royal Society of Victoria* **85**, 237–286.

Sun G, Ji Q, Dilcher DL, Zheng S, Nixon KC, Wang X (2002) Archaefructaceae, a new Basal Angiosperm family. *Science* **296**, 899–904.

Swenson U, Hill RS and McLoughlin S (2001) Biogeography of *Nothofagus* supports the sequence of Gondwana break-up. *Taxon* **50**, 1025–1041.

Talent JA and Mawson R (1994) Conodonts in relation to age and environmental framework of the Burdekin Basin (Mid-Devonian) north-eastern

Australia. *Courier Forschunginstitut Senckenberg* **168**, 61–81.

Tate R (1885) Descriptions of new species of mollusca of the Upper Eocene beds at Table Cape. *Papers and Proceedings of the Royal Society of Tasmania (1884)*, 226–231.

Tate R (1886) The lamellibranchs of the older Tertiary of Australia (Part 1). *Transactions of the Royal Society of South Australia* **8**, 91–176.

Tate R (1887) The lamellibranchs of the older Tertiary of Australia (Part 2). *Transactions of the Royal Society of South Australia* **9**, 142–189.

Tate R (1888) The gastropods of the older Tertiary of Australia (Part 1). *Transactions of the Royal Society of South Australia* **10**, 91–176.

Tate R (1889) The gastropods of the older Tertiary of Australia (Part 2). *Transactions of the Royal Society of South Australia* **11**, 116–174.

Tate R (1896) Palaeontology. In *Report on the work of the Horn Scientific Expedition to Central Australia*, volume 3. Geology and Botany. (Ed. B Spencer) pp. 97–116. Melville, Mullen and Slade, Melbourne.

Tate R (1899) A revision of the Older Tertiary Mollusca of Australia. *Transactions of the Royal Society of South Australia* **23**, 249–277.

Tate R and Dennant J (1893) Correlation of the marine Tertiaries of Australia. Part 1. Victoria. *Transactions of the Royal Society of South Australia* **17**, 203–226.

Taylor DW and Hickey LJ (1990) An Aptian plant with attached leaves and flowers: implications for angiosperm origin. *Science* **247**, 702–704.

Tedford RH (1984) The Diprotodons of Callabonna. In *Vertebrate Zoogeography and Evolution in Australia* (Eds M Archer and G Clayton) pp. 999–1002. Hesperian Press, Perth.

Tedford RH (1991) Vertebrate palaeontology in Australia: the American contribution. In *The Vertebrate Palaeontology of Australasia*. (Eds P Vickers-Rich, JM Monaghan and TH Rich) pp. 45–83. Pioneer Design Studio and Monash University Publications Committee, Melbourne.

Tedford RH and Wells RT (1990) Pleistocene deposits and fossil vertebrates from the dead heart of Australia. *Memoirs of the Queensland Museum* **28**(1), 263–284.

Tedford RH, Wells RT and Barghoorn SF (1992) Tirari Formation and contained faunas, Pliocene of Lake Eyre Basin South Australia. *The Beagle. Records of the Northern Territory Museum of Arts and Sciences* **9.1**, 173–194.

Teichert C (1954) A new Permian crinoid from Western Australia. *Journal of Paleontology* **28**, 70–75.

Tenison Woods JE (1876) On some Tertiary fossils from Table Cape. *Papers and Proceedings of the Royal Society of Tasmania (1875)*, 13–26.

Tenison Woods JE (1877) Notes on the fossils referred to in the foregoing paper. *Papers and Proceedings of the Royal Society of Tasmania (1876)*, 91–116.

Thomas DE (1960) The zonal distribution of Australian graptolites. *Journal and Proceedings of the Royal Society of New South Wales* **94**, 1–58.

Thompson RB (1991) *A Guide to the Geology and Landforms of Central Australia*. Northern Territory Geological Survey, Alice Springs.

Thomson BP, Daily B, Coat RP and Forbes BG (Eds) (1976) *Late Precambrian and Cambrian Geology of Adelaide Geosyncline and Stuart Shelf, South Australia*. Excursion Guide No 33A, 25th International Geological Congress, Sydney.

Thomson KS (1973). Observations on a new rhipidistian fish from the Upper Devonian of Australia. *Paleontographica Abteilung A* **143**, 209–222.

Thomson KS and Thomas B (2001) On the status of species of *Bothriolepis* (Placodermi, Antiarchi) in North America. *Journal of Vertebrate Paleontology* **21**(4), 679–686.

Thulborn RA (1990) *Dinosaur Tracks*. Chapman and Hall, London.

Thulborn RA, Hamley T and Foulkes P (1994) Preliminary report on sauropod dinosaur tracks in the Broome Sandstone (Lower Cretaceous) of Western Australia. *Gaia* **10**, 85–96.

Thulborn RA and Wade M (1979) Dinosaur stampede in the Cretaceous of Queensland. *Lethaia* **12**, 275–279.

Thulborn RA and Wade M (1984) Dinosaur trackways in the Winton formation (mid-Cretaceous) of Queensland. *Memoirs of the Queensland Museum* **21**(2), 413–517.

Thulborn RT (1979) A proterosuchian thecodont from the Rewan Formation of Queensland. *Memoirs of the Queensland Museum* **19**, 331–355.

Thulborn RT (1983) A mammal-like reptile from Australia. *Nature* **303**, 330–331.

Thulborn RT (1994) Ornithopod dinosaur tracks from the Lower Jurassic of Queensland. *Alcheringa* **18**, 247–258.

Thulborn T, Warren A, Turner S and Hamley T (1996) Early Carboniferous tetrapods in Australia. *Nature* **381**, 777–780.

Tidwell WD (1987) A new species of *Osmundacaulis*

(*O. jonesii* sp. nov.) from Tasmania, Australia. *Review of Palaeobotany and Palynology* **52**, 205–216.

Tidwell WD (1991) *Lunea jonesii* gen. et. sp. nov., a new member of Guareaceae from the mid-Mesozoic of Tasmania, Australia. *Palaeontographica*, **B223**, 81–90.

Tidwell WD and Ash SR (1994) A review of selected Triassic to Early Cretaceous ferns. *Journal of Plant Research* **107** (4), 417–442.

Tidwell WD and Jones R (1987) *Osmundacaulis nerii*, a new Osmundaceous species from Tasmania, Australia. *Palaeontographica* **204**, 181–191.

Tidwell WD, Kim JH and Kimura T (1987) Mid-Mesozoic leaves from near Ida Bay, southern Tasmania, Australia. *Papers and Proceedings of the Royal Society of Tasmania* **121**, 159–170.

Tidwell WD, Munzig GE and Banks MR (1991) Millerocaulis species (Osmundaceae) from Tasmania, Australia. *Palaeontographica* **223**, 91–105.

Tidwell WD and Skog JE (1992) Two new fossil matoniaceous stem genera from Tasmania, Australia. *Review of Palaeobotany and Palynology* **70**, 263–277.

Tillyard RJ (1916) Mesozoic and Tertiary insects of Queensland and New South Wales: description of the fossil insects. *Queensland Geological Survey Publications* **253**, 11–60.

Tillyard RJ (1918) Permian and Triassic insects from NSW in the collection of Mr John Mitchell. *Proceedings of the Linnean Society of New South Wales* **42**, 720–756.

Tillyard RJ (1926) Upper Permian insects of New South Wales. I. Introduction and the Order Hemiptera. *Proceedings of the Linnean Society of New South Wales* **51**, 1–30.

Tillyard, RJ and Dunstan, B (1924) *Mesozoic Insects of Queensland.* Queensland Geological Survey, Brisbane.

Tindale NB (1945) Triassic insects of Queensland and New South Wales. 1 Eoses, a probable lepidopterous insect from the Triassic beds of Mt Crosby, Queensland. *Proceedings of the Royal Society of Queensland* **56**, 37–46.

du Toit AL (1937) *Our Wandering Continents: Hypothesis of continental drifting.* Oliver and Boyd, London.

Travouillon KJ, Archer M, Hand SJ and Godthelp H (2006) Multivariate analyses of Cenozoic mammalian faunas from Riversleigh, northwestern Queensland. *Alcheringa Special Issue* **1**, 323–349.

Trewin NH and McNamara KJ (1995) Arthropods invade the land: trace fossils and palaeoenvironments of the Tumblagooda Sandstone (?late Silurian) of Kalbarri, Western Australia. *Transactions of the Royal Society of Edinburgh: Earth Sciences* **85**, 177–210.

Trinajstic K and Long JA (2009) The potential of x-ray and synchrotron CT scanning in determining soft tissue anatomy in early vertebrates. *Geological Society of Australia Abstracts* **93**: 59.

Trinajstic K, Marshall C, Long J and Bifield K (2007) Exceptional preservation of nerves and muscle tissues in Devonian placoderm fish and their phylogenetic implications. *Biology Letters* **3**, 197–200.

Trueman CNG, Field JH, Dortch J, Charles B, and Wroe S (2005) Prolonged coexistence of humans and megafauna in Pleistocene Australia. *Proceeding of the National Academy of Sciences of the United States of America* **102**(23), 8381–8385.

Turnbull WD and Lundelius EL (1970) The Hamilton fauna. A late Pliocene mammalian fauna from the Grange Burn, Victoria, Australia. *Fieldiana Geology* **19**, 1–163.

Turnbull WD, Lundelius Jr EL and McDougall I (1965) A potassium/argon dated Pliocene marsupial fauna from Victoria, Australia. *Nature* **206**(4986), 816.

Turnbull WD, Rich THV and Lundelius EL (1987a) The petaurids (Marsupialia: Petauridae) of the early Pliocene Hamilton Local Fauna, southwestern Victoria. In *Possums and Opposums: Studies in Evolution*, volume 2 (Ed. M Archer) pp. 629–638. Surrey Beatty & Sons Pty Ltd and the Royal Zoological Society of New South Wales, Mosman.

Turnbull WD, Rich THV and Lundelius EL (1987b) Pseudocheirids (Marsupialia: Pseudocheiridae) of the early Pliocene Hamilton Local Fauna, southwestern Victoria. In *Possums and Opposums: Studies in Evolution*, volume 2 (Ed. M Archer) pp. 689–713. Surrey Beatty & Sons Pty Ltd and the Royal Zoological Society of New South Wales, Mosman.

Turnbull WD, Rich THV and Lundelius EL (1987c) Burramyids (Marsupialia: Burramyidae) of the early Pliocene Hamilton Local Fauna, southwestern Victoria. In *Possums and Opposums: Studies in Evolution*, volume 2 (Ed. M Archer) pp. 729–739. Surrey Beatty & Sons Pty Ltd and the Royal Zoological Society of New South Wales, Mosman.

Turner S (1982) Saurichthys (Pisces, Actinopterygii) from the Early Triassic of Queensland. *Memoirs of the Queensland Museum* **20**, 545–551.

Turney CSM, Bird MI, Fifield LK, Roberts RG, Smith M, Dortch CE, Grun R, Lawson E, Ayliffe LK, Miller GH, Dortch J and Cresswell RG (2001). Early human occupation at Devil's Lair, southwestern Australia 50,000 years ago. *Quaternary Research* **55**(1), 3–13.

Twelvetrees WH and Petterd WF (1900) Note on the humeri of Tasmanian labyrinthodonts. *Papers and Proceedings of the Royal Society of Tasmania (1898–1899)*, 27–31.

Tyler I (1996) 'Geology and Landforms of the Kimberley'. Department of Conservation and Land Management, Perth, WA.

Tyler IM (2000) Geological map of the Halls Creek orogen, East Kimberley region. Geological Survey of Western Australia, Department of Minerals and Energy, East Perth.

Tyler PA and Hill RS (1988) Lower Gordon River and Regatta Point. In *Cainozoic Vegetation of Tasmania* (Ed. EA Colhoun) pp. 72–77. Newcastle University Special Publication, Newcastle, NSW.

UNEP-WCMC (1994/2006) 'Australian Fossil Mammal Sites Riversleigh/Naracoorte datasheet'. United Nations Environment Programme World Conservation Monitoring Centre. <http://whc.unesco.org/en/list/698>

Vandenberg AHM (2003) Graptolite zonation schemes and the Victorian sequence. In *Geology of Victoria.* (Ed. WD Birch) pp. 695–698. Geological Society of Australian Special Publication No. 23, Geological Society of Australia (Victoria Division), Melbourne.

Vandenberg AHM and Cooper RA (1992) The Ordovician graptolite sequence of Australasia. *Alcheringa* **16**, 33–85.

Vandenburg AHM, Wright AJ, Sherwin L and Rickards RB (2002) 'Ordovician–Silurian graptolite succession of southeastern Australia'. First International Palaeontological Congress (IPC-2002). Post-2 Field Excursion Guidebook, Australasia.

Vernon JEN (2008) *A Reef in Time: The Great Barrier Reef from Beginning to End*. The Belknap Press of Harvard University Press, Cambridge, Mass.

Vickers-Rich P, Constantine A and Rich TH (1999) Environmental setting of the polar faunas of southeastern Australia and adaptive strategies of the dinosaurs. *National Science Museum Monographs* **15**, 181–195.

Vickers-Rich P and Molnar RE (1996) The foot of a bird from the Eocene Redbank Plains Formation of Queensland, Australia. *Alcheringa* **20**, 21–29.

Vickers-Rich P and Rich TH (1993a) Australia's polar dinosaurs. *Scientific American* **269**(1), 50–55.

Vickers-Rich P and Rich TH (1993b) *Wildlife of Gondwana*. Reed Books, Frenchs Forest, NSW

Wade M (1969) Medusae from uppermost Precambrian or Cambrian sandstones, central Australia. *Palaeontology* **12**, 351–365.

Wade M (1972a) Hydrozoa and Scyphozoa and other medusoids from the Precambrian Ediacara fauna, South Australia. *Palaeontology* **15**, 197–225.

Wade M (1972b) Dickinsonia: polychaete worms from the Late Precambrian Ediacara fauna, South Australia. *Memoirs of the Queensland Museum* **16**, 171–190.

Wade M and Molnar R (2000) *Geology 4: Winton dinosaur trackways*. Queensland Museum, Brisbane.

Wade RT (1941) The Jurassic fishes of NSW. *Journal and Proceedings of the Royal Society of New South Wales* **75**, 71–84.

Wade RT (1953) Jurassic fishes of NSW (Macrosemiidae) with a note on the Jurassic genus Promecosomina. *Journal and Proceedings of the Royal Society of New South Wales* **87**, 63–72.

Waggoner B (2003) The Ediacaran biotas in space and time. *Integrative and Comparative Biology* **43**, 104–113.

Wagstaff BE and McEwen-Mason J (1989) Palynological dating of lower Cretaceous coastal vertebrate localities, Victoria, Australia. *National Geographic Research* **5**, 54–63.

Walkom AB (1917a) Mesozoic floras of Queensland. Part 1 continued. The flora of the Ipswich and Walloon Series. (c) Filicales etc. *Publications of the Geological Survey of Queensland* **257**, 1–67.

Walkom AB (1917b) The Mesozoic floras of Queensland. Part 1 concluded. (d) Ginkgoales, (e) Cycadophyta, (f) Coniferales. *Publications of the Geological Survey of Queensland* **259**, 1–51.

Walter MR, Elphinstone R and Heys GR (1989) Proterozoic and Early Cambrian trace fossils from the Amadeus and Georgina Basins, central Australia. *Alcheringa* **13**, 209–256.

Walter MR, Krylov IN and Preiss WV (1979) Stromatolites from Adelaidean (Late Proterozoic) sequences in central and South Australia. *Alcheringa* **3**, 287–305.

Warne MT, Archbold NW, Bock PE, Darragh TA, Dettmann ME, Douglas JG, Gratsianova RT, Grover M, Holloway DJ, Holmes FC, Irwin RP, Jell PA, Long JA, Mawson R, Partridge AD, Pickett JW, Rich TH, Richardson JR, Simpson AJ, Talent JA and Vandenberg AHM (2003) Palaeontology: the biogeohistory of Victoria. In *Geology of Victoria* (Ed. WD Birch) pp. 605–652. Geological Society of Australian Special Publication No. 23, Geological Society of Australia (Victoria Division), Melbourne.

Warren AA (1980) Parotosuchus from the Early Triassic of Queensland and Western Australia. *Alcheringa* **4**, 25–36.

Warren AA (1981) A horned member of the labyrinthodont super-family Brachyopoidea from the Early Triassic of Queensland. *Alcheringa* **5**, 273–288.

Warren AA (1985a) Triassic Australian plagiosauroid. *Journal of Paleontology* **59**, 236–241.

Warren AA (1985b) Two long snouted temnospondyls (Amphibia, Labyrinthodontia) from the Early Triassic of Queensland. *Alcheringa* **9**, 293–295.

Warren AA (1991) Australia fossil amphibians. In *Vertebrate Palaeontology of Australia* (Eds P Vickers-Rich, JM Monaghan, RF Baird and TH Rich) pp. 570–589. Pioneer Design Studio with the Monash University Publications Committee, Melbourne.

Warren AA and Black T (1985) A new rhytidosteid (Amphibia, Labyrinthodontia) from the Early Triassic Arcadia Formation of Queensland Australia and the relationships of Triassic temnospondyls. *Journal of Vertebrate Paleontology* **5**, 303–327.

Warren AA and Hutchinson MN (1988) A new capitosaurid amphibian from the Early Triassic of Queensland. *Palaeontology* **31**, 857–876.

Warren AA and Hutchinson MN (1990a) The young ones: small temnospondyls from the Arcadia Formation. *Memoirs of the Queensland Museum* **28**, 103–106.

Warren AA and Hutchinson MN (1990b) *Lapillopsis*, a new genus of temnospondyl amphibians from the Early Triassic of Queensland. *Alcheringa* **14**, 149–158.

Warren JW and Wakefield NA (1972) Trackways of tetrapod vertebrates from the Upper Devonian of Victoria, Australia. *Nature* **238**, 469–470.

Watt (1896) Report on the work of the Horn Scientific Expedition to Central Australia (ed. B Spencer). Dulau, London. <http://openlibrary.org/books/OL6589554M/Report_on_the_work_of_the_Horn_Scientific_Expedition_to_Central_Australia>

Webb GE (2001) Famennian mud-mounds in the proximal fore-reef slope, Canning Basin, Western Australia. *Sedimentary Geology* **145**, 295–315.

Webby BD (1972) Devonian geological history of the Lachlan geosyncline. *Journal of the Geological Society of Australia* **19**, 99–123.

Webby BD (1991) Ordovician stromatoporoids from Tasmania. *Alcheringa* **15**, 191–227.

Webby BD, Zhen YY and Percival IG (1997) Ordovician coral- and sponge-bearing associations: distribution and significance in volcanic island shelf to slope habitats, eastern Australia. *Boletin de la Real Sociedad Espanola de Historica Natural* **92**, 163–175.

Weidenhofer M (1976) *Maria Island – a Tasmanian Eden*. Darlington Press, Melbourne.

Wells AT (1976) 'Geology of the Late Proterozoic–Palaeozoic Amadeus Basin'. International Geological Congress 25, Excursion Guide 48A.

Wells PM and Hill RS (1989) Fossil imbricate-leaved Podocarpaceae from Tertiary sediments in Tasmania. *Australian Systematic Botany* **2**, 387–423.

Wells RT (1996) Earth's geological history – a contextual framework for assessment of World Heritage fossil site nominations. Natural Heritage Program, IUCN, Gland, Switzerland.

Wells RT and Callen RA (Eds) (1986) The Lake Eyre Basin – Cainozoic sediments, fossil vertebrates and plants, landforms, silcretes and climatic implications. *Australasian Sedimentologists Group Field Guide* Series **4**, Geological Society of Australia, Adelaide.

Wells RT, Moriarty K and Williams D (1984) The fossil vertebrate deposits of Victoria Fossil Cave: an introduction to the geology and fauna. *Australian Zoologist* **21**(4), 305–333.

Wells RT and Tedford RH (1995) *Sthenurus* (Macropodidae Marsupialia) from the Pleistocene of Lake Callabonna, South Australia. *Bulletin of the American Museum of Natural History* **225**, 1–111.

Westall F, Walsh MM, De Vries ST and Nijman W (2001) Fossil microbial biofilms from Early Archaean volcaniclastic sediments. In *4th International Archaen Symposium 2001, Geoscience Australia, Record 2001/37*. (Eds KF Cassidy, JM Dunphy and MJ van Kranendonk), pp. 266–268. Australian Geological Survey Organisation, Canberra.

Western Australian Museum (2011) Cundlego Formation Siltstone. <www.museum.wa.gov.au/research/collections/earth-and-planetary-sciences/fossil-collection/large-block-early-permian-cundlego-formation-siltstone>.

Whang SS and Hill RS (1995) Phytolith analysis in leaves of extant and fossil populations of *Nothofagus* subgenus *Lophozonia*. *Australian Systematic Botany* **8**, 1055–1065.

White ME (1961) 'Report on 1960 collections of Mesozoic plant fossils from the Northern Territory'. Record 61/146, Bureau of Mineral Resources, Canberra.

White ME (1966) 'Report on 1965 plant fossil collections'. Record 1966/111, Bureau of Mineral Resources, Canberra.

White ME (1974) 'Tertiary plant fossils from Melville Island, NT'. Record 1974/5, Bureau of Mineral Resources, Geology and Geophysics, Canberra.

White ME (1981) Revision of the Talbragar Fish Bed Flora (Jurassic) of New South Wales. *Records of the Australian Museum* **33**(15), 695–721.

White ME (1991) Time in our hands: semi-precious gemstones: keys to the geological past. Reed, Balgowlah.

White ME (1994) *The Greening of Gondwana*. 2nd edn. Reed Books, Frenchs Forest NSW.

Whitehead PW and Stephenson PJ (1998). Lava rise ridges of the Toomba basalt flow, north Queensland, Australia. *Journal of Geophysical Research, B, Solid Earth and Planets* **103**(11), 27 371–27 382.

Whitehouse FW (1926) The Cretaceous Ammonoidea of Australia. *Memoirs of the Queensland Museum* **8**, 195–242.

Whitehouse FW (1927) Additions to the Cretaceous ammonite fauna of eastern Australia. *Memoirs of the Queensland Museum* **9**, 109–120.

Willis PMA and Archer M (1990) A Pleistocene longirostrine crocodilian from Riversleigh: first fossil occurrence of *Crocodylus johnsoni* Krefft. *Memoirs of the Queensland Museum* **28**, 159–164.

Winchester-Seato T and Carey SP (2000) Chitinozoans and associated conodonts from the Early Devonian Point Hibbs Formation, Tasmania. *Australian Records of the Western Australian Museum Supplement* **58**, 163–177.

Withnall IW, Lang SC, Jell JS, McLennan TPT, Talent JA, Mawson R, Fleming PJG, Law SR, Macansh JD, Savory P, Kay JR and Draper JJ (1988) Stratigraphy, sedimentology, biostratigraphy and tectonics of the Ordovician to Carboniferous, Broken River Province, north Queensland. *Australasian Sedimentologists Group Field Guide* Series 5, Geological Society of Australia, Sydney.

Wood R (2000) Novel paleoecology of a postextinction reef: Famennian (Late Devonian) of the Canning Basin, northwestern Australia. *Geology* **28**, 987–990.

Wood R (2002) Dodging mass extinction. *Natural History* **111**(10), 58.

Woodburne MO (1967a) Three new diprotodontids from the Tertiary of the Northern Territory. *Bulletin of the Bureau of Mineral Resources, Geology and Geophysics Australia* **85**, 53–103.

Woodburne MO (1967b) The Alcoota fauna, central Australia. An integrated palaeontological and geological study. *Bulletin of the Bureau of Mineral Resources, Geology and Geophysics Australia* **87**, 1–187.

Woodburne MO, MacFadden BJ, Judd AC, Springer MS, Pledge NS, Power JM, Woodburne JM and Springer JD (1994) Land mammal biostratigraphy and magnetostratigraphy of the Etadunna Formation (Late Oligocene) of South Australia. *Journal of Vertebrate Paleontology* **13**, 483–515, for 1993.

Woodburne MO and Tedford RH (1975). The first Tertiary monotreme from Australia. *Novitates of the American Museum of Natural History* **2588**, 1–11.

Woodburne MO, Tedford RH, Archer M, Turnbull WD, Plane MD and Lundelius EL (1985) Biochronology of the continental mammal record of Australia and New Guinea. *South Australian Department of Mines and Energy, Special Publication* **5**, 347–363.

Woods JT (1961) Mesozoic and Cainozoic sediments of the Wrotham Park Area. *Geological Survey of Queensland Publication* **304**, 1–6.

Woodward AS, David TWE and Pittman EF (1895). The fossil fishes of the Talbragar Beds (Jurassic?), with a note on their stratigraphical relations. *Memoirs of the Geological Survey of New South Wales, Palaeontology* **9**, 1–27.

Wroe S and Mackness BS (2000a) A new genus and species of dasyurid from the Pliocene Chinchilla Local Fauna of southeastern Queensland. *Alcheringa* **24**, 319–326.

Wroe S and Mackness BS (2000b) Additional material of *Dasyurus dunmalli* from the Pliocene Chinchilla Local Fauna of Queensland and its phylogenetic implications. *Memoirs of the Queensland Museum* **45**(2), 641–645.

Yates AM (1996) New material of the small temnospondyl *Lapillopsis* from the Early Triassic Arcadia Formation of Australia. *Journal of Vertebrate Paleontology*, **16**(3) Suppl, 74.

Yates AM (2000) A new tiny rhytidosteid (Temnospondyli; Stereospondyli) from the Early Triassic of Australia and the possibility of hidden temnospondyl diversity. *Journal of Vertebrate Paleontology* **20**(3), 484–489.

Yeates AN (2001) 'An assessment of Australian geological sites of possible national or international significance. Volume 2: Fossils'. Report for the Australian Heritage Commission, Canberra.

Young GC (1984) Further petalichthyid remains (placoderm fishes, Early Devonian) from the Taemas–Wee Jasper region N.S.W. *BMR Journal of Australian Geology and Geophysics* **9**(2), 121–131.

Young GC (1999) Preliminary report on the biostratigraphy of new placoderm discoveries in the Hervey Group (Upper Devonian) of central New South Wales. *Records of the West Australian Museum* **57** (supplement), 139–150.

Young GC (2005) New Phyllolepids (Placoderm fishes) from the middle–late Devonian of southeastern Australia. *Journal of Vertebrate Paleontology* **25**(2), 261–273.

Young, GC (2008) Early evolution of the vertebrate eye – fossil evidence. *Evolution: Education and Outreach Special Issue: The Evolution of Eyes* **1**(4), 427–438.

Young, GC (2010) Placoderms (armored fish): Dominant vertebrates of the Devonian Period. *Annual Review of Earth and Planetary Sciences* **38**, 523–550.

Young GC and Laurie JR (Eds) (1996) *An Australian Phanerozoic Timescale.* Oxford University Press, Melbourne.

Young GC and Long JA (2005) Phyllolepid placoderm fish remains from the Devonian Aztec Siltstone, southern Victoria Land, Antarctica. *Antarctic Science* **17**(3), 387–408.

Zhen YY (1991) Devonian rugose coral faunas and biostratigraphy from the Fanning River Group, north Queensland. PhD thesis. University of Queensland, Australia.

Zhen YY and Jell JS (1996) Middle Devonian rugose corals from the Fanning River Group, north Queensland, Australia. *Palaeontographica Abteilung A* **242**, 15–98.

Zhen YY and Webby BD (1995) Upper Ordovician conodonts from the Cliefden Caves Limestone Group, central New South Wales, Australia. *Courier Forschungsinstitut Senckenberg* **182**, 265–305.

Index

Acacia bulbosa 80
Acanthostega 93
Acmopyle glabra 80
Ajax Mine archaeocyathid locality (SA) 57–8
Alcoota vertebrate fossil beds (NT) 19–21, 24
Anderson, William 15
Anglesea site, eastern view formation (Middle Eocene) (Vic) 59, 107–9
Anomalocaris 56
Araucaria readiae 80, 81
Araucarioides linearis 80, 81
Araucarioides sinuosa 80
Archer, Michael 16, 41, 93
Australian Fossil Mammal Sites 44, 69–70
Austrophyllolepis 90
Austrosequoia wintonensis 50
aves (birds) 103

Bacchus Marsh, Hines Quarry Diprotodon site (Vic) 96–7
Baldina Creek Swamp (SA) 67
Banksia strahanensis 80
Barwickia 90
Bath Tub Gorge (SA) 55
Baragwanathia longifolia 100
Baragwanathia macoflora (Yea Flora Fossil Site) (Vic) 100–1
Bauhinia Downs flora, Balbarini (NT) 21, 107
Beelarongia 90
Bell, Ken 97
Belmont insect beds (NSW) 5–6
Belubula River Valley (Cliefden Caves) (NSW) 13–14
Betula 50
Big Bend 'Leichhart Cliffs' Devonian reef complex (Qld) 32–4
Bitter Springs formation, Ross River homestead (NT) 21–3
Black Mountain Cambrian–Ordovician Boundary Section (QLD) 35–6
Blina Shale, Canning Basin (WA) 35, 88
Bluff Downs Local Fauna (Qld) 16, 36, 37, 98
Bothriolepis 8, 28, 90, 92
Boulia (Qld) 35
Bow Local Fauna (NSW) 16, 37, 98
Bowen basin (QLD) 34, 44
Bowenia 81
Brachina George (SA) 55, 58
Broken River Province (Jack Hills Gorge and Fish Hill) (Qld) 40
Broome Sandstone dinosaur trackways and macroflora, Broome (WA) 48, 61, 121–2
Buchan (Vic) 8, 9, 113
Buckley, J.H. 71
Bulldog Shale 60–1
Bullock Creek vertebrate fossil site (NT) 23–4

Cabonnichthys 8
Callitris strahanensis 80
Callytharra Formation, Callytharra Springs (Permian invertebrates and foraminifera) (WA) 121, 124
Cameron, W.E. 44
Canning Basin (WA) 33, 34, 35, 111, 121
Canowindra 8
Canowindra fossil fish site (NSW) 8–9, 90, 92, 93, 113
Cape d-Estaing, Kangaroo Island (SA) 55–6
Cape Leeuwin–Cape Naturaliste region, Devil's Lair (WA) 116–17
Cape Van Dienem flora (NT) 25
Carinda (NSW) 9
Carl Creek Limestone (Qld) 41
Carnarvon National Park (Qld) 34
Castle Rock, Flinders Ranges (SA) 56
Cetacea (whales and dolphins) 103
Cethana (Tas) 77–8
Chace Range (SA) 55
Chatsworth Limestone echinoderm and trilobite localities (Qld) 37–8
Chewton–Castlemaine railway cuttings 99
Chinchilla Rifle range (Qld) 16, 36–7
Chondrichthyes (sharkes and rays) 102
Christmas Hills (Tas) 71–2
Clarke, Rev William Branwhite 1, 2, 3, 13, 75
Cleeland, Michael 93
Cliefden Caves – Belubula River Valley (NSW) 13–14
Coleoptera (beetles) 5
Collie Basin (WA) 46
Coober Pedy (Moon Plain Cretaceous fossil site) (SA) 60–1
Cook, Alex 47
Cookson, Isabel 101
Cooksonia 101
Cooyal locality 2, 46

Crenulostomina 84
Crisp gorge (Dead goat Soak) (SA) 55
Cuddie Springs (NSW) 9–12
Cullen, Charles 15
Cundlego Formation Permian invertebrate fauna, Gascoyne Junction (WA) 124
Cyathophyllum leichhardtii 33

Dacrycarpus carpenteri 80
Dacrydium 58, 77
Darling Downs (Qld) 12
Darling Plateau, West Dale (WA) 118–20
Darlington Limestone, Maria Island (Tas) 74, 75, 76
Darwin, Charles 16–17
Darwin Meteorite Crater (Tas) 72
David, Sir Tannant William Edgeworth 57
de la Hunty, L.E. 123
Devil's Lair, Cape Leeuwin–Cape Naturaliste region (WA) 12, 116–17
Devonian 'Great Barrier Reef' and Gogo fish fossil site, Fitzroy Crossing (WA) 9, 111–16
Dickinsonia costata 55
Dickinsonia rex 51
Dicroidium 46
Dinmore Quarry (Qld) 46
Dinosaur Cove (Otway Ranges) (Vic) 93–6
Dinosaur Stampede National monument 46–50
Diplichnites 118
Diprotodon 9, 16, 20, 23, 36, 62, 66, 67, 68
  Hines Quarry, Bacchus Marsh (Vic) 96–7

Early Cretaceous (Aptain) Koonwarra site (Vic) 106–7
early life sites 125
  Ediacara sites, Nilpena, Flinders Ranges (SA) 51–5, 56
  Mount Skinner fauna (NT) 28–9
  Warrawoona group (North Pole and Strelley Pool sites), Proterozoic early life fauna (WA) 123–4
Ediacara sites, Nilpena, Flinders Ranges (SA) 51–5, 56
Elizabeth Creek (Qld) 31–2
Ellery Creek Cambrian–Ordovician trace fossil locality (NT) 14, 24–5
Ellery Creek Section (NT) 14, 25–8
Emu Bay (Cape d-Estaing to Big Gully), Kangaroo Island (SA) 55–6
Etadunna Formation (SA) 20, 24, 44, 58, 61, 62, 63, 65
Etymalbaillella yennienii 13
Eucryphia microstoma 80
Eurydesma 74–5, 76
Evans, John William 6

Fagus 50
Fennel Bay (Kurrur-Kurran) (NSW) 1–3
fish sites
  Buchan (Vic) 8, 9, 113
  Canowindra (NSW) 8–9, 90, 92, 93, 113
  Gogo Formation, Fitzroy Crossing (WA) 8, 9, 111–16
  Knocklofty Formation (Tas) 84–8
  Mt Howitt (Vic) 8, 9, 89–91, 113
  Mt Watt and Mt Charlotte (NT) 29
  Talbragar (NSW) 14–15, 107
  Wee Jasper (NSW) 8, 9 12–13, 33, 113, 116
Fish Hill, Broken River Province (Qld) 13, 40
Fitzroy Crossing, Devonian 'Great Barrier Reef' and Gogo fish fossil site (WA) 111–16
Flannery, Tim 93
Fletcher, Harold 6, 9
Fletcherview – Big Bend 'Leichhart Cliffs' Devonian reef complex (Qld) 32–4, 35
Fleurieu Peninsula (Myponga Beach) (SA) 56–7
Flinders Ranges (Ediacara sites, Nilpena) (SA) 51–5, 56, 58, 67
flora sites 46, 50, 59
Florentine Valley (Tas) 72–4
Fossil Bluff (Table Cape), Wynyard (Tas) 82–3
Fossil Cliffs, Maria Island (Tas) 74–6
fossil sites, by type 125–6

Gascoyne Junction, Cundlego Formation Permian invertebrate fauna (WA) 124
Gavinia 90
Genoa River Devonian tetrapod site (Vic) 91–3
Genyornis newtoni 9
geological timeline 127
Gill, Edmund 97
Ginkgo 44, 46, 50, 106
Gippsland Basin (Vic) 106–7
Giralia Anticline, Miria Marl (WA) 32, 36, 122–3
glosselytrodea (an extinct order of winged insects) 5
glossopterids 1–2, 46
Glossopteris 5, 44–5, 46
Gogo Formation fish fossil site, Fitzroy Crossing (WA) 8, 9, 111–16
Gogonasus andrewsae 112
Golden Grove (SA) 59
Gooloogongia 8, 90

Grange Burn Pliocene mammal site, Hamilton (Vic) 16, 37, 97–8, 105
graptolite province, Victorian Ordovician 98–100
Great Barrier Reef (Qld) 116
Green Waterhole Cave (SA) 63–4
Gregory River (Qld) 41
Griman Creek Formation (NSW) 4
Groenlandaspis 8, 9, 90
Gymnostoma 81

Hamilton, Grange Burn Pliocene mammal site (Vic) 16, 37, 97–8, 105
Hamilton Miocene–Pliocene sites 16, 37, 105–6
Harris, W.J. 98, 99
Heysen Range (SA) 55
Hine, Kerry 97
Hines Quarry Diprotodon site, Bacchus Marsh (Vic) 96–7
Homevale Station (Qld) 2, 44–6
Homoptera (cicadas, leafhoppers, aphids) 5
Howchin, Sir Walter 57
Howidipterus 90
Howittacanthus 90
Howqualepis 90
human habitation/activity 3, 4, 5, 9, 12, 32, 69, 70, 104, 105, 116–17

Inaria 55
insect fossils 5–6
Ipswich Coal Measures (Qld) 38, 46, 48
invertebrate fossil sites 125
- Ajax Mine archaeocyathid locality (SA) 57–8
- Belmont insect beds (NSW) 5–6
- Bitter Springs formation, Ross River homestead (NT) 21–3
- Black Mountain Cambrian–Ordovician Boundary Section (QLD) 35–6
- Callytharra Formation, Callytharra Springs (Permian invertebrates and foraminifera) (WA) 121, 124
- Chatsworth Limestone echinoderm and trilobite localities (Qld) 37–8
- Christmas Hills (Tas) 71–2
- Cliefden Caves – Belubula River Valley (NSW) 13–14
- Cundlego Formation Permian invertebrate fauna, Gascoyne Junction (WA) 124
- Ediacara sites, Nilpena, Flinders Ranges (SA) 51–5, 56
- Ellery Creek Section (NT) 14, 25–8
- Emu Bay (Cape d-Estaing to Big Gully), Kangaroo Island (SA) 55–6
- Fossil Cliffs, Maria Island (Tas) 74–6
- Hamilton, Grange Burn Pliocene mammal site (Vic) 16, 37, 97–8, 105
- Miria Marl, Giralia Anticline (WA) 32, 122–3
- Mount Crosby Triassic insect locality (Qld) 37–40
- Ordovician faunal succession of the Florentine Valley (Tas) 72–4
- Point Hibbs shoreline, western Tasmania 81–2
- Roe Calcarenite invertebrate fauna, Roe Plains (WA) 120
- Victorian Ordovician graptolite province 98–100

Jack, Robert Logan 40
Jack Hills Gorge, Broken River Province (Qld) 13, 40
Jago, J.B. 71
Jan Juc Formation (Vic) 101–2, 103, 104
Janjucetus hunderi 103, 104
Jimbacrinus bostocki 124

Kalbarri trace fossils (WA) 117–18
Kalbarria brimmellae 118
Kangaroo Island, Emu Bay (Cape d-Estaing to Big Gully) (SA) 55–6
Kimberley region (WA) 8, 13, 111, 113
Knight, O.L. 6
Knocklofty Formation (Tas) 84–8
Kollikodon ritchei 4
Kool, Lesley 93
Koonwarra site, early Cretaceous (Aptain) (Vic) 106–7
Konservat-Lagerstätten 5, 56
Kronosaurus 32, 50
Kurrur-Kurran (Fennel Bay) (NSW) 1–3
Kryoryctes cadburyi 93

Lagerstätten 56
Lake Callabonna (SA) 12, 62, 67–9, 97
Lake Eyre (SA) 20, 24, 44, 58, 61–3, 65, 66
Lake Frome Basin, Namba Formation (SA) 63, 65
Lake George (NSW) 3–4
Lake Ngapakaldi Leaf Locality (SA) 65–6
Lake Palankarinna (SA) 61–3
Lancefield Quarry (Vic) 99
Lancefield Swamp (Vic) 12
landscape sites 125
- Bitter Springs formation, Ross River homestead (NT) 21–3

Devonian 'Great Barrier Reef' and Gogo fish fossil site, Fitzroy Crossing (WA) 9, 111–16
Ellery Creek Cambrian–Ordovician trace fossil locality (NT) 14, 24–5
Fletcherview – Big Bend 'Leichhart Cliffs' Devonian reef complex (Qld) 32–4, 35
Jack Hills Gorge, Broken River Province (Qld) 13, 40
Taemas–Cavan–Wee Jasper (NSW) 8, 9 12–13, 33, 113, 116
Tumblagooda Sandstone, Kalbarri and Murchison River George trace fossils (WA) 117–18

Lark Quarry Dinosaur Stampede National monument (Winton Formation) (Qld) 35, 36, 46–50

Late Cretaceous (Cenomanian) Winton Formation (Qld) 50

Latrobe Valley coal measures (Vic) 109

Laurophyllum australum 80

Lea River (Tas) 77, 78

Leichhart Cliffs (Qld) 32–4

Lennard Shelf (WA) 8, 13, 111, 113, 116

Lepidocyclina 105

Lepidodendron veltheimianum 119

Lepidomena 74

Leptolepis koonwarri 107

Lightning Ridge (NSW) 4–5

Little Rapid River (Tas) 77, 78

Lune River Jurassic plant site (Tas) 78–9

Lowe, A. 14, 15

Lundelius, Ernest L. 97, 98, 116

Lyell, Charles 16, 17

McCoy, Frederick 100, 103, 105, 106

MacDonald, Robert 97

MacDonnell Ranges (NT) 14, 24, 25

Mcnamarapsis kaprios 112

Macquarie Harbour, eastern shore of (Tas) 79–81

mammal fossils xi, 4, 16, 17, 41, 44, 62, 63, 65–6, 67, 69–70, 93, 97–8, 104
biochronological scheme 20–1, 23, 66
Cretaceous 4
Grange Burn 97–8
Pliocene–Pleistocene 16, 97–8, 116–17

Mammalodon colliveri 103, 104

Mandageria 8

Maria Island, Fossil Cliffs (Tas) 74–6

Marsdenichthys 90, 91

Maslin Bay (SA) 59, 78

Maydenella 74

megafauna site 16, 67, 69, 70, 116, 117
Cuddie Springs (NSW) 9–12
Hines Quarry, Bacchus Marsh (Vic) 96–7

Megalania 16, 62

Megalasauropus broomensis 121

Melville Island (NT) 25

micro fossil sites 126
Bitter Springs formation, Ross River homestead (NT) 21–3
Darwin Meteorite Crater (Tas) 72
Lake George (NSW) 3–4

Middle Eocene Anglesea site, eastern view formation (Vic) 107–9

Miria Marl, Giralia Anticline (WA) 32, 122–3

Mitchell, John 6

Mitchell, Thomas 13, 17

mixed fossil sites 126
Lake Ngapakaldi Leaf Locality (SA) 65–6
Lightning Ridge (NSW) 4–5
Moon Plain Cretaceous fossil site, Coober Pedy (SA) 60–1
Redbank Plains Formation (Qld) 38
Table Cape – Fossil Bluff, Wynyard (Tas) 82–3
Talbragar fossil beds (NSW) 14–15, 107
Torquay Oligocene whale and invertebrate site (Vic) 101–5
Walsh River and Elizabeth Creek (Qld) 31–2

Mole Creek (Tas) 74

Moon Plain Cretaceous fossil site, Coober Pedy (SA) 60–1

Mt Charlotte (NT) 29

Mount Crosby Triassic insect locality (Qld) 38–40

Mount Gambia, Green Waterhole Cave (SA) 63–4

Mt Howitt fish fossil site (Vic) 8, 9, 89–91, 113

Mount Morgan Jurassic dinosaur footprint site (Qld) 34, 35, 36, 48

Mount Scott Range (SA) 57

Mt Skinner fauna (NT) 28–9

Mt Watt (NT) 29

Muddy Creek (Vic) 105

Murchison River Gorge trace fossils (WA) 117–18

Myponga Beach, Fleurieu Peninsula (SA) 56–7

Myrtaceaephyllum pleistocenicum 80

Namba Formation (Lake Frome Basin) (SA) 63, 65

Naraoia 56

Narracoorte Caves (SA) 12, 16, 20, 37, 44, 63, 64, 69–70

National Heritage List xi, 48

Nelly Creek fossil plant sites (SA) 58–9

Neuroptera (lacewings) 5

Newcastle Coal Measures 2, 5, 6
Nilpena, Flinders Ranges (SA) 51–5
Nothofagus 50, 77, 78, 80, 108
Nothofagus pachyphylla 80
Nullarbor Pain (WA) 16, 117, 120
Nypa 81
Nypa australis 80

odonata (dragonflies) 5
Oligocene macroflora sites of north-western Tasmania 77–8
Ordovician graptolite province, Victorian 98–100
Ordovician faunal succession of the Florentine Valley (Tas) 72–4
Orites truncata 80
Osteichthyes (bony fish) 103
Otway Ranges (Vic) 4, 93–6
Owen, Sir Richard 12, 16, 17
Oxylobium pungens 80

Palaeophragmodictya 55
Palaeoscolex 56
Palorchestes azael 69
Pandanus 15
Parasqualodon wilkinsoni 103
Parvancorina 56
Pentoxylon 15
Phyllozoon 55
Pincombe, T. 6
plant fossil sites 125
  Bauhinia Downs flora, Balbarini (NT) 21, 107
  Broome Sandstone dinosaur trackways and macroflora, Broome (WA) 48, 61, 121–2
  Dinmore Quarry (Qld) 46
  Early Oligocene macroflora sites of north-western Tasmania 77–8
  Fennel Bay (Kurrur-Kurran) (NSW) 1–3
  Homevale Station (Qld) 2, 44–6
  Lake George (NSW) 3–4, 72
  Late Cretaceous (Cenomanian) Winton Formation (Qld) 50, 61
  Latrobe Valley coal measures (Vic) 109
  Lune River Jurassic plant site (Tas) 78–9
  Macquarie Harbour (Tas) 79–81
  Silurian Baragwanathia site, Yea (Vic) 100–1
  Stuart, Poole and Nelly Creek fossil plant sites (SA) 58–9
  West Dale, Darling Plateau (WA) 118–20
Platycephalus 84
Plecoptera (stoneflies) 5
Pleistocene fossil sites 12
Point Hibbs, western Tasmania 81–2
Poole Creek fossil plant sites (SA) 58–9
Prosqualodon sp. 103
Prosqualodon davidis 84, 103
Psigraptus jacksoni 73, 74
Psocoptera (related to book lice) 5
Pterostoma hirsutus 80

Quintinia tasmanensis 80

Ranken, George 17
Redbank Plains Formation (Qld) 38
Remigolepis 8, 92
Rewan, The Crater (NT) 34–5
Rich, Thomas 66, 96, 97
Ritchie, Alex 6, 9
Riversleigh (Qld) 20, 23, 24, 37, 41–4, 63, 65, 66, 69–70
Roe Calcarenite invertebrate fauna, Roe Plains (WA) 120
Rubus nebuloides 80

Sarcophilus harrisii 116
Seal Creek (Vic) 14
Seymour, Glen 48
Seymour Quarry (Qld) 47, 48
Silurian Baragwanathia site, Yea (Vic) 100–1
Simpson, Wiliam 9
Singh, Gurdip 4
Skania 56
Smith, Thomas 99
Soederberghia 8
Stanley, M. 6
Steropodon galmani 4
Sthenurus 9, 62, 67, 117
Stirton, Ruben 16, 19, 24, 62, 63, 65–6, 67
Strzelecki Ranges (Vic) 4, 93–6
Stuart Creek fossil plant sites (SA) 58–9

Table Cape – Fossil Bluff, Wynyard (Tas) 82–3
Taemas–Cavan–Wee Jasper (NSW) 8, 9 12–13, 33, 113, 116
Talbragar fossil beds (NSW) 14–15, 107
Tarwinia australis 106
Tasmanorthis 74
Tasmaniosaurus 79, 88
Tedford, Richard 16, 62, 65
Teichert, Curt 116

Telopea strahanensis 80
Tetradium 82
Thomas, D.E. 73, 98, 99
Thulborn, Tony 121, 122
Thylacoleo (marsupial lions) 16
Tillyard, Robert John 6, 38
Tirari Desert (SA) 61–3
Toombs, Harry 116
Torquay Oligocene whale and invertebrate site (Vic) 101–5
trace fossil sites 125
  Broome Sandstone dinosaur trackways and macroflora, Broome (WA) 48, 61, 121–2
  Ellery Creek Cambrian–Ordovician trace fossil locality (NT) 14, 24–5
  Lark Quarry Dinosaur Stampede National monument (Winton Formation) (Qld) 35, 36, 46–50
  Mount Morgan Jurassic dinosaur footprint site (Qld) 34, 35, 36, 48
  Myponga Beach, Fleurieu Peninsula (SA) 56–7
  Tumblagooda Sandstone, Kalbarri and Murchison River George trace fossils (WA) 117–18
Triassic vertebrate faunas, Knocklofty Formation (Tas) 84–8
Trusler, Peter 96
Tumblagooda Sandstone, Kalbarri and Murchison River George trace fossils (WA) 117–18

vertebrate fossil sites 126
  Alcoota vertebrate fossil beds (NT) 19–21, 24
  Bullock Creek vertebrate fossil site (NT) 23–4
  Canowindra fossil fish site (NSW) 8–9, 90, 92, 93, 113
  Chinchilla Rifle range (Qld) 16, 36–7
  Cuddie Springs (NSW) 9–12
  Devil's Lair, Cape Leeuwin–Cape Naturaliste region (WA) 12, 116–17
  Devonian 'Great Barrier Reef' and Gogo fish fossil site, Fitzroy Crossing (WA) 111–16
  Early Cretaceous (Aptain) Koonwarra site (Vic) 106–7
  Genoa River Devonian tetrapod site (Vic) 91–3
  Grange Burn Pliocene mammal site, Hamilton (Vic) 16, 37, 97–8, 105
  Green Waterhole Cave (SA) 63–4
  Hines Quarry Diprotodon site, Bacchus Marsh (Vic) 96–7
  Jack Hills Gorge, Broken River Province (Qld) 13, 40
  Lake Callabonna (SA) 12, 62, 67–9, 97
  Lake Frome Basin, Namba Formation (SA) 63, 65
  Lake Palankarinna (SA) 61–3
  Middle Eocene Anglesea site, eastern view formation (Vic) 107–9
  Mt Howitt fish fossil site (Vic) 8, 9, 89–91, 113
  Mt Watt (NT) 29
  Otway Ranges (Vic) 4, 93–6
  Rewan, The Crater (NT) 34–5
  Riversleigh (Qld) 20, 23, 24, 37, 41–4, 63, 65, 66, 69–70
  Torquay Oligocene whale and invertebrate site (Vic) 101–5
  Triassic vertebrate faunas, Knocklofty Formation (Tas) 84–8
  Wellington Caves (NSW) 12, 15–18, 37, 97
Vetustovermis 56
Vickers-Rich, Patricia 66, 96
Victoria Fossil Cave (SA) 69
Victorian Ordovician graptolite province 98–100

Wakefield, Norman 91
Walloon Coal Measures (Qld) 48
Walsh River (Qld) 31–2
Warner's Bay (NSW) 5, 6
Warrawoona group (North Pole and Strelley Pool sites), Proterozoic early life fauna (WA) 123–4
Warren, James 89, 91
Wee Jasper (NSW) 8, 9 12–13, 33, 113, 116
Wellington Caves (NSW) 12, 15–18, 37, 97
Wellington River (Vic) 99
Wellsford Rifle Range (Vic) 99
West Dale, Darling Plateau (WA) 118–20
Winton Formation (Qld) 46–50
  Lark Quarry Dinosaur Stampede National monument 46–50
  Late Cretaceous (Cenomanian) Winton Formation 50, 61
Woodburne, Michael O. 19, 20, 23, 62, 65–6
Wollemia 50
Wynyard (Tas) 82–3
Wynyardia bassiana 84

Xandrella 56

Yea, Silurian Baragwanathia site (Vic) 100–1

Zygomaturus trilobus 69

www.ingramcontent.com/pod-product-compliance
Ingram Content Group UK Ltd.
Pitfield, Milton Keynes, MK11 3LW, UK
UKHW060018051026
15591UKWH00003B/270

* 9 7 8 0 6 4 3 1 0 1 7 7 7 *